Gotthilf Hempel (Ed.)

Antarctic Science

Global Concerns

With 88 Figures

Springer-Verlag
Berlin Heidelberg New York
London Paris Tokyo
Hong Kong Barcelona
Budapest

Professor Dr. Gotthilf Hempel
Institut für Polarökologie
Wichhofstr. 1–3, Gebäude 12
D-24148 Kiel, Germany

ISBN 3-540-57559-6 Springer-Verlag Berlin Heidelberg New York
ISBN 0-387-57559-6 Springer-Verlag New York Berlin Heidelberg

Library of Congress Cataloging-in-Publication Data. Antarctic science global concerns / Gotthilf Hempel (ed.). p. cm. "Papers from the Antarctic Science Conference, in 1991" – Galley. Includes bibliographical references and index. ISBN 3-540-57559-6. – ISBN 0-387-57559-6 (U.S.) 1. Antarctica – Congresses. 2. Research – Antarctica – Congresses. I. Hempel, Gotthilf. II. Antarctic Science Conference (1991) G845.5.A54 1994 509.98′9 – dc20 94-10695 CIP

Typesetting: Best-set Typesetter Ltd., Hong Kong
SPIN: 10129806 31/3130-5 4 3 2 1 0 – Printed on acid-free paper

Preface

International Science and the Antarctic Treaty System

Antarctic science is relatively young, effectively dating from the early years of this century, but the Antarctic Science Conference in 1991 showed the value of Antarctic science to the world; many papers in this volume demonstrated this fact with respect to their own fields of science, particularly the new programmes being formulated on global change and on the Antarctic sea-ice zone. However, a single example suffices, the discovery of the "ozone hole", a discovery which justified every penny that has been spent on Antarctic research by all countries. Antarctic science has a vital contribution to make to the understanding of global phenomena and it is therefore crucial that what scientific capacity is available should be used as productively as possible. This is not something that can be achieved by regulation or direction; but it is something which governments should seek to encourage by creating a climate of opinion which values Antarctic science (Handbook of the Antarctic Treaty System 1990).

SCAR – the Scientific Committee on Antarctic Research – has a clear-cut dichotomy of function: (1) to take primary responsibility, within the context of international science, for the co-ordination of national science programmes in Antarctica, for the promotion of co-operation among scientists on an international level, and for the initiation of new research projects. (2) To provide expert advice on a range of scientific, environmental and conservation matters within the Antarctic Treaty System (ATS) and to liaise and cooperate with other organizations on relevant Antarctic matters. The first function is much more complex than it used to be, because there are many more countries that are members of SCAR, but also because science is now more sophisticated and is integrally linked with global change problems; the International Council of Scientific Unions (ICSU) has given SCAR the task of co-ordinating global programmes in this region of the world. The second function is also more complex than it used to be, because of the growth and questioning role of the environmental pressure groups, the media interest stimulated by them, the development of tourism and the interest of the United Nations in Antarctic affairs. Concerning the first function, the 18 papers in this volume clearly demonstrate the great strides that Antarctic science has taken; it is now recognized as a significant component of environmental research worldwide, particularly as a major contribution to understanding global change. The authors deal with the sea-ice cover, geological issues, various aspects of oceanography and climate, the polar ice sheets, the biosphere, and environmental protection. One paper addresses conflicts of interests in the use of the Antarctic and the remainder of this extended preface deals mainly with this issue.

The current regime for the Antarctic stems from the activities of ICSU in planning and implementing the International Geophysical Year (IGY) in 1957–1958. So successful was the Antarctic element of the IGY that the opportunity was taken to set up a permanent system to promote Antarctic science, by establishing the Special – changed later to Scientific – Committee on Antarctic Research in 1958. This led indirectly to the signing of the Antarctic Treaty, which was drawn up to set aside the continent for peaceful purposes, especially for scientific use, and came into force in 1961. SCAR is not explicitly mentioned in the Treaty, but from the very first Antarctic Treaty Consultative Meeting (ATCM) SCAR has been referred to in many Treaty recommendations. The high initial regard of the Antarctic Treaty Consultative Parties (ATCPs) for SCAR – a non-governmental organization (NGO) – continued for many years and whenever they were in need of scientific advice concerning Antarctica, they came to SCAR, which has consistently supported them; the many requests for advice have involved a great deal of work.

This has contributed significantly to the success of the Antarctic Treaty System, particularly in matters relating to the environment and conservation. Over the years SCAR advice has been influential in the initial formulation and subsequent evolution of the Agreed Measures for the Conservation of Antarctic Flora and Fauna (1964), particularly in the development of the Protected Area system. There are many relevant SCAR publications and SCAR continues to review and develop the arrangements for these matters. Since 1964 SCAR has also been active in considering the control of pelagic sealing and contributed to the formulation of the Convention for the Conservation of Antarctic Seals (CCAS) which came into force in 1978. This confirmed the special role of SCAR by formally inviting it – an NGO – to provide independent scientific advice under an intergovernmental instrument.

Although the question of mineral resources is not referred to in the Antarctic Treaty, a Treaty Recommendation in 1975 invited SCAR to make an assessment of the possible impact on the Antarctic environment if mineral exploration and/or exploitation were to occur there. Also, SCAR had been concerned with Antarctic marine living resources and had been urged by the ATCPs "to continue its scientific work on these matters and to consider convening, as soon as practicable, a meeting to discuss current work and report on programmes for the study and conservation of Antarctic marine living resources." SCAR's response was a comprehensive, international research programme on Biological Investigations of Marine Antarctic Systems and Stocks (BIOMASS), formulated in 1976 and spanning 15 years, which aimed to gain a deeper understanding of the structure and dynamic functioning of Antarctic marine ecosystems, essential for the wise management of the living resources of the Southern Ocean. It ended in 1993 with a colloquium in Bremerhaven to evaluate the achievements of the programme. This research programme led to an impressive number of publications, as well as the creation and operation of the BIOMASS Data Centre, probably the first international relational database in biological oceanography.

BIOMASS also led to the negotiation of a Convention for the Conservation of Antarctic Marine Living Resources (CCAMLR) which came into force in 1982. Its objective is the conservation (including rational use) of all Antarctic marine life, appling also to waters south of the Antarctic Polar Front (Antarctic Convergence). The BIOMASS programme has been an important source of advice for CCAMLR and SCAR's role in relation to this convention is a complementary one, encouraging basic scientific research to provide knowledge which is essential for management purposes.

After the failure of governments to bring the Convention on the Regulation of Antarctic Mineral Resource Activities (CRAMRA) into effect, despite many years of negotiations, intergovernmental discussions on Comprehensive Measures for the Protection of the Antarctic Environment, rapidly led in 1991 to the adoption of a Protocol on Environmental Protection to the Antarctic Treaty. The Protocol has yet to enter into force and it will probably be a few years before a mechanism evolves to implement its provisions. Clearly, SCAR has amply demonstrated its commitment to environmental conservation over the years. However, most Antarctic scientists do not believe that there has been significant widespread environmental impact in the Antarctic due to the human activities originating there (the wider impacts, such as atmospheric pollution and ozone destruction, originating from human activities in developed regions of the world), that the very localized problems due to scientific and supportive activities are more an aesthetic than an ecological problem, and that evolutionary progress within the Protocol towards improving the existing arrangements is the best way forward. Also, many Antarctic scientists have been disconcerted to find their long sustained efforts in environmental protection belittled or ignored in the media's current focus on Antarctic conservation. The underlying concern of Antarctic scientists is that actions may be taken without adequately consulting them, ostensibly to protect the environment, some of which are unnecessarily rigorous and may seriously limit their ability to conduct basic research. Because of the key role of the Antarctic in global change research, such measures could have serious consequences, not just for Antarctica but for the world.

One experienced Antarctic diplomat put this well: In democratic societies it is an established principle that the law needs to be acceptable to the governed... The risk in the Antarctic situation is that the ATCPs may find themselves getting legislatively ahead of their real constituents, the scientific community in Antarctica. Impossible, impracticable or unnecessary laws are undesirable. Politicians should take account of the views of the scientific community, expressed through SCAR, before making new laws.

SCAR is worried about the possibility under the new protocol of a diminution of its role and influence in relation to the ATS, about legislation leading to costly environmental protection measures, drawing funds from science, and about the effect this may have on "freedom of scientific investigation and co-operation toward that end" – a keystone of the Antarctic Treaty. Specific current concerns of SCAR are fourfold: first, the inadequate resources available to maintain and expand its activities, in particular to cope with the increase in work related to an advisory role in relation to the ATS; second, the prospect of duplication of scientific effort in relation to the management of Antarctic, which compounds the first problem; third, undue diversion of funds from science to environmental protection and monitoring as a result of the disproportionate influence of environmental pressure groups; and fourth, misconceptions about the nature of geological and geophysical research in the Antarctic.

Antarctic affairs are entering a new phase, the full implications and characteristics of which are as yet unclear. In principle, the role of scientists, in addition to continuing to conduct relevant research, is to provide data and advice on the consequences of various actions; it is for administrators, and politicians, to consider this advice and to act through the legal processes. A relatively new factor is the role of environmentalist pressure groups, which have strong, but not always valid, ideas about the Antarctic environment and which attempt to persuade governments to support their objectives. A large part of the problem for the future is the very limited funding available to SCAR. Scientists are therefore posing the question: With such inadequate funding to exploit scientific opportunities of global significance, should SCAR still be trying both to promote primary research and also to advise on management, at the risk of doing both inadequately?

So what is the way forward? We have three international fora in which decisions can be made which, in one way or another, affect what is, or is not, done in Antarctica – the ATCMs, SCAR, and the Council of Managers of Antarctic Programs (COMNAP). The urgent need is to develop, in these three fora, clear ideas as to what their functions are and as to how they can work together effectively in a complementary way.

ATCMs are about the governance of Antarctica by international agreement involving intergovernmental obligations. This provides a legal umbrella for Antarctic science, but the scientific community needs to be able to influence (international) legislation, as taxpayers in a single country do. Scientific research, including its logistic support, is one of the few activities in the Antarctic to be regulated by such legislation. Thus, the question for SCAR is whether it should continue, as an act of positive policy, to seek an advisory role to the Treaty or whether it should concentrate its energies on its "primary responsibility . . . for the co-ordination of national science programmes . . . for the promotion of co-operation . . . and for the initiation of research projects"? SCAR wishes to continue to advise the Antarctic Treaty System (for the reasons already mentioned), while recognizing that the need to fulfil our "primary responsibility" has considerably increased and that the arrangements for the provision of advice have changed. Turning to COMNAP, that organization (which is federated to SCAR and incorporates the functions of the former SCAR Working Group on Logistics) is about resource allocation and sharing costs and practical "know-how" between countries. Its principal task is the support of science, which it achieves through the regular exchange of information on operational matters, seeking solutions to common problems and reviewing the major logistic requirements for international scientific programmes. In recent years co-operation between SCAR and COMNAP has greatly improved and they have jointly submitted advice to the Treaty on environmental matters.

In the light of all this perhaps it would be sensible for SCAR to confine itself to offering expert advice on purely scientific matters. Its input to environmental and conservation matters would then be confined to answering strictly scientific questions put by governments. However, it is undeniable that a likely consequence might be reduced influence on decisions (such as unduly rigorous controls on research activities) that could adversely affect scientific research. The ATCPs themselves seem to acknowledge the undesirability of this and the issue continues to be debated within Antarctic circles.

It is for these reasons that SCAR decided to hold the conference on Antarctic Science – Global Concerns, in order to draw attention to the value of Antarctic science and the dilemma it now faces. Subsequently, in order to bring the climate and global change-related research efforts in the Antarctic into the context of wider international research programmes, SCAR has set up a Group of Specialists on Global Change and the Antarctic to oversee a major, new long-term programme, with six core projects, addressing broad interdisciplinary topics which are central to global change studies. These concern: the Antarctic sea-ice zone; palaeo-environmental records from the ice sheet and marine and land sediments; the mass balance of the ice sheet and sea level change; Antarctic stratospheric ozone and ultra-violet effects; the role of the Antarctic in biogeochemical exchanges; and detection of global changes in the Antarctic. The Global Change Programme (GLOCHANT) will be the flagship programme of SCAR over the next 10 years or so, taking us into the next century. It will also be the largest single programme that SCAR has attempted to organize, equivalent in scale to several BIOMASS programmes. I believe that the contents of this volume give a foretaste of the exciting science still to come in this field.

Richard M. Laws

President of SCAR
St. Edmunds College
Cambridge CB3 0BN
United Kingdom

Table of Contents

Antarctic Science – Global Concerns: *Introduction*

Gotthilf Hempel[1]

The title of this book stems from an international conference held in Bremen, Germany, in September 1991 under the auspices of the Scientific Committee on Antarctic Research (SCAR). Its main objectives were to increase public awareness of the importance of Antarctic science, particularly in relation to global problems, and to foster the interaction of Antarctic scientists working in different disciplines. About 500 scientists, students and engineers, as well as politicians and media from all over the world attended the conference. The first 2 days of the conference were concerned with current scientific programmes and results; the last 2 days addressed policy matters and the future directions of Antarctic science.

This volume presents revised and up-dated texts of key lectures given at the conference by some of the leading Antarctic researchers, spanning a broad spectrum of Antarctic science from the "ozone hole" to the microbiology of sea ice. Nevertheless, some fields are missing, e.g. geophysics of the Antarctic continent, terrestrial ecology, and biology of seals, whales and birds.

The main focus is on the role of Antarctica and the Southern Ocean in the world climate system, a topic of great importance and heated debate. Antarctic ice, both on land and at sea, determines much of the "slow climate system" which responds to, and produces changes on time scales of years to centuries, whereas most processes on land and in the atmosphere are short-lived. The contribution by P. Wadhams deals with the interaction between sea ice and atmosphere. The formation of sea ice is a complex process, which results in unconsolidated pancake ice covering a third of the sea-ice zone in winter, whilst the solid pack ice nearer the continent has open leads and coastal polynyas where new ice is continuously forming.

Sea ice is a rich, but seasonal, habitat for ice algae and a variety of small animals. M. Spindler and G. Dieckmann emphasize its ecological significance, in particular the extent to which Antarctic krill depends on sea ice for survival and growth in winter. They predict that the disappearance of the sea-ice cover due to global warming would completely upset the Antarctic marine ecosystem. Changes in the composition and quantity of phyto-

[1] Institut für Polarökologie, Wischhofstr. 1-3, Gebäude 12, D-24148 Kiel, Germany

Hempel (Ed.) Antarctic Science

plankton would also affect other organisms in the Antarctic food web, particularly krill and its predators, and the stocks of whales, seals and penguins. The floating ice shelves extending from the continental ice cap far into the Southern Ocean, particularly in the Weddell Sea and Ross Sea, have a profound effect on the formation of the Antarctic Bottom Water which circulates across the Equator into the North Atlantic. G. Fahrbach et al. describe the latest findings from cruises of Norwegian, German and Russian research vessels into the pack ice and coastal polynya of the Weddell Sea where 70% of the Antarctic Bottom Water is formed. Ocean-atmosphere interaction in the Antarctic will rank high on the agenda of future work in global change. E. Augstein focuses on themes including ocean-atmosphere momentum, energy and gas exchanges, deep mixing, atmospheric radiation and interrelations with cloudiness and snow/ice surfaces. Modelling and remote sensing are immensely powerful tools, although scientists will have to continue to undertake fieldwork and new techniques must be developed for in situ studies.

The incorporation of the biosphere into global models is accompanied by great difficulties. As yet it is not known whether the Southern Ocean is a source or sink of atmospheric CO_2, or what effect the disappearance of sea ice might have on CO_2 uptake by phytoplankton. These changes might be due to differences in the uptake of CO_2 by phytoplankton in surface waters of the ocean and its transport to the sea bed. The importance of this "biological pump" in the oceans is described by P. Treguer. If it could run unlimited by nutrients – like nitrate, phosphate and silica – the CO_2 in the atmosphere could be lowered to about one-half of its present level, while a simulation model without biological pumping led to a 50% increase in atmospheric CO_2. The present level of atmospheric CO_2 is unprecedented over the entire period covered by the ice-core records and is an alarming indicator of the growing greenhouse effect of global warming. The increase in UV-B radiation due to the ozone hole may have adverse effects on marine productivity and H. Marchant reports on differences in the sensitivity of various kinds of plankton to UV-B, and the potential consequences, involving the organisms themselves and, potentially, wider climatic change due to the effect of relatively more tolerant *Phaeocystis* blooms [which produce large amounts of dimethylsulphate (DMS) – a cloud-condensing nucleus] on cloud formation.

The climate of the past is recorded in the sediments of the Southern Ocean, in Antarctic lakes, and in the glacial deposits in the Transantarctic Mountains. They all reveal large fluctuations in glaciation in the recent geologic past. Whether the present very cold ice sheet will remain stable under conditions of rising temperature in the next century is not only a question of the amount of temperature increase but also of the rate and distribution of precipitation. As yet it is not possible to say whether the total ice mass is presently growing or shrinking. Radar altimetry, by ESR-1 and other Earth observing satellites, is the essential tool for obtaining an overall picture of the thickness of the Antarctic continental ice sheet. J. Zwally gives a comprehensive account of detection of changes, mainly from satellites, including ozone depletion, ice stream flow and ice shelf disintegration; techniques are being developed for measuring other changes to overcome the difficulty of estimating the present overall mass balance of the Antarctic continental ice sheet.

Ice sheets in both polar regions are by no means uniform masses, but consist of different parts – resting on land, filling marine basins or floating on the sea – each with a distinct history and response to climatic change. While large parts of northern Eurasia and America were ice-covered during the Ice Age, Antarctica's ice sheet may have grown only by about one-third, because it rested on a large continent with little precipitation. The most detailed information on changes in climate and in the composition of the atmosphere over Antarctica is contained in ice cores – drilled in 3- to 4-km-thick ice sheets of central Antarctica – which permit the reconstruction of world climate over the past 230 000 years; post-glacial warming was extremely rapid after the last glaciation, ending about 12 000 years B.P. The CO_2 content of the atmosphere, as retained in the tiny air bubbles in the ice, was lower in the cold periods than in the interglacials. Past periods of warming have affected the Antarctic ice sheet to a much greater degree than hitherto expected.

H. Oeschger puts the Antarctic results into context with the most recent data obtained by European drilling at Summit on Greenland. The high resolution of the Greenland cores made it possible to detect drastic and rapid changes in temperature during the interglacial period 100 000 years ago.

Climate is just one of the various interconnections between the Antarctic and the rest of the world. Under the general heading "The Antarctic in the global scene" M. J. Rycroft points out that Antarctica provides a window to outer space, where it comes closest to planet Earth, because of the characteristics of the Earth's magnetic field and its interaction with charged particles streaming from the Sun.

The high southern latitudes, particularly the Pole station itself, are uniquely suitable for studying Sun, stars and planets, leading to the new research field of helioseismology, that will yield a better understanding of the ultimate source of energy for all processes on the Earth's surface. Another, more unexpected application of Antarctic research to studies in extraterrestrial space is described by D. J. Lugg: biomedical research on small confined groups wintering in the Antarctic provides valuable clues for the preparation of outer space missions.

The role of Antarctica in biological evolution is analyzed by J. A. Crame in a wide ranging account based on the fossil record of plants and animals over the past 400 million years; Antarctica may well have been a site of significant evolutionary innovation. The past two decades have seen a rapid change in geological knowledge about Antarctica, although only less than

2% of Antarctica's surface is ice-free and hence directly accessible to geological surveys. Antarctica and the seas around it are major pieces in the puzzle of plate tectonics and of Godwana in particular, hence Antarctica is a continent which cannot be left out of studies of global geological problems, as emphasized by F. Tessensohn.

Examples for Antarctic research in its own rights are given by S. Nicol on krill, and by W. Arntz and V. A. Gallardo on Antarctic benthos.

S. Nicol deals with the changing perceptions of the Antarctic krill, both as a potential resource and in terms of its role in the Antarctic ecosystem; far from being a just "small shrimp-like planktonic animal" it is larger than most plankton organisms and has a long life span; sea ice is very important as a habitat for krill. In fact, much of the krill population undergoes a seasonal shift from a filterer of phytoplankton to a grazer of ice algae. It is the staple food not only of baleen whales, and of most of the Antarctic seals and penguins, but also of fish and squid. The present annual harvest of 200 000–400 000 tons is of little significance to the krill stock and its consumers as a whole.

The Antarctic benthos is very rich in both abundance and number of species in certain groups. Arntz and Gallardo provide a comprehensive review of the state of knowledge of the major groups and their geographical and bathymetric distribution. They also point to needs in future benthos research, which has so far concentrated on small, relatively shallow areas near biological research stations and on some transects across the deeper parts of the shelf. Therefore, more work in the deep sea is needed as well as experimental studies in both the field and the aquaria of research stations and vessels. The future of research in the Antarctic biosphere in general is discussed by G. Hubold, who presents new programmes, particularly on the ecology of the Antarctic sea-ice zone. In regard to biological aspects of global change, the study of the sea-ice zone will be one of the most promising fields of Antarctic research in the years to come in terms of biogeochemical cycles.

Research in Antarctica has financial, political and environmental implications: work in the Southern Ocean and on the Antarctic continent and islands is expensive. It is benefitting from the climate of peace and international co-operation under the umbrella of the Antarctic Treaty and within the framework of programmes of SCAR. Any presence of man in the Antarctic has a limited impact on the pristine environment. Overwintering research stations shared by two countries and open to others represent a means to reduce costs and environmental pressure. An example of the effective common use of ship facilities is the European Polarstern Study EPOS 1988/1989, which involved 130 scientists from 11 countries in 3 cruise legs on the *R.V. Polarstern*, provided by the German government.

"The conflict of interests in the use of Antarctica" is the title of the paper by D. Drewry, Director of the British Antarctic Survey. According to the Antarctic Treaty, science is the foremost reason for man's presence in the Antarctic, but there are also other legitimate reasons for being there: the political will of national presence results in the fact that some overwintering stations are still much bigger and more crowded than necessary for their scientific programme. The economic interest in Antarctic resources is presently reduced to some limited fishing operations, while mineral exploration is prohibited. Tourism, on the other hand, is increasing and requires more regulation in order to protect the Antarctic environment and allow undisturbed implementation of scientific research. "How to harmonize the interests of Antarctic scientists and environmentalists" is the topic of the closing paper by N. Bonner, former Chairman of the SCAR Group of Specialists on Environmental Affairs and Conservation.

The Madrid protocol of the Antarctic Treaty declares Antarctica a nature reserve dedicated to peace and science. Therefore, it must be the aim of the scientists to control research activities while preserving the value of the Antarctic to science and the world. On the other hand, science need protection against impracticable and unnecessary laws which might unduely interfere with the carrying out of research activities. The President of SCAR in his extended preface to this book made this point with respect to the growing importance of Antarctic science, on the one hand, and to the increase in legislation in the Antarctic Treaty System, on the other.

Environmental Protection and Science in the Antarctic

W. Nigel Bonner[1]

During the past 5000 years, man has wrought more change on the global environment than has any other species, living or extinct. The last two centuries have seen the rapid development of sophisticated technologies which have provided the means greatly to accelerate these changes, with results that are not always beneficial, either to man or to the environment. Concurrently, advances in medical science, coupled with new systems of food production and generally more stable social conditions, have led to an astonishing increase in the number of people living on this planet. Only recently have people come to realize that if the world is to continue to support the demands made on it by increasing populations, hungry for resources and space, active steps must be taken to protect the environment. As a consequence, we have seen the introduction of national, and more significantly, international legislation to protect animals and plants, their habitats and those environmental features that contribute to the quality of human life.

In this respect the Antarctic is a very special region. The last two centuries, the time of the development of industrialization, have encompassed the whole of human impact on the Antarctic, with most of this being concentrated in the last 50 years. Early interest in the Antarctic was concentrated on its potential for resources. The most obvious of these were the astonishingly abundant seals and whales, and most of the early activities around the fringes of Antarctica were conducted by sealers. Commercial seals stocks were soon exhausted, but at the beginning of this century an enormous whaling industry was initiated in Antarctic waters. The Antarctic whaling industry, pioneered by the Norwegian C. A. Larsen, at its peak in the 1930s provided about 15% of the world's total marine harvest. In the absence of proper regulatory agreements, the whales declined to a level which made hunting uneconomic. Marine harvesting, based on a small stock of fish and the abundant but not very valuable shrimp-like krill, continues in Antarctic waters but scarcely impinges upon the continent itself.

Activities on land were initially largely concerned with exploration, but even from the earliest voyages, such as those of Wilkes or Ross, scientific

[1] Scientific Committee on Antarctic Research, c/o Scott Polar Research Institute, Lensfield Road, Cambridge CB2 0ET, UK

Hempel (Ed.) Antarctic Science

observations were made. The geographical position of Antarctica made it of especial interest for many geophysical studies which are best performed at high latitudes. Cooperative "Polar Years" had been organized by geophysicists in 1882–1883 and 1932–1933. The first of these saw a German expedition working in South Georgia. For the second, although some Antarctic activities had been planned, no stations were operated in the Antarctic. However, a third Polar Year was planned for 1957–1958, a year of maximum sunspot activity. This was a much more extensive operation than the two previous Polar Years, and was formalized as the International Geophysical Year or IGY. Sixty-six nations took part, of which 12 established stations in the Antarctic and on many peri-Antarctic islands. It was a remarkable collaboration between the scientists of these 12 nations, some of which had no official contact other than that involved in planning the IGY, and the programme was scientifically highly productive.

The IGY saw the establishment not only of geophysical programmes in the Antarctic. Research in geology, glaciology, biology and medicine was conducted as part of the IGY programme, and exploration, with accompanying cartography and mapping, continued.

To co-ordinate all these activities, the International Council of Scientific Unions, ICSU, set up the Special Committee on Antarctic Research, later renamed the Scientific Committee on Antarctic Research, SCAR. This provided, and has continued to provide, a forum in which scientists can discuss and coordinate their work and plan mutually supportive programmes.

As important as the development of coordinated science in the Antarctic, was the opportunity provided by the IGY for representatives of the 12 nations concerned to develop a common political structure for Antarctica. Prior to the IGY, disputed territorial claims in the Antarctic had provided a source of friction which might have resulted in serious destabilization of diplomatic relationships. The common accord of scientists from different countries provided a basis for diplomatic advantage that was too valuable to discard when the IGY ended. In 1958 the USA took the initiative to call a conference to "seek an effective joint means of keeping Antarctica open to all nations to conduct scientific research or peaceful activities there".

The outcome of this conference was the Antarctic Treaty, signed at Washington in December 1959, and which came into force in 1961.

The Antarctic Treaty provided a legal code regulating activities in the Antarctic. The Treaty itself had little to say about conservation, though it named the preservation and conservation of living resources as one of the subjects of common interest about which Treaty Parties should consult together. No mention at all was made of environmental protection. However, at subsequent meetings of Parties to the Treaty, a series of recommendations were adopted which form the basis of environmental protection in the Antarctic today.

Regulations exist to protect fauna and flora, to establish protected areas, to control the disposal of wastes in the Antarctic, or pollution at sea, and to ensure that the environmental consequences of planned scientific or logistical activities are adequately assessed before action is taken.

These regulations were assembled in rather an ad hoc manner and in 1990 and 1991 meetings were held in Viña del Mar, Chile, and Madrid, Spain, at which a Protocol on Environmental Protection was drafted. This develops the concepts contained in the earlier recommendations through a series of annexes and provides a formal structure for their implementation. The Protocol was signed in Madrid in 1991 and was endorsed and amplified with an additional Annex at the Sixteenth Antarctic Treaty Consultative Meeting in Bonn, FRG in October of that year.

The negotiation of the Protocol has been seen as an environmentalist triumph. It will indeed provide a comprehensive code for environmental protection that is sufficiently flexible, through its annex structure, to be easily revised when that appears necessary. However, it can still be questioned whether the introduction of the Protocol will result in all the general benefits expected of it.

From what does the Antarctic environment need protection? Two sorts of pressures impact on the Antarctic: those arising from activities outside the area, and those arising from activities internal to the Antarctic.

The first of these are environmentally the more serious. They are characterized by being generally distributed about the continent. Examples are the occurrence of traces of chlorohydrocarbon insecticides, such as DDT, first identified in Antarctic penguins in 1966. These compounds, used elsewhere in the world, had reached the Antarctic food chain presumably through aerial routes. Equally pervasive are the radionucleides that can be identified in snow samples from inland Antarctic areas. These are derived from nuclear explosions many thousands of kilometres from the Antarctic. Perhaps neither of these contaminants can be said to be likely to have serious consequences to the Antarctic environment. The same is not true for the increasing concentration of carbon dioxide in the atmosphere, derived largely from the burning of fossil fuels in the northern hemisphere, and which is predicted to lead to global warming. Global warming may have one of its most serious consequences in the Antarctic, where the stability of some of the ice shelves may be disrupted. The ozone hole, about which so much public concern has been expressed, and which we now know to be caused by traces of chlorofluorocarbons in the upper atmosphere, is centred over Antarctica and the destructive effects of the increased ultra-violet radiation reaching the earth's surface may be first noticed on Antarctic fauna and flora.

These impacts cannot be controlled by legal codes applying only to the Antarctic. They are global problems and call for global solutions. Unfortunately, in most cases the damage is done, and all that can be hoped for now is a measure of restraint from the rest of the world to avoid aggravating the present situation. Depressingly, a realistic assessment is that this restraint is unlikely to be applied. International agreements, such as those on nuclear testing or on the use of chlorofluorocarbons, are environmentally inadequate and there has been little international response to the most significant and far-reaching environmental problem of all, global warming.

The other category of impacts, those arising from activities conducted within the Antarctic, are generally highly localized to the vicinity of research stations, field camps and traverse lines. An exception to this may be the products of combustion of fuels used at stations, or in ships, over-snow vehicles or aircraft, which are released into the atmosphere, and which may be widely distributed, but in negligible concentrations. Since nearly all the persons conducting activities within the Antarctic are governed by the Antarctic Treaty, the Treaty regulations on environmental protection should provide the means to control these impacts.

The major activity carried on in the Antarctc is scientific research, and its associated logistic support. Tourism, though frequently cited as a significant impact on the Antarctic, is in fact a very minor operation, though its local impact may be high in places.

How has science and its associated logistics impacted on the Antarctic? In the early years of Antarctic science there was rather little concern for active environmental protection. Penguin eggs were collected for food and seals were used to feed sledge dogs. Rubbish was generally disposed of by dumping in a convenient gully or in the sea. Inevitably, the oldest established stations built up some impressive garbage heaps and in one, or perhaps two places local seal populations were depleted. But the absolute magnitude of these impacts on the 14 million km^2 Antarctic Continent was very small.

The question of the scale of the Antarctic continent is often overlooked by those who are critical of activities within it. Its extent is unfamiliar to all but a few. It is larger than Australia (7.7 million km^2) or the USA and Mexico together (9.7 million km^2). Even if consideration is restricted to the ice-free area of Antarctica (about 200 000 km^2), this is still larger than the size of a state such as Nebraska (198 531 km^2) or a small country such as Uruguay (186 876 km^2).

With increasing awareness of the need for environmental protection, as well as the introduction of the protective measures of the Antarctic Treaty, standards in the field have greatly improved. Some of the newer stations are models of environmental responsibility. The impacts that are discernible are almost exclusively those associated with logistics, rather than research. Very few scientific programmes have any measurable impacts on the environment. Some of those that might seem to have a major impact, such as seismic shooting, in fact have very transitory and minor effects. The collection of living specimens of animals or plants for scientific purposes has not resulted in significant, or even observable, changes in populations. This is because species in the Antarctic are generally abundant, if often very localized, and because scientists have practised restraint in collecting.

Logistic impacts have been more severe. The establishment and maintenance of a land station in the Antarctic inevitably has a destructive effect on the local environment. In some regrettable cases damage has been caused by unnecessary road building, or the leakage of fuel. However, these effects are invariably highly localized. Much has been made of the effects of sewage discharges. However, at one of the largest Antarctic bases, divers report that within a few hundred metres of the outfall no effects are visible. In many cases what are seen as environmental problems are in fact no more than aesthetic eyesores.

It is of course right that people should be vigilant over the protection of the Antarctic environment. Environmental protection is a priority, but it is not the only priority. Other papers in this volume refer to scientific research carried out in Antarctica that is of fundamental importance to the rest of the world. The understanding of ice-sheet dynamics will be of vital significance as global warming proceeds. The record of atmospheric carbon dioxide contained in bubbles trapped in ice cores can provide us with baseline data essential to understanding atmospheric chemistry. Unique aspects of our links with geospace, our contact with the rest of the universe, are best studied from Antarctic observatories. We are all aware of the role Antarctica played in the discovery of the ozone hole.

In order to come to terms with the global environment, and our place in it, we need to understand the whole of its structure and functioning. Antarctica is too large a part of the world – some 10% – to be ignored, and this is the overwhelming justification of Antarctic science.

Conservation measures inevitably restrict activities. It is important that in the Antarctic conservation measures are not unnecessarily carried to the point where they impede scientific research. By this I do not mean to suggest that environmental protection should be weakened, but rather that it should be restricted to that which is necessary to achieve its objectives. Within the last decade we have seen the development of a movement of well-meaning persons whose objective is to preserve the wilderness status of Antarctica. This movement has highlighted the environmental scars associated with some research stations and, presenting themselves as the guardians of what is often called "the last remaining wilderness", have generated what has been described as "an inflated media-hyped campaign about damage to the Antarctic environment". Many Antarctic scientists who have for several decades been promoting environmental conservation in the Antarctic have felt that their efforts have been ignored by this campaign which has perhaps led governments to go further than is either necessary or desirable in environmental protection.

An example of this is to be found in the field of the avoidance of marine pollution. Both the current measure controlling this under the Antarctic Treaty (Recommendation XV-4) and Annex IV to the Protocol on Environmental Protection prohibit the discharge of sewage from vessels within 12 nautical miles of land or ice shelves. Bearing in mind the small amount of shipping operating in the Antarctic, the unrestricted and highly dynamic nature of much of the coastal waters, not to mention the fact that the discharge of sewage from research stations to the sea is permitted by another Antarctic Treaty measure and Annex III, this prohibition may seem to be based more on political or aesthetic requirements than on scientifically based environmental protection.

Enforcement of a legal code in the Antarctic will always be problematical. Much reliance will have to be placed on self-regulation, and this will be most effective where restrictions are confined to those seen to be necessary. It is the scientists with experience of the Antarctic who are in the best position to give advice on what is needed in the way of environmental protection legislation in the Antarctic. The adoption of regulations tailored to other, and always more populous, regions of the world, will rarely be appropriate in the Antarctic.

SCAR is the organization which brings together collective experience of the Antarctic environment and expertise in all branches of science conducted there. Governments should take account of the views of Antarctic scientists, expressed through SCAR, before drafting environmental protection measures.

However well-drafted, no legal codes can be effective without the support of those involved. Compliance, in the absence of an extensive Antarctic police force, which no one would wish to see, depends largely on responsible individual behaviour. How can one ensure this support? The media, and particularly television, have made most people in developed communities aware of the Antarctic. Scientists must devote far more effort to explaining the significance of their researches in this region to the public at large. An effective way to do this is through one or other of the quasi-scientific and educational channels which have proved so popular on television. More general education is also needed. Schools, even at the earliest levels, can introduce the concept of the importance and value of the Antarctic environment to those who will become the decision makers of the future. Tourists, though often resented by environmentalists and Antarctic operators, can also perform a valuable task. Almost invariably, they return from Antarctica filled with enthusiasm for this beautiful and impressive region, and they can act as environmental ambassadors for the Antarctic. The great costs of getting to the Antarctic control the number of tourists that can visit this remote area, but this also implies that those who do so are usually wealthy and often influential figures in their communities, so their contribution can be very effective.

Our ultimate aim in the Antarctic must be to control our activities so as to preserve the values of this region – scientific values, wilderness values, aesthetic values – for the use and enjoyment both by the present and the future generations of mankind. These values are not exclusively Antarctic – they are integral to the whole of our environment.

Conflicts of Interest in the Use of Antarctica

David J. Drewry[1]

1 Introduction

1.1 Scientific Achievements

The International Geophysical Year of 1957–58 demonstrated the effective use of the Antarctic for peaceful international scientific activity, and the Antarctic Treaty of 1961 acknowledges the important contribution of science. The pre-eminent position accorded to science has been vindicated: 30 years of intensive research have shown the intimate connections and controlling influences of Antarctica on the principal environmental systems of planet Earth (climate, ocean circulation and sea level). The achievements of and challenges arising from research in Antarctica whilst representing fundamental scientific contributions also possess political and environmental dimensions:

- the discovery of the ozone "hole". This is arguably one of the most fundamental scientific and environmental revelations of the last 50 years, which led from scientific discovery to political action, in the form of the Montreal Protocol, in only 2 years, demonstrating that science can, at times, influence rapidly the political process;
- the detection of background levels of global pollution in Antarctic snow (e.g. heavy metals, PCBs, pesticides), drawing attention to the extreme end of the transboundary pollution spectrum;
- the climate archive contained in Antarctic ice cores. Careful chemical analyses of deep cores reveal details of changes in a wide range of environmental parameters over ice age time-scales. The analysis of past atmospheric composition contained in gas bubbles (such as CO_2, H_2O_2, CH_4) has been a fundamental, if not unique, contribution to the understanding and study of climate change;
- the role and importance of the Antarctic region, especially the Southern Ocean, to global climate change and the study of the "greenhouse effect". The Southern Ocean plays a very significant part in the carbon cycle, the

[1] British Antarctic Survey, Cambridge, CB3 0ET, UK

Hempel (Ed.) Antarctic Science

uptake of excess, anthropogenic, CO_2 and its behaviour is thus critical to climate predictions;
- the ice sheet contains sufficient ice to raise world sea levels by a total of 55–60 m. A change in ice volume of only a fraction of a percent is sufficient to alter sea level by many centimetres over the next decades. The growing knowledge and modelling of ice sheet response to climate warming is, therefore, critical to predictions of future sea level change on a variety of time scales (decades, centuries, millennia).

1.2 Political Achievements

Issues of science have to be viewed within the overarching framework of the Antarctic Treaty which stands as one of the greatest achievements of humankind in the latter part of the twentieth century – setting aside almost one-tenth of the planet to be free of conflict, confrontation and terrorism which seem pervasive elsewhere; establishing freedoms of travel, scientific research, and exchange of personnel and data; creating an area free from nuclear explosions and nuclear dumping, and from military activities; and endorsing rights of inspection – all of which have set out the path to the gradual diminution of sovereignty (Vicuna 1986; Beeby 1991).

It would seem a paradox, therefore, that in this region of the planet already devoted to peace the subject of this chapter should focus on *conflict* in the uses of Antarctica. Conflict results from, among other factors, unacceptable change; of the result of tensions created by the alteration, too quickly or too dramatically, in the state of systems.

It is 30 years since the entry into force of the Antarctic Treaty. In that time there have been profound changes to society and politics worldwide which could not have been imagined in the late 1950s – humankind has conquered space flight, discovered DNA, black holes, quarks and quasars, learnt to harness microchip technology and created terraflop computing capacity. The political climate has gone from cold to at least warm. The map of Europe has changed irrevocably. Drought, famine, homelessness and diseases like AIDS stalk the planet and cast ever lengthening shadows. It is no wonder that even in remote Antarctica there should also be change driven by these external forces and that changes in perception of the appropriate use of Antarctica should take place. Established values have come steadily under pressure creating a nexus of tension fed by the increasingly diverse nature, demands, aspirations, perceptions, attitudes and methodologies of an expanding Antarctic community.

2 Uses of Antarctica

In recent years the diversity and number of activities in Antarctica has enlarged and become better defined (Drewry 1986), together with a range of legislative provisions which regulate several of them (e.g. Antarctic Treaty Recommendations, Conventions and Protocols). In addition, new coordinating machinery has become established [e.g. Council of Managers of National Antarctic Programmes (COMNAP)]. These activities and elements are illustrated in Fig. 1 in which the emphasis has been placed on the interactions with science. The figure is not intended to be comprehensive but shows the principal areas (national endeavour, the environment, international political processes and commercial exploitation) for which possible tensions with science already exist or may develop in the future.

3 Science in Conflict with Commercial Exploitation

3.1 Non-Renewable Resources

The provisions of the 1991 Protocol for the Protection of the Environment to the Antarctic Treaty have banned mining for 50 years and there can be

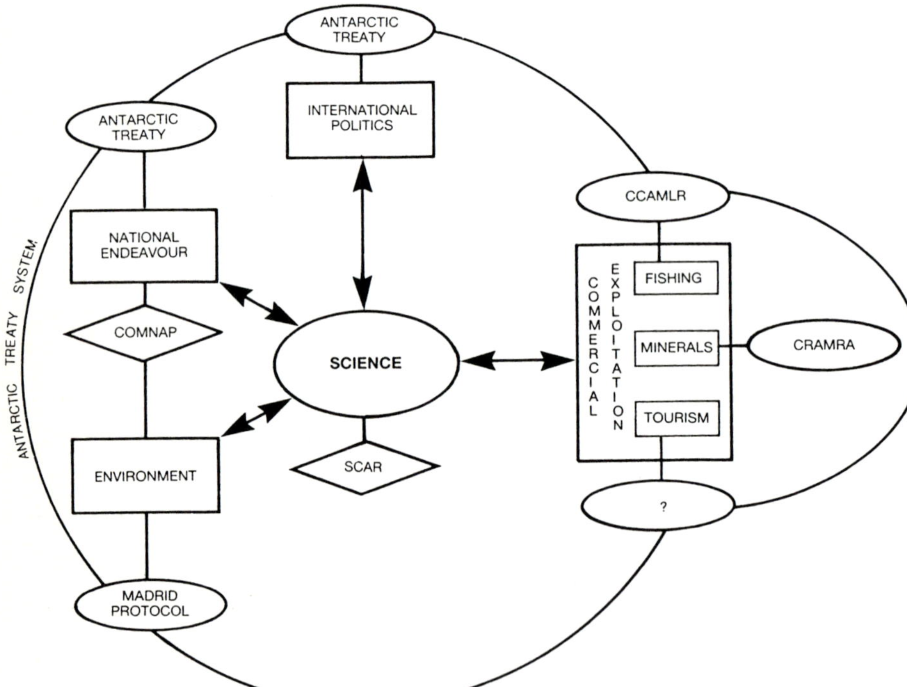

Fig. 1. Interactions between Science activity (*centre*) and various other uses of Antarctic (*boxes*) related elements of the Antarctic Treaty system (*ellipses*), and international coordinating bodies (*diamonds*) (see text for discussion). NB. CRAMRA was never ratified and lies, therefore, outside the principal circle of the Antarctic Treaty System

few who have not welcomed this development. Nevertheless, there have already been conflicts with science and problems may be encountered in the future. These stem from the perception that geo-scientific research may somehow constitute a Trojan horse for mineral prospecting or economic exploration. This linkage is a myth substantiated only in the minds of the ignorant or suspicious. The aim of the earth sciences is to study processes and patterns in time and place in order to understand the evolution of the Antarctic lithosphere. Its rationale and modus operandi are quite different from those of prospecting (Willan et al. 1991). Today it is recognized that geoscience research contributes to global change studies and has high palaeoclimate relevance: the paradigm of uniformitarism: "the present is the key to the past" is now extended to "the past is the key to the future".

Links to mineral prospecting are specious and fostered principally by groups with a political agenda: to discredit geoscience is to gain short-term political kudos and the support of an uncritical public. Perspective and balance have been casualties in the skirmishes and guerilla warfare surrounding the slow retreat from the Convention for the Regulation of Antarctic Mineral Resource Activities (CRAMRA) (Laws 1991). This is well illustrated by the following statements:

"The *local* (my underline) pollution created as a normal result of mining operations would jeopardize *Antarctica's* (my underline) uniqueness and importance as a global laboratory" (Barnes 1991 p. 196).

". . . the generation of power and heat (at Antarctic stations) has resulted in the emission of gases, waste heat, and dust which can be borne on air currents and negatively impact terrestrial flora and fauna many miles away". (ibid. p. 207)

"Inland dust and pollution (from mineral extraction activities) on the ice's surface could cause a diminished reflection of sunlight that could lead to melting of the polar ice cap". (ibid. p. 210)

Some non-governmental environmental organizations have wished to paint geoscience, sometimes *all* activity, as the tip on an exploration iceberg. This has been damaging and has threatened the core of the freedoms of intellectual pursuit. It has precipitated a reluctance amongst scientists to divulge or share their results and data or to seek non-prejudicial sponsorship from commercial companies – an essential element to the funding of science projects at a time when in many countries access to financial resources is becoming more difficult. Prior to the Madrid Protocol only British Petroleum had espoused the forgoing of all commercial interests in the Antarctic (Horton 1990).

3.2 Renewable Resource Exploitation

Renewable resource exploitation is confined to marine species (Fig. 2). Recognizing the need, albeit at an advanced stage of depletion of certain stocks (whales and fin fish), to provide regulatory mechanisms the Antarctic

Fig. 2. Japanese squid fishery vessel off South Georgia. On average, over the last 7–10 years, the catch of fin fish has been of the order of 130 k tonnes per year and that of krill 300 k tonnes per year. (Photo: courtesy BAS)

Treaty negotiated the Convention (Commission) for the Conservation of Antarctic Marine Living Resources (CCAMLR) in 1982. This instrument was novel in that it took the ecosystem approach to living resource management, with consideration of dependent as well as target species. This has, however, led to unwelcome developments for the scientists who have promoted the efficacy of this view. With the establishment of the Scientific and Technical Committee of CCAMLR scientists are now required to provide the necessary continuing expert input to these meetings – basic research, detailed and extensive analysis of data, synthesis, modelling and prediction of stocks and impacts (see Bassen and Beddington 1991). These tasks can be considered a heavy burden on the resources of research institutes and on individual scientists. It is legitimate to enquire whether the demands of CCAMLR are warping or creating unwelcome and unnecessary direction of scientific research in the Southern Ocean. Is basic scientific research productivity as good as it could be without the existence of CCAMLR? Is there conflict between the aims of CCAMLR and basic science in terms of choice, priority, flexibility and demand on limited resources for science?

An issue related to renewable resource exploitation which creates a dilemma for life scientists is the balance between the need for exploitation and the needs of conservation. It is my view that the Southern Ocean will continue to be the focus of interest for a protein-hungry world: if not now, within two or three decades. World population will increase by the middle of the next century by some 4 billion people, and will have doubled to 10 billion from present numbers certainly by the end of the twenty-first century. How will all these mouths be fed? Can advances in agriculture – genetic manipulation of crops to give enhanced yields and improve their tolerance to environmental stress, the opening up of new areas for crop production, the synthesis of new materials and substances for human consumption – achieve the necessary foodstuff levels?

At present there is a retreat by Russian fishing in the Southern Ocean due to internal political and organizational reasons. Problems also continue with the processing and acceptable presentation of krill, the overall economics of the fishery and with the development of satisfactory regulatory measures (Nicol 1991). Nevertheless, for the first decades of the next century I predict the inevitability of major harvesting, and conflict between establishing effective catch limits to sustain a conservation policy for the species of the Southern Ocean and harvesting for humankind. Scientists will be caught directly in the central and dangerous ground of this debate. Will there be sufficient understanding of the Southern Ocean ecosystem to model separately the effects of this impact and provide sound management advice?

3.3 Tourism

Tourism constitutes the final strand in the thread of commercial exploitation of Antarctica and has been a fast-growing industry in the last decade. Some 6300 tourists visited the Antarctic in 1992–93, the overwhelming majority by commercial cruise liners to the Antarctic Peninsula and a few to the Ross Sea (Polar Research Board 1993). In 1989–90 there were 21 tour-ship visits to Antarctica; in the 1991–92 season the number had risen to 49 (Enzenbacher 1991, 1993). Airborne tourism is buoyant but constitutes only some 2–3% of the annual total (ibid.) (Fig. 3).

There are clear conflicts already emerging with science. Tourists are drawn, magnetically, to Antarctic bases and research stations where their visits can be disruptive to science programmes. They are difficult to control and can have an expensive and unacceptable impact. Thin-hulled ships present obvious dangers (Anonymous 1989; Penhale 1989). Uncontrolled and inappropriate aircraft movements can give rise to potential disasters and there have been several air tourism fatalities in recent years. It is not clear who provides the search and rescue for lost, stranded or injured tourists. This burden falls, at present, upon national operators (and thus national budgets) who will not refuse assistance on humanitarian grounds. Yet such actions put at risk unnecessarily the limited and valuable resources already

Fig. 3. DC-4 aircraft of Antarctic Airways part of Adventure Network International at the blue ice runway, Patriot Hills, in support of tourist visits to the continent. [Courtesy C. W. M. Swithinbank, Polar Record 24(151):315 (1988)]

committed to science – aircraft, ships and over-snow vehicles, as well as the time of logistics and science staff. More importantly, it exposes innocent people to danger and risk. The formation in 1991 of an industry-internal coordinating body: International Association of Antarctic Tour Operators (IAATO) (Stonehouse 1992) may improve both the safety of tourist activities and environmental conduct; but affiliation is not yet mandatory. The British Antarctic Survey has stated that in agreeing quotas for tourist vessels to visit certain of its stations it will only negotiate with IAATO members.

Tourists place an additional burden and pressure on certain sites and locations in Antarctica – accessible colonies of birds and mammals come under unnecessary pressure. There may be trampling of mosses and lichens, and the taking of biological and geological souvenirs. Research sites may be unwittingly compromised (SCAR 1984). Although the Antarctic Treaty Consultative Parties failed to reach agreement on detailed or special provisions for the regulation of tourism at their meeting in 1992, framework arrangements are in place (Nicholson 1991) and a number of ATCM Recommendations already focus upon the control of tourist activities. Nevertheless, with a continuing increase in Antarctic visitor numbers and concern being expressed by national operators, the issue of tourism is likely to remain an area of active debate for some time.

4 Science and National Endeavour

Twenty-six countries work in Antarctica, each with research programmes – at 49 research stations or by ships and aircraft. The motives underlying involvement in the Antarctic which give rise to a national presence are complex. Sovereignty issues have exercised, to varying degrees, the minds of the seven claimant states (Beck 1986). In other respects, presence may be based on the seeking of national prestige, to compete with neighbouring or related states, to ensure there is no loss of political or commercial opportunity of even as a digression from difficult economic problems back home (Drewry 1986). Rarely, however, has the reason been purely scientific. Ironically, the financial bill for science, which is seen as the ticket to Antarctica, is underwritten by a web of other interests, and conflict may, therefore, arise over divergent priorities.

The above points should not be overstated. More importantly, there are interesting developments to national presence in Antarctica which have arisen as the numbers and diversity of countries involved in Antarctic work have expanded over the last few years. Some of these overlays, polarizations and developing axes are:

- Claimant – Non-Claimant states
- Eastern – Western block
- A Latin American axis
- A growing Asian presence
- Developed – Developing nations
- Traditionally active countries – New entrants
- An emerging European grouping

This diversity is richly rewarding yet has the potential to create tension and there is a need in scientific debate to be alert to cultural background. Traditionally, Antarctic science has been drawn from Euro-American traditions, embodying certain value judgements and national aspirations which affect the approach to research. Just because two people undertaking fieldwork in Antarctica wear typical and similar polar clothing, and perhaps prepare, in related fashion, their scientific results for publication in international journals, it should not be assumed that they hold the same values on the worth and nature of Antarctic endeavour. There is a great danger, stemming from the pervasiveness of the western intellectual tradition, that there can be an assessment of progress and contribution on the basis of the degree to which those traits and traditions are followed. It is salutary to recall the diverse elements that different cultures offer:

- The role of tradition (walking in the wake of one's ancestors)
- A range of different spiritual values (some focus on the past and present; others on the value of the future)
- Arbitrariness is manifest in caste traditions

- There are many differences in the value placed on the individual
- Concepts of honour and saving face can be essential ingredients to many cultures
- Ethical considerations of varying shades are frequently important in decision making.

Such issues give rise to latent problems for the continued trajectory of what is frequently and unequivocally assumed to be a homogenous, holistic entity: science in Antarctica.

A further element which has arisen in recent years is the "minimalist" approach (Drewry 1988). This purports to seek to obtain benefit from a presence in the Antarctic accruing from the minimum commitment – governed not by *what* science is of value to be undertaken, but *how* to establish a bridgehead in Antarctica for political purposes. The deployment of this minimalist approach has, of course, been exacerbated by the limited resources available to some nations. The net result, however, has been the unacceptable concentration of stations and duplication of scientific effort both geographically and by discipline. Approximately 20% of the Antarctic research stations are concentrated on one small area of Antarctica less than 1/100th of 1% of the continent (Harris 1991). This has created serious problems for the credibility of the Antarctic community when viewed from outside and is an issue still requiring attention (Nature 1990).

5 Science and Environmentalism

There is a need to reduce the perceived differences between science and environmentalism. Almost all scientists working in the Antarctic are in some way "environmental scientists". The real concern is the manner in which environmental ideas may have political context, or may be exploited for political gain.

The range of environmental issues in the Antarctic falls into two categories, external and internal. In the global context it is transboundary pollution from outside the region which is to be judged the most critical – transportation by atmospheric circulation of, for instance, heavy metals from industrial processes in other continents, pesticides, PCBs, and CFCs. These are found in ice and snow, at low concentrations, as well as in the tissues of birds and seals. Recent studies have shown that the magnitude and extent of the impact of research stations in Antarctica, however, is actually very small (Wolff 1990) and reducing as new measures to protect the environment are steadily introduced.

Environmental lobby groups have been prominent in seeking protection of the Antarctic environment, reduction in human input in the region (including that of science) and the designation of Antarctica as a "World Park"

(May 1988). The status and range of scientific activity within the "World Park", as proposed by environmental groups, have remained confused and ambiguous. Some statements consider the Antarctic should remain a zone of "limited scientific activity" (Greenpeace International 1988). This concept has been explained by Rigg (1991): "An Antarctic World Park would be a conservation zone where wilderness values are paramount; . . . and a region recognized for its role as a place to monitor global environmental pollution and atmospheric degradation. All human activities undertaken south of sixty degrees South should be judged against these values" (p. 69). It is germane to ask how much basic science would be subject to exclusion under such a definition. On the other hand, Mayer (1991) has made it clear that "Greenpeace has never felt that World Park designation would conflict with scientific research if it is sensitively executed with respect for the Antarctic continent". The Antarctic and Southern Ocean Coalition (ASOC) have listed a series of points for the comprehensive environmental protection of an Antarctic World Park which include science and defines this as the primary reason for human activity on the continent, but should be limited to that which is benign, unique to Antarctica, and/or globally significant (ASOC 1989; Rigg 1991).

On balance, it does appear that Greenpeace and other environmental pressure groups have come some way to accepting the value of science in Antarctica (Phillips 1992). Nevertheless, there remains a clear conflict which threatens to reduce intellectual advance between the full range of potential science to be admitted in the World Park and its substitution by "baseline" and "monitoring" studies only. Some extreme views are still held:

> "Even the conversion of Antarctica to a World Park that allows scientists to study various problems would have to be viewed as too lenient . . . SCAR has set aside certain areas of special scientific interest that are closed to anyone except to those obtaining special permits to enter and work there. An extension of this procedure to all Antarctica would presently be the optimal solution" (Hirsch 1991 p. 33).

Some may consider that the real objective of the environmental lobby is to gain political leverage and maintain the vitality of corporate funding. More insidious is a desire to acquire power and a pivotal position in decision making on what may or may not be permitted in the Antarctic World Park. Environmental groups have provoked direct conflict with Antarctic science inasmuch as their propaganda has alienated the public of many countries to Antarctic scientists by casting them in the role of uncaring polluters and despoilers of a pristine environment. This has been an unfortunate, unreasonable, and unnecessary policy: the support of many scientists would have been received by a more moderate and less confrontational approach.

Notwithstanding these very real challenges to scientific freedoms there is a range of measures now enshrined in the Protocol on Environmental Protection to the Antarctic Treaty and Recommendations of the XIV and XV ATCMs which place a high burden on national operators in respect of environmental impact assessment, monitoring, direct protection from pollution, waste management and conservation (Fig. 4). Many, if not most, of the detailed measures are welcome, but scientists have to recognize that the resources of the national operators and funding agencies are finite and money for environmental activities is competing already for funds for science.

6 Science and International Politics

The Antarctic Treaty has set aside Antarctica for peaceful and scientific purposes. These provisions are embodied in the original 1961 Treaty and have been reaffirmed in the 1991 Madrid Protocol. It is stated firmly in this Protocol that the region shall constitute "the natural reserve devoted to peace and science". There is acknowledgement in the preamble to the Protocol of "... the unique opportunities Antarctica offers for scientific monitoring and research on processes of global as well as regional importance". Furthermore, Article 2 states "Human activities shall be planned and conducted in Antarctica so as to accord priority to scientific research and to preserve the value of Antarctica as an area for the conduct of such research, including research essential to understanding the global environment". These statements are to be applauded; science and its value are acknowledged, if not necessarily understood.

It is highly relevant that the ATCM actively seeks scientific advice. The Scientific Committee on Antarctic Research (SCAR) considers this as welcome but recognizes that it creates a new set of problems. The ATCM does not provide additional funds to underpin the gathering and provision of advice and information that it requests, thus placing a considerable burden upon the limited resources of such organizations as SCAR (SCAR 1991). Furthermore, the ATCM does not identify SCAR as the only source of scientific comment and judgement. In the past the Treaty turned to SCAR almost de facto. Over the last decade, however, there has been the establishment, inter alia, of alternative channels including the creation of "special committees" of the Treaty such as those set up for the evaluation of environmental impacts of mineral exploration, the Scientific and Technical Committee of CCAMLR and the proposed Committee on Environmental Protection, albeit with SCAR as an observer. Other organizations have also been engaged for scientific comment on Antarctic matters such as the International Union for the Conservation of Nature (IUCN) and the World Meteorological Organization (WMO).

Thus, the Treaty has not failed science by not seeking comment or advice. Mechanisms and pathways have been established which may be quite satisfactory for science in general, but are not necessarily satisfactory for SCAR in particular. Inasmuch as SCAR was set up by the International Council of Scientific Unions in 1958 to coordinate Antarctic science and

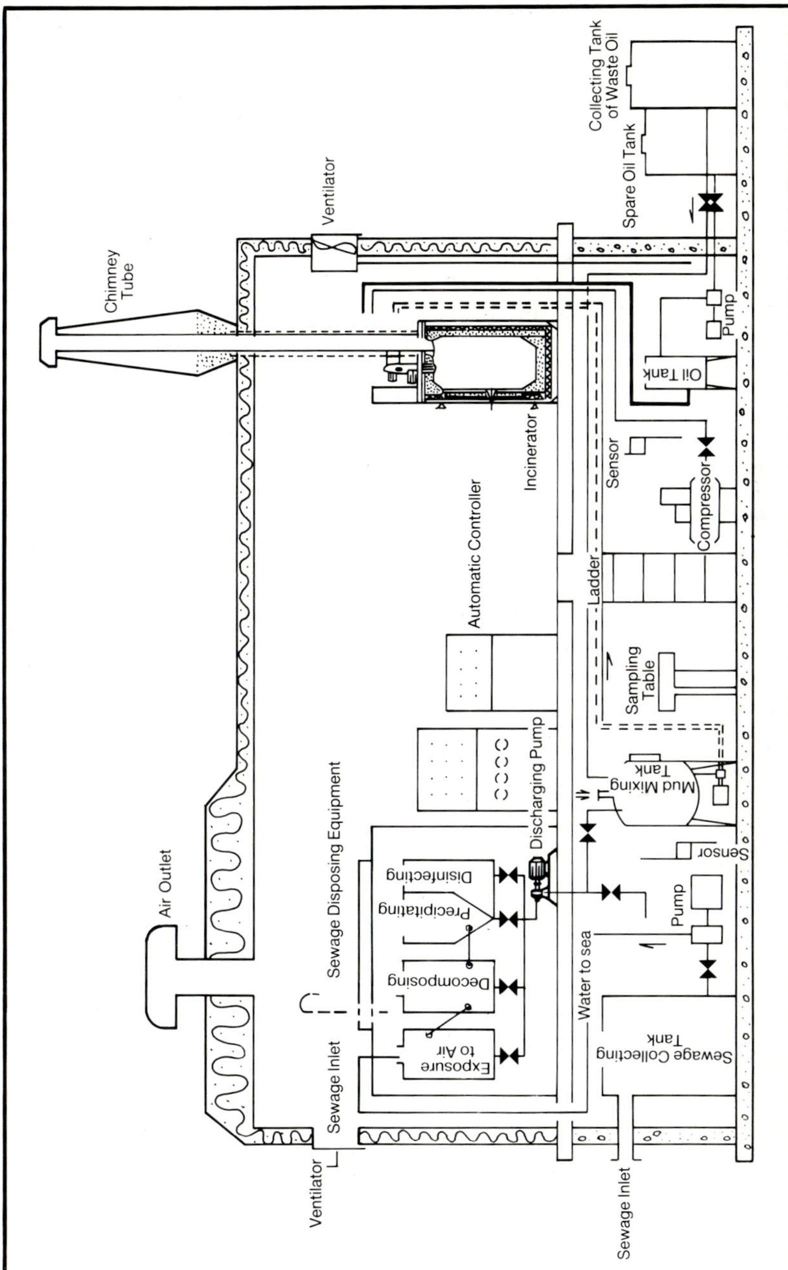

Fig. 4. Diagram of the sewage treatment plant installed at the Great Wall station of China. Using a biochemical method the plant has a capacity of 26 tonnes per day. (Courtesy Chinese National Antarctic Research Expedition)

scientists and is the pre-eminent NGO scientific body, such action by the ATCM may appear a snub. From the diplomatic angle the wide cast of the net may seem entirely reasonable and fair: the ATCM has simply cut the cake of scientific advice a different way which has led inevitably to unfortunate, and possibly avoidable, tensions with SCAR.

An internal momentum in ATCM affairs has become discernible which has, in large measure, allowed, ineluctably, the ATCM to become increasingly divorced from an understanding of the very region for which it was established to provide stewardship. This lack of understanding is manifest in what I term the "Four I's":

- *I*gnorance of Antarctica's extreme environment and the physical and psychological challenge to the individual of the harshest and most unforgiving climate of the planet;
- *I*ndifference to the values and culture developed over a period of 150 years amongst those who voyage to and undertake their enterprises in Antarctica;
- *I*nattention to the heritage of Antarctica and the lack of historical perspective;
- *I*ncomprehension of the questions, methods, rules, judgements and value of science.

These shortcomings have resulted in those qualities which make Antarctica so special being largely lost in the political arena. It is salutary to ask how many of the officials of the many nations involved in Antarctic diplomatic affairs have been to Antarctica, even briefly and how many are able, confidently to represent the interests of science in ATCM and other fora; how many can understand and defend those indefinable qualities so highly valued by those whose endeavour takes them to Antarctica? There is a burden on national operators and even individual scientists to encourage and improve the understanding and direct experience of diplomats and politicians of high southern latitudes, and prevent them from acquiring the Cyclops' vision of Antarctica.

Indifference to the consequences of ATCM actions on the special Antarctic dimension can be illustrated readily. Embedded within the Annex on the Conservation of Flora and Fauna of the Protocol on Environmental Protection is a measure which requires the removal of sledge dogs (Fig. 5). In 1991 there were less than 100 dogs remaining in Antarctica, extremely localized to two or three stations (in the Antarctic Peninsula and in East Antarctica). The exploration of Antarctica, the very essence of the opening up of the continent is a dedication to the role and value, not least in terms of human companionship, that sledge dogs have provided. They are the quintessential heritage of Antarctica. Elsewhere the destruction of heritage such as the razing of buildings, the loss of amenities and the elimination of traditional customs is considered abhorrent. This example underscores the lack of empathy with Antarctica and its spirit that can arise from the separation of the political and executive functions.

Fig. 5. A lost Antarctic heritage? A sledge dog team running on Adelaide Island, Antarctica. Dogs must be removed by 1 April 1994. (Photo: courtesy BAS)

7 Science

There are questions which relate to conflicts within science itself. Science is subject to change and evolution which present new challenges for the style, volume and significance of research activity (Drewry 1993). The programmes at present being developed by SCAR have latent conflicts. The concentration of effort on promoting an Antarctic component of the International Geosphere-Biosphere Programme (IGBP) (SCAR 1989, 1993), and termed GLOCHANT (Global Change in Antarctica), raises the question of the balance of priority and support for important regional or systematic studies and the global context.

Certain areas of science are not included in the theme of the IGBP. These include elements of geoscience not focused on palaeoenvironmental studies (such as areas of geophysics, palaeontology, petrology and geochemistry) and the study of the upper atmosphere and space from Antarctica including astronomy (Fig. 6). These scientific areas are nevertheless important and demand respect. They are disciplines (e.g. astronomy) frequently dominated by Nobel Prizewinners and command formidable budgets in their own right. The US budget for Astronomy in 1990, for instance, was 5.4 times greater than the funding for the entire US Antarctic Program (Schneider

Fig. 6. Part of the 16-antennae array at Halley V station, of the Polar Anglo-American Conjugate Experiment (PACE) for studies of plasma irregularities in the ionosphere. (Photo: courtesy BAS)

1990; US National Academy of Science 1991). It will be essential to maintain a balance and perspective in the forward strategy of SCAR if first and second leagues of science are not to be established indiscriminately, and which will be used by bureaucrats to guide funding.

The nature of the development of Antarctic science has shown progressive deepening of an understanding of Antarctic systems from the early geographical exploration of the region through the phase of reconnaissance studies, (essentially around the IGY and later) to the problem-orientated approach today. There has been an exponential increase in the number of programmes being conducted in Antarctica from an enlarged pool of nations. These points lead inevitably to questions such as "is more better?" Conflict can arise between quantity and quality of output. The solution to the question of maintaining standards is not easy and depends upon defined objectives for research, but I would have to reaffirm the need for scientific output in the context of publication in the fully reviewed international science literature (Drewry 1989). As comprising at least the kernel of most science policy there is a second question: "is the best relevant?" This invites scientists and programme managers and indeed even SCAR Working Groups and Groups of Specialists to consider carefully priorities and related factors in the development of projects. Scientists should be seeking relevance, but in which context? It must surely be the role of SCAR to establish an international scientific agenda of agreed areas for framing and developing national research. Points which need to be taken into account in both of these contexts would include:

- Excellence
  - consolidation of outstanding current programmes;
- Timeliness and relevance
  - to issues of a regional, global or fundamental nature;
- Strategic science elements
  - renewable resources-related research (to support CCAMLR);
- Environmental management and conservation (to support the Protocol on Environmental Protection);
- Minor criteria
  - balance between field and laboratory studies
  - training and career development of scientists.

Priority also relates to cost. There are only limited resources for science in Antarctica now, and this situation is unlikely to change in the future. Furthermore, a very high proportion of budgets is spent on logistics in Antarctica (Fig. 7). Neither individual nations nor international bodies (such as SCAR) can afford to protect low priority science or science of questionable quality. This issue is especially important in areas where there may be

Fig. 7. The advanced research-logistics vessel *Polarstern* of the FRG which cost approx. DM 200 million. (Photo: courtesy BAS)

collaboration between nations and it will become an increasingly relevant theme in the future as the need to cooperate to maximize limited funds becomes imperative.

8 Conclusions

1. There is an inexorable evolution of the Antarctic system involving science and politics to meet change. It is likely that this will result in a higher profile for diplomatic and legal processes. Scientists must work hard to establish and maintain their central scientific role and credibility on the diplomatic agenda. The persistent demonstration of the pursuit of science of high relevance may well be considered a lien on the continued freedoms of prosecution and more especially the adequate funding of science in Antarctica by governments.
2. Consensus will become more difficult as the number of nations with varied backgrounds are involved and this will, without doubt, result in more dilute, less effective decisions by the ATCM. The marginalizing of science and the increase in entropy in the Antarctic Treaty system may be inescapable conclusions.
3. There will be a lessening of Antarctic values. There will be loss of heritage, weakening of the positive elements of tradition. The "spirit of Antarctica" will be threatened by faceless bureaucrats and diplomats unless the active Antarctic community engage national political representatives and embue them also with the qualities and value of the region.
4. Science will share an uneasy future, but has great, indeed outstanding challenges and opportunities in the fields of global environmental research and many areas of basic science where it is recognized that this least spoiled region of our planet has a unique and virtually limitless contribution.
5. There will be pressure, inevitably, on resources for science. There is likely to be more directed research as priorities are set to maximize the funds available. There will undoubtedly need to be more coordination and sharing, and SCAR and COMNAP will be central in this matter.

This paper commenced with the issue of conflict. It is quite clear that it should conclude by stressing accord. The SCAR Antarctic Science Conference in Bremen in September 1991 attested to the foremost and crucial contributions of Antarctica in a global context. Furthermore, through frequent personal interchange, sustained development of collaborative initiatives and the shared experiences in this most challenging of regions Antarctic scientists are outstanding ambassadors for peace. There are few places in this small planet where humankind has achieved so high a spirit of cooperation and so deep a mutual understanding of each other.

References

Anonymous (1989) Argentine ship sinks near Palmer Station. Antarct J US 14(2):3–9

ASOC (1989) Permanent protection for Antarctica: a conservation convention is urgently needed. ASOC information paper No 2, PREP XV Antarctic Treaty Consultative Meeting. Inf 2, 13 pp

Barnes JN (1991) Protection of the Environment in Antarctica: are present regimes enough? In: Jorgensen-Dahl A, Ostreng W (eds) The Antarctic Treaty System in world politics. Macmillan, London, pp 186–228

Basson M, Beddington JR (1991) CCAMLR: the practical implications of an eco-system approach. In: Jorgensen-Dahl A, Ostreng W (eds) The Antarctic Treaty System in world politics. Macmillan, London, pp 54–69

Beck PJ (1986) The international politics of Antarctica. Croom Helm, London, 332 pp

Beeby CD (1991) The Antarctic Treaty System: goals, performance and impact. In: Jorgensen-Dahl A, Ostreng W (eds) The Antarctic Treaty System in world politics. Macmillan, London, pp 4–21

Drewry DJ (1986) International scientific cooperation in Antarctica. In: Millar TB (ed) Australia, Britain and Antarctica. Australian Studies Centre, Institute of Commonwealth Studies, University of London, London, pp 25–36

Drewry DJ (1988) The challenge of Antarctic science. Oceanus 31(2):5–10

Drewry DJ (1989) Science in Antarctica – a matter of quality. Antarct Sci 1(1):2

Drewry DJ (1993) The future of Antarctic scientific research. Polar Rec 29(168):37–44

Enzenbacher DJ (1991) Tourists in Antarctica: numbers and trends. Polar Rec 28(164): 17–22

Enzenbacher DJ (1993) Antarctic tourism: 1991/92 season activity. Polar Rec 29(170):240–242

Greenpeace International (1988) Greenpeace Antarctic expedition: background information 1987–1988, Lewes, Stichting Greenpeace Council

Harris CM (1991) Environmental effects of human activities on King George Island, South Shetland Islands, Antarctica. Polar Rec 27(162):193–204

Hirsch P (1991) Microbial life in Antarctic rocks and soils: a case for World Park type protection of a rare extreme environment. In: Patel J, Mayer SJ (eds) Antarctica: the scientists case for a World Park. Greenpeace International, Amsterdam, pp 29–33

Horton R (1990) Letter to the editor. The Independent, London, 23 November

Laws RM (1991) Unacceptable threats to Antarctic science. New Sci 129(1762):4

May J (ed) (1988) The Greenpeace book of Antarctica. Dorling Kindersley, London, pp 158–159

Mayer SJ (1991) Introduction. In: Patel J, Mayer SJ (eds) Antarctica: the scientists case for a World Park. Greenpeace International, Amsterdam, p 3

Nature (1990) Antarctic wilderness. 348:268–269

Nicholson IE (1991) Antarctic tourism – the need for a legal regime? In: Jorgensen-Dahl A, Ostreng W (eds) The Antarctic Treaty System in world politics. Macmillan, London, pp 415–427

Nicol S (1991) CCAMLR and its approach to management of the krill fishery. Polar Rec 27(162):229–236

Penhale P (1989) Research team focuses on environmental impact of oil spill. Antarct J US 14(2):9–12

Phillips C (1992) The management of science in the newly protected Antarctica – the environmental angle. Circumpolar J 7:139–146

Polar Research Board (1993) Science and stewardship in the Antarctic. National Academy Press, Washington, DC, 107 pp

Rigg K (1991) Environmentalists' perspectives on the protection of Antarctica. In: Cook G (ed) The future of Antarctica: exploitation versus preservation. Manchester University Press, Manchester, pp 68–80

SCAR (1984) A visitor's introduction to the Antarctic and its environment. British Antarctic Survey, Cambridge, 36 pp

SCAR (1989) The Role of Antarctica in global change. Scientific priorities for the International Geosphere-Biosphere Programme (IGBP). ICSU Press/SCAR, Cambridge, 28 pp

SCAR (1991) SCAR perspectives on Antarctic research and management. SCAR Inf Pap, XVI Antarctic Treaty Consultative Meeting, Bonn, October 1991

Schneider C (1990) Funding research in Antarctica: a look at NSF support to scientists. Antarct J US 15(2):9–11

Stonehouse B (1992) IAATO: an association of Antarctic tour operators. Polar Rec 28(167):322–324

US National Academy of Sciences (1991) The decade of discovery in astronomy and astrophysics. US National Academy of Sciences, Washington, DC

Vicuna FO (1986) Antarctic conflict and international cooperation. In: Antarctic Treaty System: an assessment. National Academy Press, Washington, DC, pp 55–64

Willan R, Macdonald D, Drewry D (1991) The mineral resource potential of Antarctica: geological realities. In: Cook G (ed) The future of Antarctica: exploitation versus preservation. Manchester University Press, Manchester, pp 25–43

Wolff EW (1990) Signals of atmospheric pollution in polar snow and ice. Antarct Sci 2(3):189–205

Antarctica – Where Space Meets Planet Earth

Michael J. Rycroft[1]

1 Introduction

A significant and ever increasing proportion of the human population of planet Earth is becoming informed – and concerned – about changes to the Earth's environment; the atmosphere is rightly the focus of particular interest nowadays. Such changes may be either natural, such as due to volcanoes (both El Chichon in 1982 and Mount Pinatubo in 1991 put vast quantities of dust and sulphur into the lower stratosphere), or anthropogenic. Enhancement of the so-called greenhouse effect is caused by increasing the concentrations of infrared active trace gases in the atmosphere. This may be due to increased evaporation from the oceans (water vapour, H_2O), deforestation or the burning of fossil fuels (carbon dioxide, CO_2), increases of the cattle population or the area of paddy fields (methane, CH_4), increases of chlorofluorocarbons (CFCs), or changed agricultural practices involving fertilizers (nitrous oxide, N_2O).

For studies of such phenomena and of the physical, chemical and biological processes underlying them, coordinated observations are required. The World Climate Research Programme (WCRP), the International Geosphere- Biosphere Programme (IGBP), the International Decade for Natural Disaster Reduction (IDNDR) and the Solar-Terrestrial Energy Programme (STEP) are just four examples of the many scientific research programmes that are agreed through the International Council of Scientific Unions (ICSU) to investigate different, yet related, aspects of the Earth's environment. The observational parts of these programmes are complemented by theoretical investigations and by numerical simulations using computers ranging, as appropriate, from personal computers (PCs) to supercomputers.

Observations made using diverse sets of sophisticated instruments aboard satellites orbiting the Earth have to be complemented by observations made from the ground to obtain a deeper understanding of the phenomenon being investigated. Examples are conventional surface meteorological obser-

[1] College of Aeronautics, Cranfield Institute of Technology, Cranfield, Bedford MK43 0AL, UK

Hempel (Ed.) Antarctic Science

vations, observations made from balloon-borne sondes throughout the troposphere and into the stratosphere, or radar investigations of the properties of the ionosphere at heights ranging from 70 to over 300 km. The observations made from the hostile environment of Antarctica are crucial components of these global data sets.

The continent of Antarctica offers several unique features for such studies (see Weller et al. 1987). Examples here are:

1. the situation of a coastal station in summer (~0 °C) and an essentially inland station in winter (~−40 °C) that is separated from the southern ocean by up to 1000 km or more of sea ice which prevents moisture from entering the atmosphere locally;
2. the vast expanse of smooth, cold terrain at high altitude which leads to intense katabatic winds and to a very stable, non-turbulent boundary layer during the long winter night;
3. the clear, relatively unpolluted atmosphere that enables several days of continuous observations of the Sun to be made for helioseismological purposes from the South Pole in summer;
4. the existence in the lower stratosphere over Antarctica in winter of a stable, cold polar vortex within which tiny ice crystals form – in spring, the returning sunlight detaches chlorine atoms from these to destroy ozone catalytically and lead to the springtime Antarctic ozone depletion, the so-called ozone hole, and
5. a large land mass on which manned or unmanned observatories can be strategically placed to investigate the Earth's magnetic field and its extension into space – geomagnetic field lines emanate from Antarctica and go down the comet-like tail of the magnetosphere into space, or return to Earth in the Arctic and northern middle to high latitude regions.

In the limited space available here, attention is concentrated first upon 5 and, secondly, on 4.

2 Solar-Terrestrial Physics from Antarctica

Antarctic research is of major significance in solar-terrestrial physics today. This is because geomagnetic field lines from Antarctica thread all the interesting regions of the magnetosphere and the near-Earth space environment (see Rodger and Smith 1989). The supersonic solar wind plasma flowing towards and then past the geomagnetic field creates the magnetosphere, the three-dimensional, time varying plasma physics laboratory within which much action takes place (see Russell 1991; Lockwood and Coates 1992). When the interplanetary magnetic field has a southward component it connects to geomagnetic field lines on the dayside. Field lines are pulled by the solar wind through two indentations on the magnetopause each of which, north or south, is termed the cusp. The field lines are then carried downstream, as illustrated in Fig. 1, and their feet in the ionosphere are pulled across the polar caps (see Rycroft 1990a, 1991a). They can reconnect in the plasma sheet on the nightside; they then catapult the electrons and ions on them towards the Earth. Accelerated by the Fermi process, by wave-particle interactions and/or electrostatic fields, electrons of up to 10 keV are precipitated into the upper atmosphere. At altitudes ~110 km these electrons excite atmospheric oxygen atoms which emit their characteristic visible radiation at wavelengths of 557.7 nm (green) and, at ~240 km altitude, 630 nm (red). In this way the aurora australis (south) and aurora borealis (north) are formed in rings around the two magnetic poles. These auroral ovals (see Fig. 2) are well observed (in the ultraviolet part of the spectrum) by instruments aboard the Dynamics Explorer (DE-2) and Viking satellites. Electrons gyrating in tight helical paths around the geomagnetic field lines explain the rayed structure of the aurora seen from the Earth's surface. The upper atmosphere thus acts like the phosphor screen of a television set. Observations of the aurora lead to information on the electron

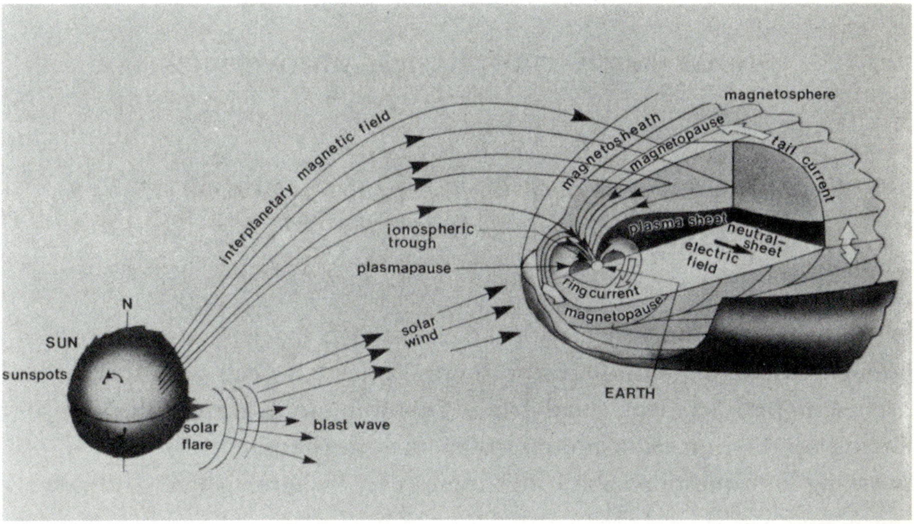

Fig. 1. Diagram (Shapland and Rycroft 1984) illustrating the Sun-Earth system, linked via interplanetary field lines. Disturbances in the Sun's photosphere associated with the reconfiguration of magnetic fields in regions near sunspots can create a blast wave which catches up with the quiescent solar wind ahead. If the interplanetary shock front so formed strikes the magnetosphere, this is compressed. The magnetosphere, a three-dimensional plasma physics laboratory, becomes more active, with enhanced electric fields and current systems. The polar regions are the windows on space through which such phenomena are studied, in ways that are well complemented by observations made from satellites orbiting the Earth

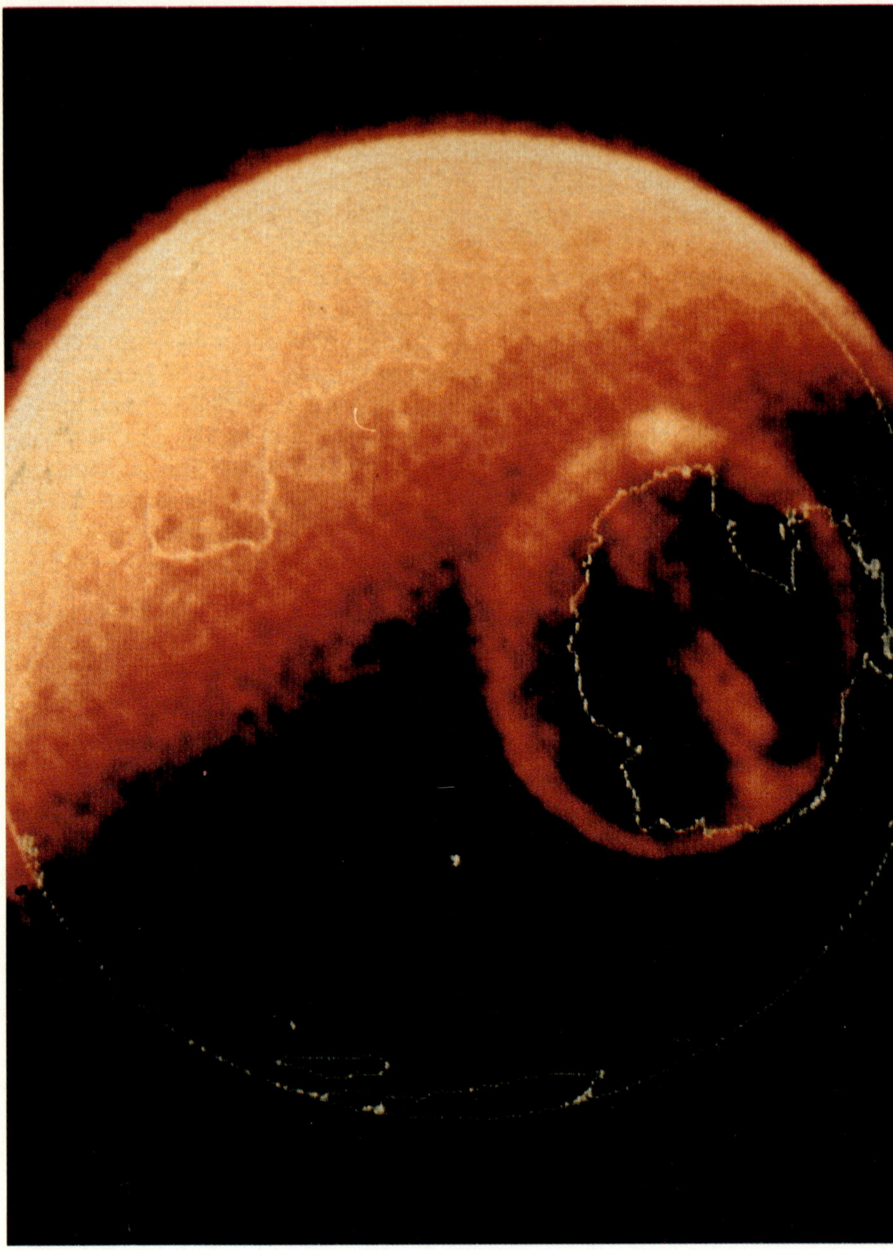

Fig. 2. False colour image of the Earth's atmosphere obtained in the ultraviolet (Frank and Craven, pers. comm. 1988) showing backscattered solar radiation on the dayside (*upper left*) and the auroral oral, with a transpolar arc, over Antarctica on the nightside. The Earth's circumference and the continents are created by computer graphics

gun at the back of the set, down the geomagnetic tail, and which transmits the latest magnetospheric programmes.

The studio in which the plasma physics being investigated takes place is vast – the Sun-Earth distance is 150 million km, and the dayside magnetopause is typically 70 thousand km from the centre of the Earth ($\sim 11\,R_E$, Earth radii). The range of spatial scales of interest is therefore enormous, from the dimensions of the magnetopause to an auroral electron's gyro radius (~ 7 m), that is a range of 10^7, 7 orders of magnitude. The time scales of interest range from less than a millisecond, the period of interesting plasma waves on auroral flux tubes, to a few seconds which is the time for a ~ 10 keV electron to bounce from hemisphere to hemisphere and cause geomagnetically conjugate phenomena, to diurnal and seasonal cycles, to the 27-day period of solar rotation and to long-term trends associated with the 11-year solar cycle. The temporal range of interest is thus a factor of 10^{12}, 12 orders of magnitude.

The number of sunspots changes over the 11-year cycle, and the sunspots themselves migrate towards the solar equator as each cycle develops. The brightness of the Sun increases in the visible part of the spectrum by $\sim 0.1\%$ from solar minimum to solar maximum. In the ultraviolet and X-ray regions, the Sun is markedly spottier, the solar output varying by a few percent over the solar cycle at a wavelength of 300 nm, and more at shorter wavelengths (see Lean 1987).

Associated with magnetic field reconfigurations in the vicinity of sunspots are high speed blast waves which propagate through the interplanetary medium. These catch up with the slower solar wind ahead, creating interplanetary shocks. If these encounter the Earth's magnetosphere they trigger geomagnetic storms, that is geomagnetic field disturbances due to enhanced electric currents flowing in the auroral ionosphere and in the ring current (see Fig. 1).

The polar ionosphere acts as an electrical load on the solar wind – magnetosphere dynamo which creates a potential difference ~ 100 kV. The electric field maps down geomagnetic flux tubes into the polar cap ionosphere; field-aligned electric currents flow (see Lin et al. 1990). These currents close as electrojets across field lines in the auroral ionosphere; their magnitudes are typically 10^6 A. Thus the power dissipated in the auroral ionosphere is $\sim 10^5$ MW, more than civilized mankind creates in power stations at the Earth's surface. This Joule dissipation heats the high latitude atmosphere at and above ~ 110 km altitude by ~ 200 °C. The dynamics of this region of the atmosphere, the thermosphere, are shown diagrammatically in Fig. 3.

Information is carried along geomagnetic flux tubes by charged particles or by waves. Of especial interest for Antarctic studies are whistler-mode waves, electromagnetic waves of audio frequency. These are guided by field-aligned ducts of enhanced plasma density which act like fibre optics light pipes. Following a geomagnetic storm, recent results (Smith and Clilverd

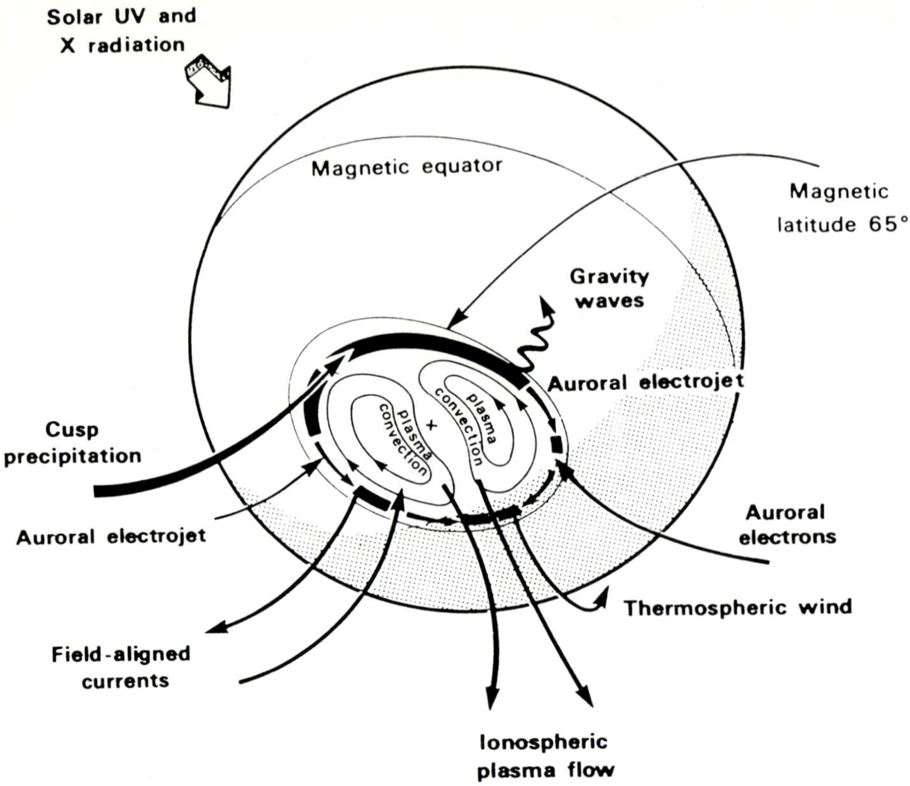

Fig. 3. Diagram from a similar perspective to that of Fig. 2 illustrating both charged particle and neutral gas processes at work in the Earth's upper atmosphere. Gravity waves carry energy from the auroral thermosphere, at ~110 km altitude, to middle and low latitudes. The *dotted region* indicates the dark nightside and the *cross* the magnetic pole near the centre of the auroral oval (where the auroral electrojet flows)

1991) have indicated that whistler-mode travel times are reduced. Plasma density changes can be inferred and the enhanced electric fields studied (Saxton and Smith 1991). Longitudinal variations have been investigated by Smith et al. (1991) recently.

A man-made transmitter of whistler-mode waves has operated at Siple station, Antarctica, for almost 20 years. This has led to many unexpected observational results (see Helliwell 1988) such as the triggering of rising frequency emissions, the characteristics of which have been modelled (see Omura et al. 1991). Balloons and rockets have been launched from Siple to investigate the X-ray bremsstrahlung radiation produced by electrons precipitating into the atmosphere from the van Allen radiation belts (see Sheldon et al. 1988).

Signals from the Siple transmitter have been recorded by Hurren et al. (1986) at Halley, South Pole and Palmer stations. Trimpi events indicative of the effects of energetic electron precipitation on the atmosphere at heights ~80 km have been detected in this way (see Rycroft 1991b).

Magnetohydrodynamic (MHD) waves with periods ~1 min have been studied at pairs of geomagnetically conjugate stations such as Halley and St. Anthony in Newfoundland (Green and Hamilton 1981), Siple and stations in Quebec, Canada (Cahill et al. 1984), and Syowa and stations in Iceland (Ishizu et al. 1981; Tonegawa et al. 1984). The field has been comprehensively reviewed by Arnoldy et al. (1988). Morrison (1990) has reported results derived from simultaneous observations of MHD waves and whistler-mode waves.

The polarization and modes of propagation of MHD waves have been studied so that these geomagnetic pulsations can be used as well as whistler-mode waves to diagnose the magnetospheric plasma distribution (see Paschal et al. 1990).

In the last few years, poleward-looking radars operating at several MHz have been installed at Halley and at Goose Bay, Labrador, Canada, as the Polar Anglo-American Conjugate Experiment (PACE). They have viewed the projections of the magnetospheric cusps in the south and north onto the high-latitude ionosphere.

They have, for the first time, detected changes of the ionospheric plasma flows in response to the changing orientation of the interplanetary magnetic field (IMF), particularly its dawn-to-dusk (B_y) component (see Greenwald et al. 1990; Dudeney et al. 1991).

At Vostok, the Soviet station on the Antarctic plateau, the IMF has been demonstrated to change the distribution of auroral light at 557.7 and 630 nm (see Troshichev et al. 1988). Mende et al. (1990) have noted magnetic impulses at South Pole station and its conjugate following 630 nm emissions on cusp field lines at higher latitudes.

The ionospheric absorption of cosmic radio noise near 30 MHz has been interestingly investigated in the Antarctic and elsewhere since the International Geophysical Year (IGY 1957–58). However, it is only recently that it became possible to devise a narrow beam, phased array antenna to examine the fine spatial structure, on a scale of ~10 km, of the absorption which is produced by the precipitating electrons. An instrument installed at South Pole station, close to the latitude of the cusp, is shedding new light on this phenomenon (see Detrick and Rosenberg 1990).

It is thus evident that Antarctic observations are playing a crucial part in elucidating the complex time and space variations of magnetospheric and ionospheric phenomena, and their coupling. They will continue to be most important during the STEP programme when new satellite missions, such as NASA's Global Geospace Science mission (see Rycroft et al. 1984) and ESA's SOHO and Cluster missions (see Domingo et al. 1988), are carried out.

3 Ozone over Antarctica

The discovery by Farman et al. (1985) of the springtime depletion of ozone over Antarctica was a bombshell. Whilst Molina and Rowland (1974) had predicted that chlorine atoms produced by the action of solar ultraviolet radiation on man-made chlorofluorocarbon (CFC) molecules that diffused up into the stratosphere would catalytically destroy ozone, no one expected the effect to be first noted over Antarctica. The careful observations of the total ozone content of the atmosphere made above Halley since the IGY demonstrate the great value of long-term observations.

Energy from the Sun drives the Earth's climate system, the result of complex interactions and feedbacks between radiation, chemistry and dynamics in the atmosphere (O'Neill 1990). Rycroft (1990a,b) has outlined not only the processes which determine the ozone concentration in the atmosphere but also the circulation of the atmosphere, with five convection cells (in latitude) in the troposphere, and one giant cell in the stratosphere. The latter has upward motions over the equator and the summer hemisphere, and a descending motion over the winter pole. Thus CFC molecules, wherever they are produced and used, leak into the atmosphere and are split up in the midlatitude or tropical stratosphere. The chlorine atoms so released participate in the stratospheric circulation, to cool and descend over the winter pole. They fill – from above – the isolated polar vortex, or cold cauldron, that develops over Antarctica in winter and attach themselves to tiny ice crystals (or Polar Stratospheric Clouds, PSCs) that form there at temperatures below −80 °C. In the spring, they are detached and proceed very effectively to destroy the ozone in situ. In the process, chlorine monoxide is formed, and much enhanced concentrations of this have been observed within the polar vortex (see Rycroft 1990b).

The spatial extent of the ozone hole, and also its temporal evolution, from day to day and month to month, are most clearly seen from the observations made from polar orbiting satellites. Figure 4 shows NASA results from the Nimbus 7 satellite taken on 10 October 1986. The scale of the phenomenon is undoubtedly continental. Herman et al. (1991) and Brasseur (1991) have commented on the 3% decrease in the globally averaged ozone content of the atmosphere from late 1978 to late 1989 that the Nimbus 7 data reveal.

The vertical profile of the ozone distribution is well investigated by sonde instruments carried aloft by hydrogen-filled balloons. Two examples, obtained from Halley station on 15 August and 13 October 1987, are shown from Shanklin and Gardiner (1989) and Rycroft (1990a,b) in Fig. 5. In the late winter the ozone profile is well behaved, with one molecule in a million in the lower stratosphere being an ozone molecule. However, 2 months later, most of the ozone at altitudes between 15 and 20 km altitude has been destroyed, in situ. Where there should be most ozone in the atmosphere, there is, in fact, least.

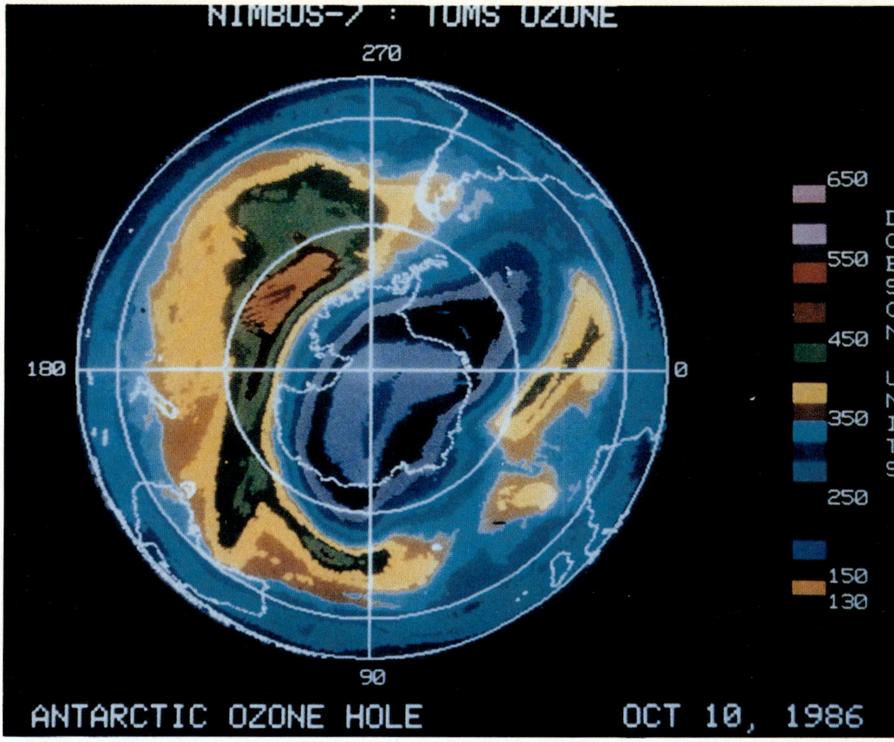

Fig. 4. Satellite observations (Total Ozone Mapping Spectrometer) of the ozone depletion over Antarctica observed on a particular day, 10 October 1986. Values less than 300 Dobson Units (corresponding to a thickness of the ozone layer, if it could be brought from the stratosphere to sea level, of 3 mm) are below the values of the 1960s

Public pressure, and some scientific arguments (see Rycroft 1990b,c), have led many governments to seek to ban the production of CFCs in 1995, a most commendable action. This will give the stratosphere the chance to heal itself, a process which will probably take ~100 years, the typical lifetime of a CFC molecule in the atmosphere.

Stephenson and Scourfield (1991) have suggested that ozone could also be destroyed by increased nitric oxide amounts due to energetic solar protons following some large solar flares observed in March 1989. Brasseur (1991) has reported that the effect should only be ~20%, at an altitude near 40 km, well above the ozone maximum. This disagreement requires further research.

With the reduction of atmospheric ozone content comes the enhancement of transmitted UV flux from the Sun. Schnell et al. (1991) and Thompson (1991) have discussed the destruction – by photodissociation – of tropospheric ozone over and around Antarctica which this causes. It is unclear what effects will ensue from that.

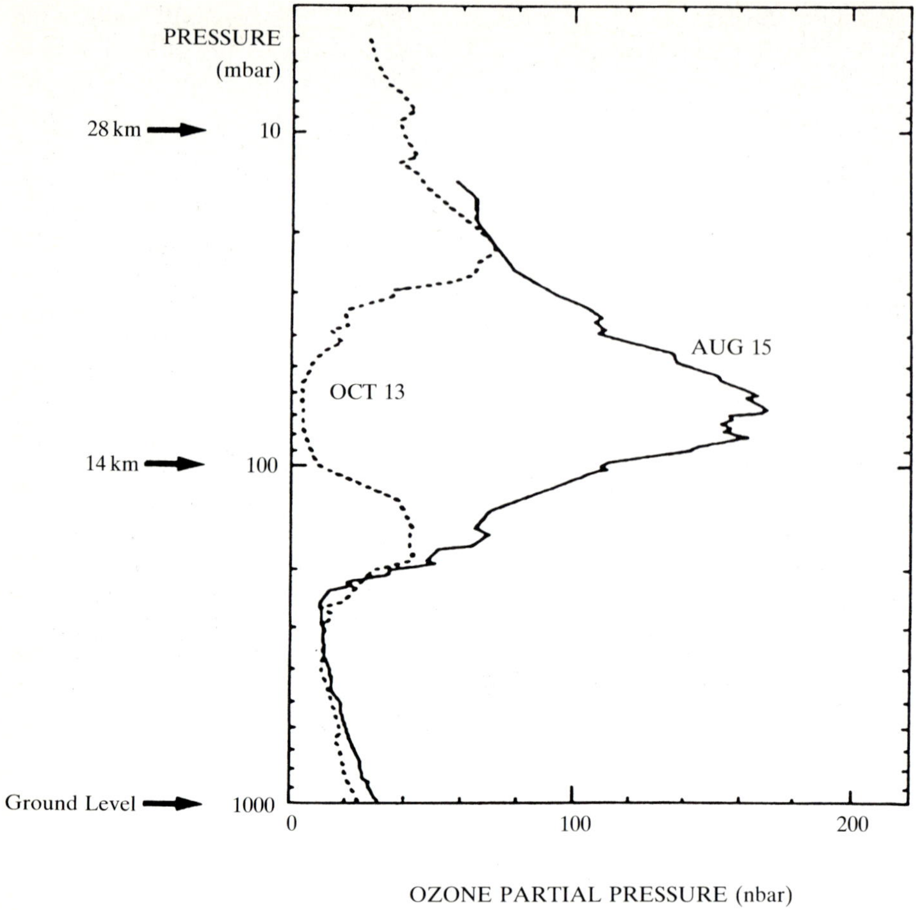

Fig. 5. Ozone profiles obtained over Antarctica on dates 2 months apart. Near 14 km altitude, where the atmospheric pressure is 100 mbar, the ozone partial pressure is 100 nbar on 15 August 1987. Here, therefore, one molecule in a million is an ozone molecule

Rycroft (1990b) and Crutzen (1992) have summarized some other effects that follow the loss of stratospheric ozone – enhanced UV radiation near 300 nm at the Earth's surface, increasing skin cancer rates, damaging DNA molecules in the cells of living organisms, and decreasing phytoplankton production rates are but four.

Smith et al. (1992) have compared phytoplankton productivity inside and outside the ozone hole, and concluded that it was reduced by ~10% below the ozone hole during the austral spring of 1990. Marchant et al. (1991) considered substances which absorb UV radiation that are produced in algae that proliferate in sea ice, the marginal ice zone and the uppermost 10 m of seawater after the break up of sea ice. Such substances will protect the organisms from the increased flux of UV radiation. This effect gives an interesting and important twist to the interactions between physical and biological phenomena in the Antarctic environment.

Figure 6 summarizes in diagrammatic form some of the changes that may occur in the Earth's atmosphere, and the driving mechanisms – both natural and man-made – that may force such changes. They involve many physical, chemical and biological processes at work in the atmosphere and many interactions between these processes. Ways of mitigating man-made environmental changes have been discussed by Cicerone et al. (1992).

One example is that the increasing greenhouse effect observed at the Earth's surface (global warming, ~0.07 °C per decade) is expected to be linked to stratospheric cooling. Stratospheric cooling is clearly observed above Antarctica (~3 °C per decade) due to the lower concentrations of ozone (which absorb solar ultraviolet radiation so warming the stratosphere). In this sense global warming and ozone depletion act together (see Kiehl 1992; Ramaswamy et al. 1992). Also, increasing concentrations of CFC molecules in the troposphere currently contribute significantly (~15%) to global warming, with CO_2 increases contributing ~50%.

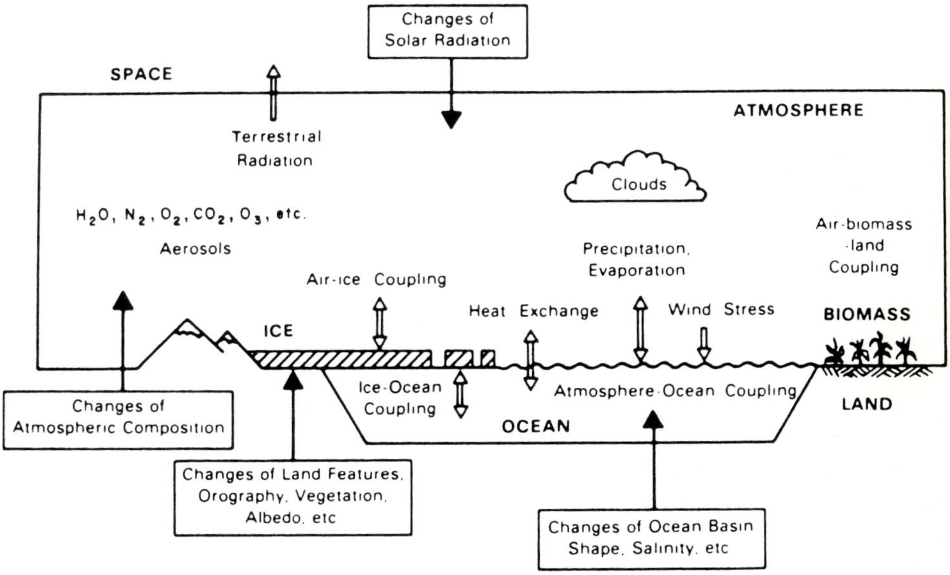

Fig. 6. Diagram illustrating processes which affect the climate, occurring at the interface between the Earth's solid or liquid surface and the gaseous atmosphere, and changes of other factors which could cause climate changes

4 Conclusions

This paper has given some examples of topical problems relating to Antarctic research where measurements made from satellites are well complemented by observations made from the Earth's surface. The field has been restricted to studies of the atmosphere and near-Earth space environment, concentrating upon the ozone hole and solar-terrestrial physics.

The several unique aspects of Antarctica for such research should be further capitalized upon, in future studies, for the benefit of mankind.

References

Arnoldy RL, Cahill LJ, Engebretson MJ, Lanzerotti LJ, Wolfe A (1988) Review of hydromagnetic wave studies in the Antarctic. Rev Geophys 26:181–207

Brasseur GP (1991) A deepening, broadening trend. Nature 352:668–669

Cahill LJ, Sugiura M, Lin NG, Arnoldy RL, Shawhan SD (1984) Observations of an oscillating magnetic shell at three locations. J Geophys Res 89:2735–2744

Cicerone RJ, Elliot S, Turco RP (1992) Global environmental engineering. Nature 356:472

Crutzen PJ (1992) Ultraviolet on the increase. Nature 356:104–105

Detrick DL, Rosenberg TJ (1990) A phased-array radio wave imager for studies of cosmic noise absorption. Radio Sci 25:325–338

Domingo V, Schmidt R, Poland AI, Goldstein ML (1988) Soho and Cluster – the scientific instruments. ESA Bull 56:25–33

Dudeney JR, Rodger AS, Pinnock M, Ruohoniemi JN, Baker KB, Greenwald RA (1991) Studies of conjugate plasma convections in the vicinity of the Harang discontinuity. J Atmos Terr Phys 53:249–263

Farman JC, Gardiner BG, Shanklin JD (1985) Large losses of total ozone in Antarctica reveal ClO_x/NO_x interaction. Nature 315:207–210

Green CA, Hamilton RA (1981) An ionospheric effect on the conjugate relationship of Pi2 magnetic pulsations. J Atmos Terr Phys 43:1133–1141

Greenwald RA, Baker KB, Ruohoniemi JM, Dudeney JR, Pinnock M, Mattin N, Leonard JM, Lepping RP (1990) Simultaneous conjugate observations of dynamic variations in high-latitude dayside convection due to changes in IMF B_y. J Geophys Res 95:8057–8072

Helliwell RA (1988) VLF stimulated experiments in the magnetosphere from Siple Station, Antarctica. Rev Geophys 26:551–578

Herman JR, Hudson R, McPeters R, Stolarski R, Ahmad Z, Gu X-Y, Taylor S, Wellemeyer C (1991) A new self-calibration method applied to TOMS and SBUV backscattered ultraviolet data to determine long-term global ozone change. J Geophys Res 96:7531–7545

Hurren PJ, Smith AJ, Carpenter DL, Inan US (1986) Burst precipitation induced perturbations on multiple VLF propagation paths in Antarctica. Ann Geophys 4:311–318

Ishizu M, Saka O, Kitamura T-I, Fukunishi H, Sato N, Fujii R (1981) Polarization study of Pc1 and Pc1-2 band pulsations at conjugate stations. Mem Natl Inst Polar Res Spec Issue, Japan 18:118–126

Kiehl JT (1992) Cold comfort in the greenhouse. Nature 355:773

Lean J (1987) Solar ultraviolet irradiance variations: a review. J Geophys Res 92:839–868

Lin ZM, Benbrook JR, Bering EA, Byrne GJ, Friis-Christensen E, Liang D, Liao B, Theall J (1990) Observations of ionospheric flux ropes above South Pole. Geophysical Monograph 58. American Geophysical Union, Washington, DC, pp 581–590

Lockwood M, Coates A (1992) When the solar wind blows. New Sci 1811:25–30

Marchant HJ, Davidson AT, Kelly GJ (1991) UV-B protecting compounds in the marine alga *Phaeocystis pouchetii* from Antarctica. Mar Biol 109:391–395

Mende SB, Rairden RL, Lanzerotti LJ, Maclennan CG (1990) Magnetic impulses and associated optical signatures in the dayside aurora. Geophys Res Lett 17:131–134

Molina MJ, Rowland FS (1974) Stratospheric sink for chlorofluoromethanes: chlorine atom-catalysed destruction of ozone. Nature 249:810–812

Morrison K (1990) Quasi-periodic VLF emissions and concurrent magnetic pulsations seen at L = 4. Planet Space Sci 38:1555–1565

Omura Y, Nunn D, Matsumoto H, Rycroft MJ (1991) A review of observational, theoretical and numerical studies of VLF triggered emissions. J Atmos Terr Phys 53:351–368

O'Neill A (ed) (1990) Dynamics, transport and photochemistry in the middle atmosphere of the southern hemisphere. NATO ASI Series C 321. Kluwer, Dordrecht, 257 pp

Paschal EW, Lanzerotti LJ, Maclennan CG (1990) Correlation of whistler mode phase delay with transient hydromagnetic waves. J Geophys Res 95:15059–15071

Ramaswamy V, Schwarzkopf MD, Shine KP (1992) Radiative forcing of climate from halocarbon-induced global stratospheric ozone loss. Nature 355:810–812

Rodger AS, Smith AJ (1989) Antarctic studies of the coupled ionosphere – magnetosphere system. Philos Trans R Soc Lond 328A:271–287

Russell CT (1991) The magnetosphere. Annu Rev Earth Planet Sci 19:169–182

Rycroft MJ (1990a) Antarctic research on the atmosphere and solar-terrestrial physics. Geodät Geophys Veröff Reihe I Berlin 15:211–238

Rycroft MJ (1990b) The Antarctic atmosphere: a hot topic in a cold cauldron. Geograph J 156:1–11

Rycroft MJ (1990c) Quo vadimus re the springtime Antarctic ozone depletion? Adv Space Res 10:(10)275–(10)277

Rycroft MJ (1991a) Atmospheric and space research from the polar regions. Phys Educ 26:153–158

Rycroft MJ (1991b) Interactions between whistler-mode waves and energetic electrons in the coupled system formed by the magnetosphere, ionosphere and atmosphere. J Atmos Terr Phys 53:849–858

Rycroft MJ, Smith AJ, Jones D, Dudeney JR, Shawhan SD (1984) Global geospace study. Nature 310:449

Saxton JM, Smith AJ (1991) Electric fields at L = 2.5 during geomagnetically disturbed conditions. Planet Space Sci 39:1305–1320

Schnell RC, Liu SC, Oltmans SJ, Stone RS, Hofmann DJ, Dutton EG, Deshler T, Sturges WT, Harder JW, Sewell SD, Trainer M, Harris JM (1991) Decrease of summer tropospheric ozone concentrations over Antarctica. Nature 351:726–729

Shanklin JD, Gardiner BG (1989) The Antarctic ozone hole. British Antarctic Survey, Cambridge, 9 pp

Shapland D, Rycroft M (1984) Spacelab – research in Earth orbit. Cambridge University Press, Cambridge, 192 pp

Sheldon WR, Benbrook JR, Bering EA (1988) Rocket investigations of electron precipitation and VLF waves in the Antarctic upper atmosphere. Rev Geophys 26: 519–533

Smith AJ, Clilverd MA (1991) Magnetic storm effects on the mid-latitude plasmasphere. Planet Space Sci 39:1069–1079

Smith AJ, Carpenter DL, Corcuff Y, Rash JPS, Bering EA (1991) The longitudinal dependence of whistlers and chorus characteristics observed on the ground near L = 4. J Geophys Res 96:275–284

Smith RC, Prézelin BB, Baker KS, Bidigare RR, Boucher NP, Coley T, Karentz D, MacIntyre S, Matlick HA, Menzies D, Ondrusek M, Wan Z, Waters KJ (1992) Ozone depletion: ultraviolet radiation and phytoplankton biology in Antarctic waters. Science 255:952–959

Stephenson JAE, Scourfield MWJ (1991) Importance of energetic solar protons in ozone depletion. Nature 352:137–139

Thompson AM (1991) New ozone hole phenomenon. Nature 352:282–283

Tonegawa Y, Fukunishi H, Hirasawa T, McPherron RL, Sakurai T, Kato Y (1984) Spectral characteristics of Pc3 and Pc4/5 magnetic pulsation bands observed near L = 6. J Geophys Res 89:9720–9730

Troshichev OA, Gusev MG, Nickolashkin SV, Samsonov VP (1988) Features of the polar cap aurorae in the southern polar region. Planet Space Sci 36:429–439

Weller G, Bentley CR, Elliot DH, Lanzerotti LJ, Webber PJ (1987) Laboratory Antarctica: research contributions to global problems. Science 238:1361–1368

The Antarctic Sea Ice Cover

Peter Wadhams[1]

1 Introduction

The sea ice cover in the Antarctic is one of the most climatically important features of the Southern Hemisphere. Its seasonal variation in extent is from 4 million km^2 in summer (February) to 20 million km^2 in winter (September) (Zwally et al. 1983). This enormous amplitude puts Antarctic sea ice second only to Northern Hemisphere snow extent as a seasonally varying cryospheric parameter. The cycle is reasonably stable from year to year; the most recent examination of filtered monthly anomalies (Folland et al. 1992) suggests that the extent diminished significantly (by about 2 million km^2) between 1973 and 1980, then increased by 1 million km^2 up to 1982 and has remained fairly constant since then (Fig. 1). The main features of the seasonal cycle are shown in Fig. 2 (after Comiso and Zwally 1989) for a typical recent year.

The presence of a sea ice cover has an enormous effect on the exchanges of heat, moisture and momentum between ocean and atmosphere; and the motion of the sea ice leads to water mass modification both in the region of generation (where fresh water is removed) and in the region of melt (where fresh water is injected into the ocean), which may be hundreds or thousands of kilometres away. Since the Antarctic sea ice and its huge seasonal cycle are of central importance to the climate and energy budgets of the Southern Ocean, it is vital that we try to understand the mechanisms leading to ice formation, transport and melt.

2 Factors Affecting Sea Ice Development

The development of sea ice in the Antarctic, both in extent and thickness, is strongly related to:

1. oceanic heat flux
2. atmospheric temperature

[1] Scott Polar Research Institute, University of Cambridge, Cambridge CB2 1ER, UK

Hempel (Ed.) Antarctic Science

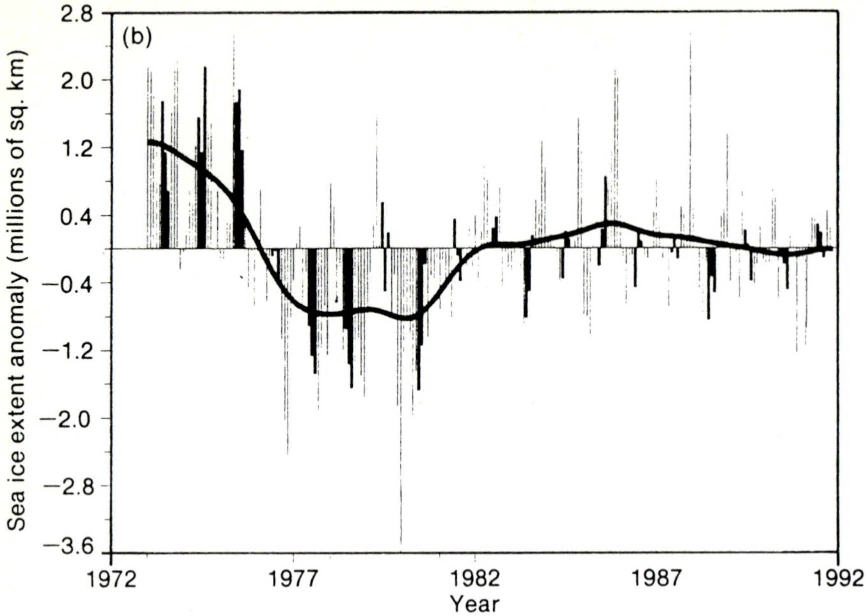

Fig. 1. Southern Hemisphere sea ice extent anomalies relative to the average for 1973–1991. The trend line is generated from a 39-point binomial filter applied to the monthly anomalies. *Heavy bars* represent winter months (June, July, August). (After Folland et al. 1992)

3. wind-driven and thermohaline circulations
4. the wave field.

Of these factors, oceanic heat flux is probably the major determinant of ice growth rate (Bagriantsev et al. 1989), and a relevant parameter is the temperature and depth of T_{max}, the temperature maximum in the warm deep water underlying the polar surface water. In the Weddell Sea the value of T_{max} is lower, and the depth at which it occurs is greater in the western part of the sea than in the Maud Rise region, and this is reflected in the thicknesses achieved by first-year sea ice in these regions.

Atmospheric temperature is important in the difference between, for instance, the southern end of the Weddell Sea and the low-latitude winter circumpolar ice edge, and also because of the effect of cold katabatic winds in enhancing growth rates in coastal polynyas. Here, the wind is also important in driving newly formed ice away from the coast, permitting new ice to form in coastal waters; the resulting polynyas are often important "ice factories", generating a considerable fraction of the ice in, for instance, the Ross Sea (Jacobs and Comiso 1989). The growth rate in coastal polynyas is enhanced not only by the cold winds but also by the fact that ice grows as frazil which does not cut off the surface water from the atmosphere.

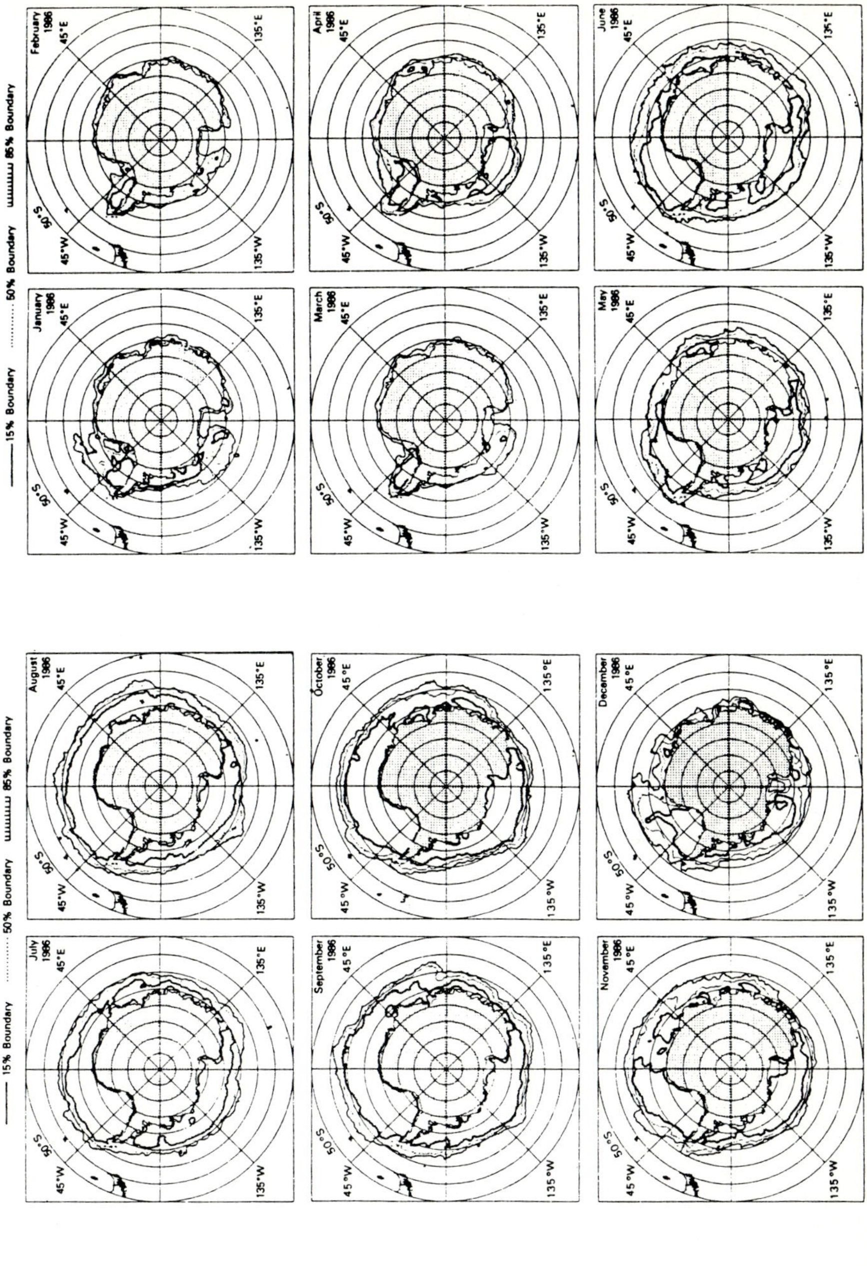

Fig. 2. Seasonal advance and retreat of Antarctic sea ice (Comiso and Zwally 1989)

In other ways the surface circulation affects ice distribution through the presence of gyres south of the Antarctic Circumpolar Current. The long-term average wind field generates a westward-flowing current near the coast and an eastward-flowing current, the Antarctic Circumpolar Current, over the deep ocean. The presence of embayments such as the Ross and Weddell Seas causes partially closed gyres to be generated from the combination of these two current systems (Deacon 1984). Within the Weddell Gyre, for instance, the ice on the western side of the Weddell Sea is thicker and older than ice on the eastern side, because it represents conditions further downstream.

Finally, the wave field influences ice characteristics mainly by the fact that it causes the advancing winter ice edge to exist as a high concentration of pancake ice rather than as a continuous sheet, thus enhancing the growth rate. This effect extents across a belt of width more than 250 km (Wadhams et al. 1987; Lange et al. 1989).

3 Some Characteristics of Antarctic Sea Ice

Antarctic sea ice differs in many ways from sea ice in the Arctic. Its thickness, composition, structure and mode of formation are all quite different, leading to differences in its behaviour and in its appearance when examined by satellite sensors. Some of these differences are described below.

3.1 The Pancake-Frazil Cycle

In the Arctic Ocean most sea ice forms within the existing pack limits, in calm water. It therefore grows as congelation ice, passing through a well-known sequence of stages. Initially a suspension of small crystals called frazil ice forms in the water column, but these soon freeze together to form a thin coherent skin called nilas. When only a few centimetres thick this is transparent (dark nilas) but as the ice grows thicker the nilas takes on a grey and finally a white appearance. Further growth yields first-year ice, which in a single season reaches a thickness of 1.5–2 m. The crystals near the top of a first-year ice sheet are small and randomly oriented, bespeaking their origin as frazil, or are oriented with c-axis vertical, which occurs if they grew as flat floating crystals in very calm water. Deeper down, long vertical columnar crystals dominate, with horizontal c-axes, since ice grows more readily at right angles to the c-axis. This columnar structure is a key identifier of congelation ice (i.e. ice which has grown thermodynamically by freezing onto an existing ice bottom).

In the Antarctic early studies of the sea ice structure (Gow et al. 1982) showed that most of the thickness of any ice core consists not of columnar ice but of small randomly oriented crystals characteristic of frazil origin. The mechanism by which the ice is generated was not elucidated until the first expedition was able to work in the pack ice zone during early winter, the time of ice edge advance. This was as recently as 1986, in the Winter Weddell Sea Project, using F.S. *Polarstern*. In transiting the ice margin region a careful study of the ice conditions and characteristics was carried out, and the pancake-frazil cycle was identified as the source of most of the first-year sea ice seen further inside the pack (Wadhams et al. 1987; Lange et al. 1989). The ice forming at the extreme edge cannot pass through the Arctic sequence because of the high energy and turbulence in the Southern Ocean wave field, so it forms a dense suspension of frazil ice out of which pancake ice congeals by wave-induced compression of the suspension, a mechanism described by Martin and Kauffman (1981). At the ice edge the pancakes are only a few centimetres in diameter, but they gradually grow in diameter and thickness with increasing distance from the ice edge, until they reach 3–5 m diameter and 50 cm thickness. The growth occurs by accretion from the frazil, which continues to form because the open water surface permits high ocean-atmosphere heat flux. The pancakes begin to freeze together in groups, but the wave field is strong enough to prevent overall freezing until a penetration of some 270 km is reached. Here the pancakes coalesce to form a continuous sheet of first-year ice. At this point, with the open water surface cut off, the growth rate drops to a very low level (estimated at 0.4 cm/day by Wadhams et al. 1987) and the ultimate thickness reached by first-year ice is only a few centimetres more than the thickness attained at the time of consolidation of the pancakes.

Ice formed in this way has a different bottom morphology from Arctic ice. The pancakes at the time of consolidation are jumbled together and rafted over one another, and freeze together in this way with the frazil acting as "glue". The result is a very rough, jagged bottom, with rafted cakes doubling or tripling the normal ice thickness, and with the edges of pancakes protruding upwards to give a surface topography resembling a "stony field" (Wadhams et al. 1987). The rafted bottom provides a large surface area per unit area of sea surface, providing an excellent substrate for algal growth and a refuge for krill. The thin ice permits much light to penetrate, and the result is a winter ice ecosystem which is more fertile than that seen in the Arctic.

Because of the large amount of free water on and around the pancakes, the zone containing this type of ice had formerly been characterized from passive microwave data as consisting of first-year ice at low concentration (Zwally et al. 1983). Thus a region which, if assumed to be circumpolar and 270 km deep, occupies an area of 6 million km^2 during the season of ice advance, is revealed as having a quite distinct ice morphology, one which requires a new type of parametrization in ice-ocean models.

Deep within the pack, as divergence occurs, open water areas become filled with new ice which forms in the traditional Arctic way. This is charac-

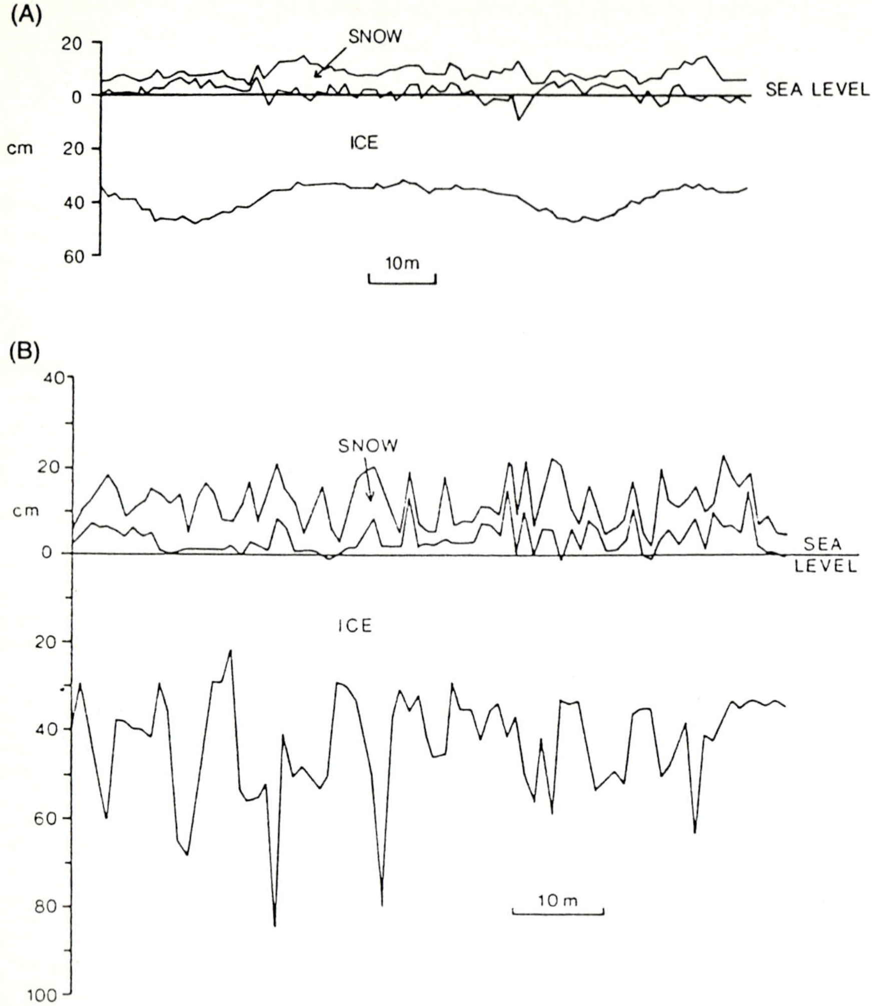

Fig. 3. First-year sea ice profiles from Antarctic Ocean, showing contrast between ice formed within calm polynyas **(A)** and consolidated pancake ice ("stony fields") **(B)**

terized by much smoother bottom and upper surfaces. The contrast between the two types of ice is seen in Fig. 3, which shows some results of profiles drilled during the 1986 experiment, with holes at 1-m intervals.

3.2 First and Multi-Year Ice

In the Arctic, sea ice commonly takes several years to either make a circuit within the closed Beaufort Gyre (7–10 years) or else be transported across the Arctic Basin and expelled in the East Greenland Current (3–4 years). More than half of the ice in the Arctic is therefore multi-year ice, which has survived at least one summer melt season and which eventually reaches thicknesses of 3–4 m. In contrast, Antarctic sea ice normally has a northward component to its motion, so that it is constantly spreading from its source area into lower latitudes, where it eventually melts. Usually this process takes less than a year, so almost all Antarctic ice is first-year ice. Only in the Weddell Sea, the Ross Sea, and to a lesser extent the Bellingshausen Sea, do significant quantities of pack ice survive the summer to become multi-year ice. Usually this occurs because of the existence of a semiclosed gyre. In the Weddell Sea, for instance, ice formed in the eastern part of the Weddell-Enderby Basin is carried west in the southern part of the Weddell Gyre (the Antarctic Coastal Current) into the southern Weddell Sea, and then westward and finally north along the east side of the Antarctic Peninsula (the route followed by the drift of Shackleton's *Endurance*). This track has been found by the use of drifting ice buoys to take about 18 months. Thus second-year ice may be found in the western Weddell Sea within the zone of northward motion. The only very old multi-year ice to be found is the small amount that has broken out from the fast ice which grows for many years in embayments along the fringes of ice shelves (Wadhams et al. 1987).

Again, the first opportunity to study such ice during the winter was a comparatively recent cruise, the 1989 Winter Weddell Gyre Study of F.S. *Polarstern*, which involved a transect across the Weddell Sea from west (Antarctic Peninsula) to east (Kap Norvegia) (Wadhams and Crane 1991). Earlier work was done in summer (Ackley 1979) and early spring (Lange and Eicken 1991). Multi-year (i.e. second-year) ice was found only west of 40° W, in the region of northward ice drift, and when sampled by drilling had a mean thickness of 1.17 m in regions free of ice deformation. Undeformed first-year ice in the same regions had a mean thickness of only 60 cm. The fact that second-year ice is twice as thick as first-year ice, despite the fact that the growth rate slows radically as ice thickness increases, is evidence of more extreme conditions in the southern Weddell Sea, through which the second-year ice has passed, than in the lower-latitude circumpolar Antarctic Ocean. These conditions may consist purely of lower mean air temperatures, but it is more likely that lower ocean heat fluxes are also important.

3.3 Snow Loading

The Arctic Ocean is a cold desert, and snow on sea ice makes only a small relative contribution to the overall thickness or mass. In the Antarctic snow is much more important. The annual snowfall is greater, and in coastal regions snow is blown onto the sea ice by katabatic winds off the tops of ice shelves. During the July–September 1986 experiment in the eastern Weddell Sea we found a mean snow thickness of 14–16 cm on the surface of first-year ice. Since the ice itself is so thin, this was sufficient to bring the ice surface below sea level in 15–20% of cases, leading to the infiltration of sea water into the overlying snow and the formation of either a wet slushy layer on top of the ice or, in the case of freezing, the formation of a "snow ice" layer between the unwetted snow and the original ice upper surface. In September–October 1989 we also found a mean snow thickness of 16 cm over undeformed first-year ice (23 cm over deformed ice, because of piling up of snow against ridges), but in multi-year ice in the western Weddell Sea the snow thickness was much greater. The average was 0.63 m over un-

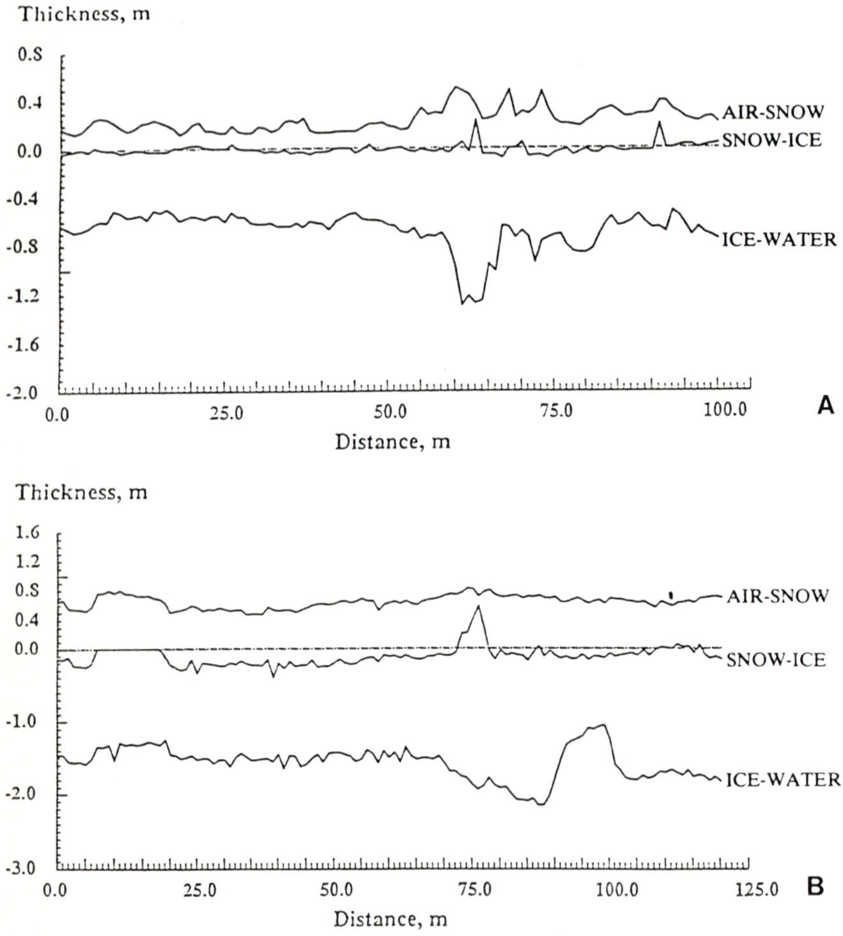

Fig. 4. First-year (**A**) and multi-year (**B**) ice profiles from the Weddell Sea (Wadhams and Crane 1991). Note that the snow-ice interface lies below sea level in the multi-year case

deformed ice and 0.7 m over deformed ice. This was sufficient to push the ice surface below sea level in almost every case. Figure 4 shows the contrast between the two types of ice cover in this respect. The resulting flooded layer will have an enormous effect on passive microwave signatures, making it difficult to unequivocally identify multi-year ice using current algorithms, and will also have an impact on mass and energy balances of sea ice which must be taken into account in modelling efforts.

3.4 Pressure Ridging

In the Arctic most pressure ridges are formed by the crushing of young ice in leads between two thicker, converging floes. Individual ridges can be up to 50 m deep, with 20 m common and 30 m attained at least once every 100 km (Wadhams 1978). Ridges make a large contribution to the overall mass of sea ice; probably about 40% on average and more than 60% in coastal regions where the mean ice draft reaches 7 m. In contrast, Antarctic ridging is much less intense. Individual ridges are much shallower. Every ridge sampled during the 1986 and 1989 experiments was less than 6 m in draft. Figure 5 shows a typical ridge cross-section, which resembles that of an Arctic ridge in everything except total depth. Examination of the structure of ridges showed that in most cases the block thickness in the ridge was similar to the thickness of the floes on either side, that is, the ridge has formed by buckling of the floe itself, or by buckling due to the collision

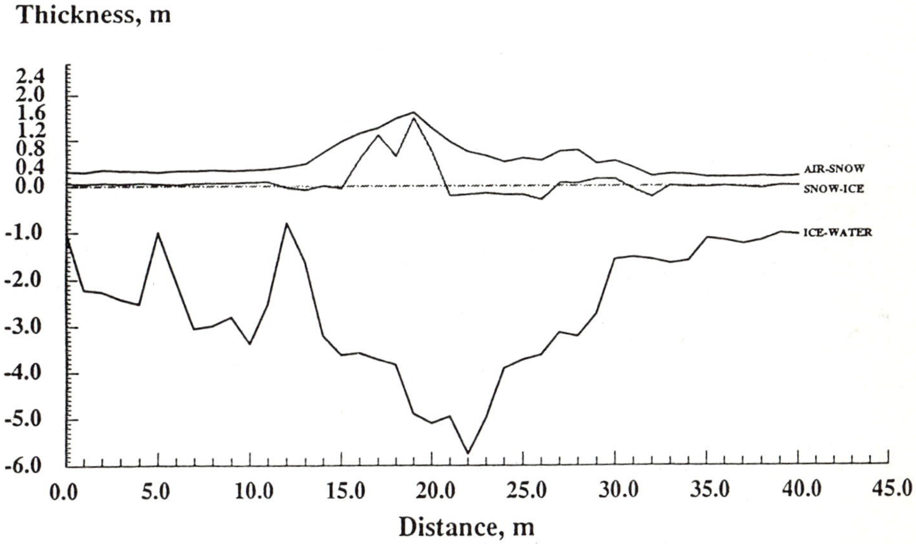

Fig. 5. Cross section through a typical Antarctic pressure ridge, from 67° S, 28° W

between two floes, rather than by the crushing of young ice. This means that the stress due to convergence can be relieved by a large number of individual buckling events rather than by a small number of refrozen lead closures, so that a given ridge need not grow to a great height. There ought to be more frequent ridges than in the Arctic, on this argument, but once again this does not appear to be the case, except possibly in coastal regions. On average, Antarctic sea ice is in a state of divergence as it moves northward into wider spaces of the Southern Ocean, and so it is likely that ridge-building events occur less frequently than in the Arctic. Furthermore, ridges formed in this way may well disintegrate and revert to brash ice when the pressure is relieved.

It is difficult to assess the overall contribution made by ridging to the mean thickness of Antarctic sea ice. Drilled profiles offer an inadequate data set for such large-scale averages. Nevertheless, the contribution does appear to be important. In 1989, for instance, the mean ice thickness in profiles containing no deformed ice, and some deformed ice, respectively, were 0.60 and 1.03 m (first-year) and 1.17 and 2.51 m (multi-year ice). In future we may expect further insight from the extension of laser profiling surveys over the ice cover (Weeks et al. 1989).

3.5 Coastal Polynyas

As we have already noted, coastal polynyas play an important role in Antarctic sea ice production in winter. They tend to exist at well-defined locations where the coastal topography leads to katabatic winds. In the case of the polynya in Terra Nova Bay (Kurtz and Bromwich 1985) the Drygalski Ice Tongue also plays a role in shielding the bay region from coastal pack ice drifting northwards in the Antarctic Coastal Current. This particular polynya has been estimated to produce 10% of the ice generated over the entire Ross Sea shelf, so polynyas are clearly important as locations of enhanced ice production. They are also important, however, in their ocean interactions. Such a large amount of ice production is also associated with a large amount of brine rejection, and so if the polynya is located in a critical region where mixing is occurring leading to bottom water generation, the additional brine production may have a strong effect enhancing the rate of bottom water production (Zwally et al. 1985).

4 Current Research Problems

Much remains to be learned about air-sea-ice interaction in the Antarctic. Four important questions which are at present receiving attention are described below.

4.1 The Interaction Between Climate Change and Antarctic Sea Ice

General circulation models (GCMs), which attempt to predict the magnitude of global warming, show very large discrepancies in their predictions for the Antarctic. In equilibrium models, such as that of Hansen et al. (1988), it is predicted that the warming in the ice regime of the Antarctic Ocean will be almost as great as in the Arctic and much greater than that occurring at low latitudes. Circumpolar rises of 1–2 °C are predicted at all longitudes, with rises of 3 °C or more in the region north of the Ross and Weddell Seas, all within 60 years. In time-dependent models, such as that of Stouffer et al. (1989) or the more recent efforts reviewed by Gates et al. (1992), the Antarctic sea ice zone is predicted to have the lowest temperature rise of any latitude in the globe, amounting to less than 1 °C in 60 years. The difference between the two types of model arises from the way in which oceanic heat transport is parametrized. Equilibrium models assume that a step change (e.g. a doubling) is given to the CO_2 concentration in the atmosphere and that the climate is allowed to attain a new equilibrium state in response to this. Time-dependent models are more realistic, in that CO_2 concentration is increased at the observed rate (about 1% per year) and account is taken of the role of the ocean in absorbing much of the warming which occurs. The difference between time-dependent and equilibrium models is thus greatest in the "ocean hemisphere" and transforms the Antarctic from being a zone of anomalously high predicted warming into a zone of anomalously low warming. The generation and interpretation of fresh oceanographic data, particularly in winter (e.g. Gordon and Huber 1990), will do much to improve our understanding of the role of Antarctic sea ice in climate change, assisted by the use of models which deal with air-sea-ice interaction in the absence of ice dynamics or ocean advection and which in this way enable sensitivity studies to be carried out. An example of such a model is that of Martinson (1990), who demonstrates that a complex set of feedback mechanisms comes into play if a parameter such as air temperature is changed. The balance of lead concentration, upper ocean structure, and pycnocline depth will adjust itself to minimize the impact of changes, and he concludes that Antarctic first-year sea ice, apparently such a thin and delicate skin that may easily be removed by a modest warming, is more likely to be quite resilient and resistant to the impact of warming. By contrast, some coupled sea ice-mixed layer models, such as that of Lemke et al. (1990), show responses of ice edge position to climate change which are essentially linear – the edge advances or retreats approximately in proportion to changes in air temperature.

4.2 Monitoring Antarctic Sea Ice by Remote Sensing

Problems with passive microwave algorithms due to ice surface flooding and to the unusual ice types found near the advancing ice edge, and problems due to disagreement between algorithms, cause real uncertainty in the delineation of total ice area (i.e. extent × concentration) and multi-year ice distribution. It is important that these quantities be reliably monitored in order to give us warning of any trend indicating that global warming may be having an impact on the area of Antarctic sea ice. Furthermore, it would be very desirable for passive microwave interpretations to be reliable enough to take over from advanced very high resolution radiometer (AVHRR) visible imagery as the prime means of generating ice charts and ice extent interpretations. More direct validation measurements are needed before this can happen.

The future of Antarctic sea ice monitoring is an exciting one, because of the advent of synthetic aperture radar (SAR) aboard the ERS-1 satellite in 1991. This permits ice motion vectors to be determined, and also ice concentration and (to some extent) ice type, all with high resolution. The problem is that retrievals from the Antarctic may not offer a dense enough coverage to be of synoptic value, both because of a lack of complete coverage by receiving stations and because SAR switch-on time is limited. Further SAR satellites such as the Japanese JERS-1 and Canadian Radarsat will ease this problem. An alternative instrument which is also aboard ERS-1 and which operates continuously is the radar altimeter. This shows ice edge position and also the depth of penetration of ocean wave energy, hence the likely width of the pancake ice zone. However, the relationship between altimeter return and true ice surface topography (from which ice thickness may possibly be inferred) has not yet been demonstrated.

4.3 The Role of Sea Ice in Bottom Water Production

We have mentioned how brine rejection from ice production in coastal polynyas may enhance bottom water production. The physics are actually more complex when the coastal polynya occurs at the edge of an ice shelf, for here the flow of water under the shelf with associated cooling is also a factor. The complex interaction of these two mechanisms for water mass transformation in winter is a problem which awaits adequate in situ field measurement and interpretation. The likely consequences of climate warming on bottom water production can then be assessed.

4.4 The Biological Role of Sea Ice

It has been found in many field experiments that the winter sea ice cover harbours an entire ecosystem, starting from nutrients and working up from bacteria to zooplankton. Energy pathways in the ecosystem remain to be investigated, including the role of physical parameters such as sea ice roughness (for bottom-dwelling biota); the geometry of brine drainage channels;

and ice thickness (defining light levels). A critical question awaiting an answer is to what extent the sea ice ecosystem in winter is responsible for "seeding" the spring plankton bloom, which occurs with greatest intensity in the region from which the sea ice has most recently retreated (Comiso and Sullivan 1986, Comiso et al. 1993). Eicken (1992) has drawn attention to the contrasting roles of sea ice in controlling primary production under ice through reduction of irradiative fluxes by the ice and its snow cover, and in stimulating a "kryohaline" mode of life for organisms within the pores of the sea ice, controlled by salinity and ambient temperature as well as irradiance.

5 The Future Course of Research

It is certain that the pace of research on Antarctic sea ice will increase, because of its importance in the global climate system.

We can expect to see an increase in *buoy programmes*, since these offer a means of monitoring ice motion and of providing the basic temperature, air pressure and wind speed data needed to model ice thermodynamics and provide better synoptic pressure fields for modelling ice dynamics. In regions which are difficult or dangerous of access by ship, e.g. multi-year ice regions, "smart" buoys, which possess a large suite of ocean current, temperature and salinity sensors, can provide vital data at less than the cost of a manned survey. In the Arctic, a buoy programme has been in existence for more than a decade, funded until recently by the USA but now with international input. In the Antarctic a variety of national buoy programmes have existed (UK, German, USA, Australian, Japanese), but there would be great scientific value in a true collaboration between nations, with sharing of data, if the entire circumpolar zone is to be covered. The WCRP has proposed an International Antarctic Drifting Buoy Project, to be coordinated by the University of Tasmania, which will accomplish this aim.

We can expect improved *satellite monitoring* in future years, as ERS-1 is supplemented and replaced by SAR satellites with greater capabilities, such as ERS-2 (1995) and ENVISAT-1 (1999) of ESA, JERS-1 (Japanese) and Radarsat (Canadian). It will become routine to monitor the entire field of ice motion, ice extent, ice type and ice concentration.

There will be more *field programmes*, involving shipborne surveys, especially in winter, and also drifting ice camps. A camp in the western Weddell Sea was successfully deployed and recovered in 1992 in the so-called Anzone programme (LDGO 1989) involving the US and Russia. Such programmes are exceedingly expensive and can only flourish under conditions of genuine international co-operation.

There will be new ways of *acoustic monitoring* of changes in the ocean and possibly the ice. Munk and Forbes (1989) carried out in 1991 a preliminary trial for a Global Acoustic Transmission Experiment, based on a source deployed off Heard Island. The aim is to monitor changes in the average temperature and structure of the upper ocean through changes in acoustic transmission time. Receiving systems on the Antarctic coastline (e.g. at Davis Station) allowed the particular properties and variability of the Antarctic Circumpolar Current to be measured. Tomography-type experiments may also be carried out, and it has been predicted (Guoliang and Wadhams 1989; Guoliang et al. 1993) that acoustic tomography can be used to measure mean ice thickness within a region by travel time changes.

We can expect to see a great improvement in the adequacy of *models*, especially general circulation models, in their applicability to the Antarctic.

The most important aspect of all of these research strategies will be international co-operation. The Antarctic sea ice zone is a vast region encompassing every longitude and therefore of interest to every Antarctic Treaty nation and every nation in SCAR. International co-operation in data collection is essential for achieving fundamental advances in understanding.

References

Ackley SF (1979) Mass-balance aspects of Weddell Sea pack ice. J Glaciol 24:391–405

Bagriantsev NV, Gordon AL, Huber BA (1989) Weddell Gyre: temperature maximum stratum. J Geophys Res 94(C6):8331–8334

Comiso JC, McClain CR, Sullivan CW, Ryan JP, Leonard CL (1993) Coastal Zone Color Scanner pigment concentrations in the Southern Ocean and relationships to geophysical surface features. J Geophys Res 98(C2):2419–2451

Comiso, JC, Sullivan CW (1986) Satellite microwave and in situ observaions of the Weddell Sea ice cover and its marginal ice zone. J Geophys Res 91(C8):9663–9681

Comiso, JC, Zwally HJ (1989) Polar microwave brightness temperatures from Nimbus 7 SMMR: time series of daily and monthly maps 1978–1987. NASA Ref Publ RP1223. NASA, Washington, DC, 102 pp

Deacon G (1984) The Antarctic Circumpolar Ocean. Cambridge University Press, Cambridge, 180 pp

Eicken H (1992) The role of sea ice in structuring Antarctic ecosystems. Polar Biol 12:3–13

Folland CK, Karl TR, Nicholls N, Nyenzi BS, Parker DE, Vinnikov KYa (1992) Observed climate variability and change. In: Houghton JT, Callander BA, Varney SK (eds) Climate change 1992. The Supplementary Report to the IPCC Scientific Assessment. Cambridge University Press, pp 135–170

Gates WL, Mitchell JFB, Boer GJ, Cubasch U, Meleshko VP (1992) Climate modelling, climate prediction and model validation. In: Houghton JT, Callander BA, Varney SK (eds) Climate change 1992. The Supplementary Report to the IPCC Scientific Assessment. Cambridge University Press, pp 97–134

Gordon AL, Huber BA (1990) Southern Ocean winter mixed layer. J Geophys Res 95(C7):11655–11672

Gow AJ, Ackley SF, Weeks WF, Govoni JW (1982) Physical and structural characteristics of Antarctic sea ice. Ann Glaciol 3:113–117

Guoliang J, Wadhams P (1989) Travel time changes in a tomography array caused by a sea ice cover. Prog Oceanog 22:249–275

Guoliang J, Lynch JF, Pawlowicz R, Wadhams P, Worcester P (1993) Effects of sea ice cover on acoustic ray travel times with applications to the Greenland Sea Tomography Experiment. J Acoust Soc Am 94(2):1044–1056

Hansen J, Fung I, Lacis A, Rind D, Lebedeff S, Ruedy R, Russell G (1988) Global climate changes as forecast by Goddard Institute for Space Studies three-dimensional model. J Geophys Res 93(D8):9341–9364

Jacobs SS, Comiso JC (1989) Sea ice and oceanographic processes over the Ross Sea continental shelf. J Geophys Res 94:18195–18211

Kurtz DD, Bromwich DH (1985) A recurring, atmospherically forced polynya in Terra Nova Bay. In: Jacobs SS (ed) Oceanology of the Antarctic Continental Shelf. Antarctic Res Ser 43, American Geophysical Union, Washington, pp 177–201

Lange MA, Ackley SF, Wadhams P, Dieckmann GS, Eicken H (1989) Development of sea ice in the Weddell Sea. Ann Glaciol 12:92–96

Lange MA, Eicken H (1991) The sea ice thickness distribution in the northwestern Weddell Sea. J Geophys Res 96(C3):4821–4337

LDGO (1989) AnZone: International coordination of oceanographic studies of the Antarctic Zone. 23–25 May 1989. In: Gordon A (ed) Meeting report. Lamont-Doherty Geological Observatory, Palisades NY

Lemke P, Owens WB, Hibler WD III (1990) A coupled sea ice-mixed layer pycnocline model for the Weddell Sea. J Geophys Res 95(C6):9513–9525

Martin S, Kauffman P (1981) A field and laboratory study of wave damping by grease ice. J Glaciol 27:283–313

Martinson DG (1990) Evolution of the Southern Ocean winter mixed layer and sea ice: open ocean deepwater formation and ventilation. J Geophys Res 95(C7):11641–11654

Munk WH, Forbes AMG (1989) Global ocean warming: an acoustic measure? J Phys Oceanogr 19:1765–1778

Stouffer RJ, Manabe S, Bryan K (1989) Interhemispheric asymmetry in climate response to a gradual increase of atmospheric CO2. Nature 342:660–662

Wadhams P (1978) Characteristics of deep pressure ridges in the Arctic Ocean. In: Muggeridge DB (ed) Proc 4th Int Conf Port & Ocean Engng under Arctic Condns, St. John's, 26–30 Sept. 1977. Memorial Univ Newfoundland, St John's 1:544–555

Wadhams P, Crane DR (1991) SPRI participation in the Winter Weddell Gyre Study 1989. Polar Rec 27(160):29–38

Wadhams P, Lange MA, Ackley SF (1987) The ice thickness distribution across the Atlantic sector of the Antarctic Ocean in midwinter. J Geophys Res 92(C13):14535–14552

Weeks WF, Ackley SF, Govoni J (1989) Sea ice ridging in the Ross Sea, Antarctica, as compared with sites in the Arctic. J Geophys Res 94(C4):4984–4988

Zwally HJ, Comiso JC, Parkinson CL, Campbell WJ, Carsey FD, Gloersen P (1983) Antarctic sea ice 1973–1976: Satellite passive-microwave observations. NASA, Washington, DC, Rept SP-459, 206 pp

Zwally HJ, Comiso JC, Gordon AL (1985) Antarctic offshore leads and polynyas and oceanographic effects. In: Jacobs SS (ed) Oceanology of the Antarctic Continental Shelf. Antarctic Res Ser 43. American Geophysical Union, Washington, DC, pp 203–226

Ecological Significance of the Sea Ice Biota

Michael Spindler[1] and Gerhard Dieckmann[2]

The Antarctic sea ice cover and biota associated with it play a key role in the marine ecosystems of the Southern Ocean. Due to the high biomass of autotrophic algae living at the ice/water interface and within the brine channel system of the sea ice, a large proportion of the total primary production in these regions can be attributed to these autotrophs. During winter the algae survive the harsh conditions of low temperatures, high salinities and low irradiance levels. It is during this time that they become an important food source for a number of pelagic animals particularly the southern krill *Euphausia superba*. During ice melt, algae are released to the water column where they may contribute to the initiation of the phytoplankton spring bloom, while sedimentation of sea ice organisms provides the benthos with food.

The above statements on the ecological role of sea ice biota are based on recent findings which will be summarized in the following pages.

Sea ice is an extremely heterogeneous environment, the result of a number of different processes of growth and development which lead to various ice types termed fast ice, pack ice, multi-year ice to name but a few (e.g. Maykut 1985; Weeks and Ackley 1986; Eicken and Lange 1989; Lange et al. 1989; Wadhams, this Vol.). In his paper, Wadhams (this Vol.) has discussed the importance of sea ice for the climate regime because it affects the exchange of energy between ocean and atmosphere. Thermodynamic and radiative properties (e.g. the albedo) of sea ice regulate the heat flux, which in turn plays a role in sea ice growth, final thickness, ice melt and areal extent of the sea ice cover. A possible warming of the global atmosphere will be amplified by changes in the sea ice regime at high latitudes, as current models demonstrate (e.g. Augstein, this Vol.). Also of importance is the influence of sea ice formation on oceanography. Haline convection due to brine rejection into the water column is one example. Sea ice undoubtedly affects the rate of gas exchange between ocean and atmosphere and therefore plays a role in the transfer of CO_2.

[1] Institute for Polar Ecology, University of Kiel, Wischhofstr. 1-3, Geb. 12, D-24148 Kiel, Germany

[2] Alfred Wegener Institute for Polar and Marine Research, Columbusstr., D-27568 Bremerhaven, Germany

Hempel (Ed.) Antarctic Science

Sea ice, however, also plays a key role in polar marine ecosystems because of the many organisms associated with and dependent on it in one way or another. These include animals living on the ice, a variety of microalgae, microbiota and invertebrates living within the ice as well as algae and larger organisms associated with the underside of the ice (Horner 1985a; Garrison 1991).

Warm blooded animals such as seals and penguins are the conspicuous inhabitants of sea ice. They depend on a stable sea ice cover during certain times of their ontogeny. Weddell seals give birth to their pups on fast ice near the ice shelf coast. Their offspring are dependent on stable sea ice conditions only for a short time. After 2 to 3 weeks the pups enter the water for the first time and are weaned after 4 weeks. Thereafter sea ice is frequently used as a resting place (Stonehouse 1972).

In contrast, emperor penguins are entirely dependent on stable sea ice conditions for more than 7 to 9 months where they assemble into large colonies, often consisting of more than 10000 birds. After courtship on the ice, the male emperor incubates the single egg laid by the female for 2 months. After hatching, the chicks are fed regularly by their parents. They remain on the sea ice for about another 6 months before the chicks fledge and develop their plumage which enables them to safely enter the water (Stonehouse 1972).

Other penguins, such as adelie and chinstrap, feed near the marginal ice zone which is a biologically productive area (Smith and Nelson 1986; Garrison and Buck 1989; Comiso et al. 1990). Pack ice is also the floating home of thousands of crabeater and leopard seals which profit from the krill associated with the sea ice, as is discussed later.

Less conspicuous but highly diverse and abundant is the life within the sea ice. During ice formation, fresh water crystals are formed from seawater and the concentrated brine is rejected into the water below. However, some brine remains within the ice matrix where it is enclosed in brine channels and pockets which serve as a habitat for the organisms. The matrix of interconnected brine channels differs with ice type which is determined by the developmental history of the ice. This is demonstrated by casts of the brine channel system (Weissenberger et al. 1992; Figs. 1, 2). The channels are strongly branched and are connected to each other in sea ice of frazil origin, while in congelation ice, less branching occurs and the main orientation of the channels is vertical. In the interior of the sea ice, channel diameters are in the order of a few hundred microns. Connections between channels may be very delicate and only a few microns thick (Weissenberger et al. 1992). Towards the underside of the ice the diameter of the channels increases from the millimetre range up to some centimetres.

As the ice temperatures increase with the onset of summer, brine pockets and channels will enlarge and thus provide more space for microorganisms to colonize.

Fig. 1. The cast shows the brine channels of a small piece of frazil ice which is composed of smaller granular ice crystals. The channels are highly branched and connected to each other and have diameters of about 200 to 300 μm. Scale = 1 mm. (Weissenberger et al. 1992)

The most common inhabitants of Antarctic sea ice are the diatoms (Fig. 3) which are represented by some 200 to 300 species (Horner 1985b). Apart from diatoms, the sea ice assemblages comprise a heterogeneous group of organisms ranging from bacteria and lower fungi to protozoans, such as ciliates and foraminiferans, as well as metazoans such as turbellarians, nematodes, copepods and amphipods.

These assemblages of organisms impart the often observed characteristic brown colouration to the ice. These are sometimes distinctly visible as dark striations. Cell counts in these layers have yielded up to 10^8 diatom cells per litre (up to 2000 μg Chl a per litre) (Garrison et al. 1983; Spindler and Dieckmann 1986; Bartsch 1989; Spindler et al. 1990). Such high concen-

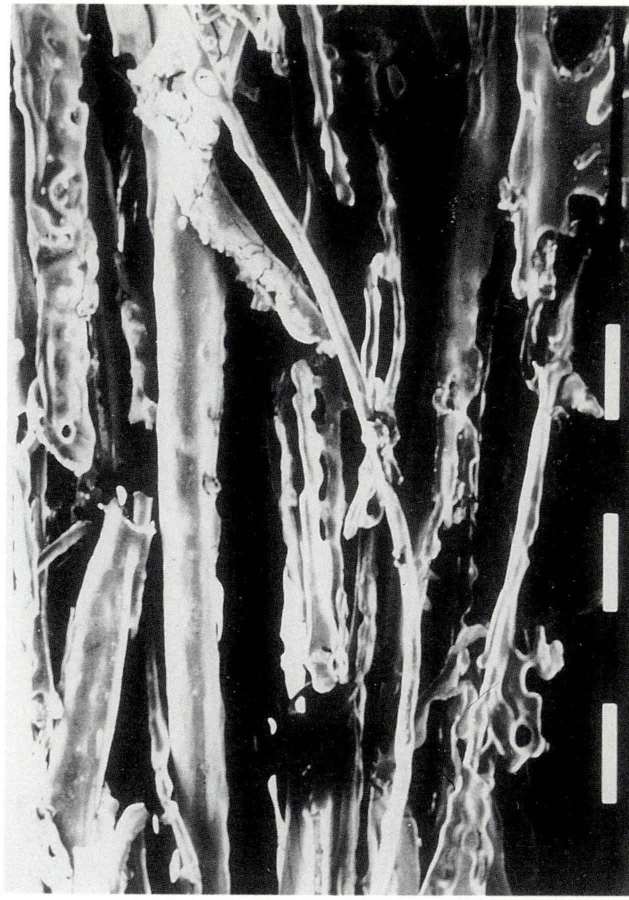

Fig. 2. In congelation ice, which is composed of elongated columnar ice crystals, the brine channels are less branched and oriented more or less vertically. Scale = 1 mm. (Weissenberger et al. 1992)

trations are among the highest ever recorded in sea water in any ocean. Thus the diatoms form the basis of an intricate food web comprising protozoans and metazoans. Interactions between these organisms are not yet well understood, mainly because of the difficulty in studying sea ice in an undisturbed form.

Not only does sea ice provide an internal habitat for many organisms, but its peripheries, in particular the ice/water interface, serve as an area for attachment of microalgae. Here algae may form dense mats or strands which are accessible to larger grazers such as krill (*Euphausia superba*) or amphipods. Krill which form immense swarms during the ice-free summer months disperse during winter to feed on the underside of sea ice where

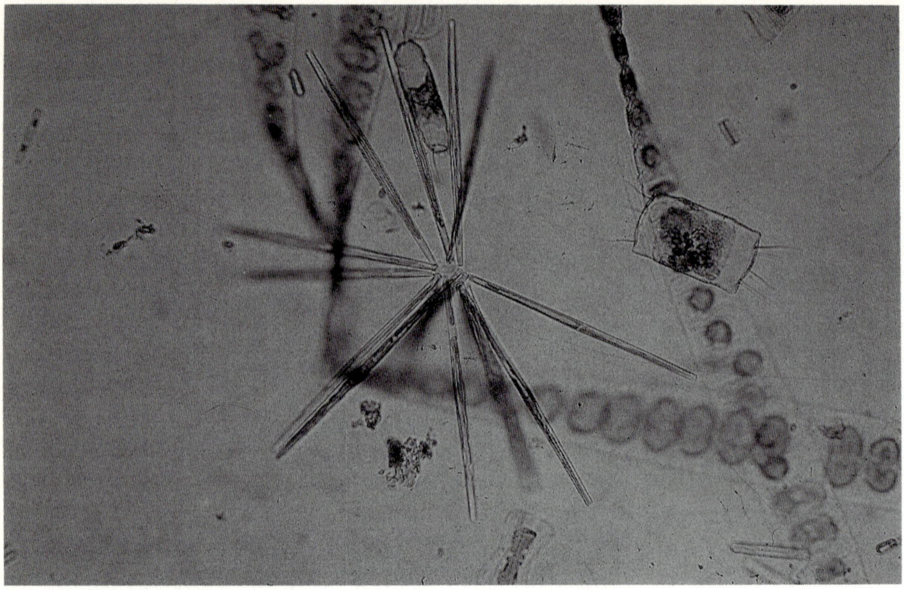

Fig. 3. Diatoms from the interior of Antarctic sea ice. Species shown include *Amphiprora kufferathii*, *Odontella weissflogii* and *Nitzschia* spp. Width of picture 100 μm

Fig. 4. Krill (*Euphausia superba*) actively scraping off algae from underneath Antarctic pack ice. (Courtesy of P. Marschall)

there is an ample supply of food while the water column is virtually devoid of phytoplankton (Marschall 1988; O'Brien 1988; Stretch et al. 1988; Fig. 4). Krill are particularly well adapted to feed both in the water column and from sea ice where they are capable of scraping off algae, as was demonstrated in experiments by Marschall (1988). Counterpart to the Antarctic krill are in the Arctic the gammarid amphipods which have successfully occupied a similar niche as the krill, with regard to sea ice (Gulliksen 1984; Lønne and Gulliksen 1991).

Both the krill under the ice and the gammarid amphipods are preyed upon by larger predators such as jellyfish, fish, penguins, seals and minke whales.

A more localized yet important secondary sea ice habitat, is the "platelet layer" which usually accumulates under fast ice in the coastal regions up to 50 km off the Antarctic continent. The ice platelets are formed in supercooled water which flows from under the ice shelves. They rise to the surface and accumulate under fast or pack ice to form layers several metres thick (Dieckmann et al. 1986). These aggregations of platelets provide a habitat for algae which grow preferentially at the fast ice/platelet layer interface where they have an optimal light environment. However, nutrient replenishment here is restricted, so that nutrients may become limiting for growth. On the other hand, the algae are protected from larger grazers and, under optimal conditions, may attain a standing stock of up to 13 mg Chl a per litre (Dieckmann et al. 1992).

The three major ice systems discussed above all contribute to the overall productivity of the Southern Ocean. Estimates of the proportion of primary production in sea ice lie as high as 30% of total Antarctic production (Legendre et al. 1992). However, these figures are based on several assumptions and extrapolations. One of the major uncertainties is the limited data available, due to the large logistical constraints on the sampling of sea ice. Furthermore, data were collected randomly and during different seasons. It is a well-established fact that sea ice organisms have a patchy distribution. Replicate cores taken less than 30 cm apart yielded organism numbers which varied by an order of magnitude (Spindler and Dieckmann 1986).

Patchiness is also the consequence of different ice formation processes which lead to distinct ice types such as congelation or granular ice. Analyses of ice cores revealed that most organisms are associated with the granular ice (congealed frazil ice crystals) which confirms that the organisms are incorporated into the ice when it forms from frazil ice (Spindler et al. 1990).

Although organism numbers in the ice are lowest during winter, algae apparently still grow during this period. They are shade adapted and thus are able to photosynthesize at very low light levels. Furthermore, they are also less sensitive to low temperatures and high salinities which occur in the ice during winter. It was shown experimentally that growth of ice algae occurs at temperatures down to −5.5 °C and at a corresponding salinity of 95 psu (practical salinity unit). Even lower temperatures

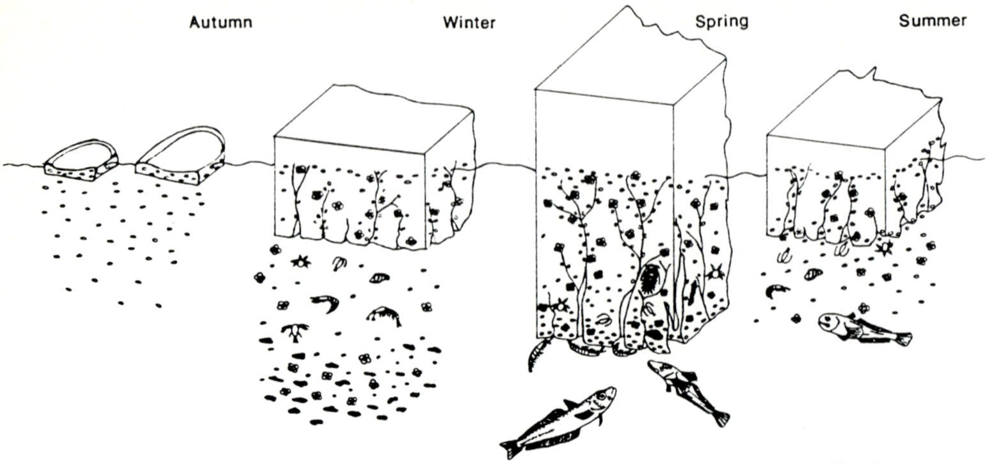

Fig. 5. Generalized seasonal cycle of sea ice development and biological activity. During ice formation in autumn and winter algae and smaller zooplankton are incorporated into the ice. They survive during winter and are fed on by pelagic organisms such as krill. After thawing most organisms are released into the water column. Reprinted by permission of Kluwer Academic Publishers. (Spindler 1990)

and higher salinities, often found within the sea ice during winter, are tolerated. At −7 °C and 115 psu and at −10 °C and 145 psu the algae cease to grow but survive these conditions for at least 6 weeks. If they are subsequently transferred to more normal conditions they resume growth (Bartsch 1989).

Despite our limited knowledge of the sea ice system and its associated organisms, it has become clear that sea ice is a key Southern Ocean marine habitat for the following reasons:

1. A sea ice cover, stable for more than 7 months is crucial for the survival of some warm-blooded animals such as emperor penguins.
2. About a quarter to one-third of the overall primary production of the Southern Ocean is provided by algae associated with sea ice.
3. The ice with its abundant primary producers supports marine life during winter, e.g. krill and other central organisms of the Antarctic marine food web depend on this source during winter in particular.
4. During ice melt the ice organisms are released into the water column. There they may play a significant role in seeding the phytoplankton spring bloom. An additional effect during ice melt may be a strong food pulse to the benthos (Fig. 5).

What are the potential consequences of global warming on this unique sea ice system? Needless to say there is a strong interaction between a complexity of processes which result in the constitution of this habitat. For instance, ice formation, growth and decay all have a strong influence on the hydrographic and meteorological regimes in the polar regions. Salt rejected during ice formation results in convective processes which contribute to deep water formation. During ice melt, the upper water column becomes stabilized, favouring the development of phytoplankton blooms. In case of meteorology, changes in humidity result in different cloud cover and subsequent changes in snow accumulation on the ice. Both in turn have effects on the light availability for the ice algae. In addition, a fragile system such as the Antarctic marine ecosystem is likely to be sensitive to small changes which will eventually have devastating effects.

Due to these complicated interactions, straightforward answers and predictions regarding the question of the influences of global warming on the sea ice biota are not possible. Nevertheless, it is perhaps interesting to contemplate some worst case scenarios:

A less extensive and stable fast ice cover may seriously affect emperor penguins. Early melting and breakup of the sea ice may result in the loss of the entire year's brood due to one of several factors: adults will no longer find their juveniles while drifting on a floe and thus these chicks will not be fed; juveniles can no longer withstand low temperatures because the large colonies break up into smaller units, protection by huddling is less effective; direct contact with sea water or by spray will be deleterious to the chicks with their down feathers. They will lose some of their insulation and suffer from cold temperatures or even become soaked and drown.

Any change in the areal extent of the annual sea ice cover will affect the overall productivity of the Southern Ocean. As a consequence, the food web will become destabilized for most consumers, in the extreme case resulting in the vanishing of entire stocks.

References

Bartsch A (1989) Die Eisalgenflora des Weddellmeeres (Antarktis): Artenzusammensetzung und Biomasse sowie Ökophysiologie ausgewählter Arten. Ber Polarforsch 63:1–110

Comiso JC, Maynard NG, Smith WO, Sullivan CW (1990) Satellite ocean color studies of Antarctic ice edges in summer and autumn. J Geophys Res 95:9481–9496

Dieckmann G, Rohardt G, Hellmer H, Kipfstuhl J (1986) The occurrence of ice platelets at 250 m depth near the Filchner Ice Shelf and its significance for sea ice biology. Deep-Sea Res 33:141–148

Dieckmann G, Arrigo K, Sullivan CW (1992) A high resolution sampler for nutrient and chlorophyll *a* profiles of the sea ice platelet layer and underlying water column below fast ice in polar oceans: preliminary results. Mar Ecol Prog Ser 80:291–300

Eicken H, Lange MA (1989) Development and properties of sea ice in the coastal regime of the southeastern Weddell Sea. J Geophys Res 94:8193–8206

Garrison DL (1991) Antarctic sea ice biota. Am Zool 31:17–33

Garrison DL, Buck KR (1989) Protozooplankton in the Weddell Sea, Antarctica: abundance and distribution in the ice-edge zone. Polar Biol 9:341–351

Garrison DL, Ackley SF, Buck KR (1983) A physical mechanism for establishing algal populations in frazil ice. Nature 306:363–365

Gulliksen B (1984) Under-ice fauna from Svalbard waters. Sarsia 69:17–23

Horner RA (1985a) Sea ice biota. CRC Press, Boca Raton, 215 pp

Horner RA (1985b) Ecology of sea ice microalgae. In: Horner RA (ed) Sea ice biota. CRC Press, Boca Raton, pp 83–103

Lange MA, Ackley SF, Wadhams P, Dieckmann G, Eicken H (1989) Development of sea ice in the Weddell Sea, Antarctica. Ann Glaciol 12:92–96

Legendre L, Ackley SF, Dieckmann GS, Gulliksen B, Horner R, Hoshiai T, Melnikov IA, Reeburgh WS, Spindler M, Sullivan CW (1992) Ecology of sea ice biota. 2. Global significance. Polar Biol 12:429–444

Lønne OJ, Gulliksen B (1991) On the distribution of sympagic macro-fauna in the seasonally ice covered Barents Sea. Polar Biol 11:457–469

Marschall H-P (1988) The overwintering strategy of Antarctic krill under the pack-ice of the Weddell Sea. Polar Biol 9:129–135

Maykut GA (1985) The ice environment. In: Horner RA (ed) Sea ice biota. CRC Press, Boca Raton, pp 21–82

O'Brien DP (1988) Direct observations of the behaviour of *Euphausia superba* and *Euphausia crystallorophias* (Crustacea: Euphausiacea) under pack ice during the Antarctic spring of 1985. J Crust Biol 7:437–448

Smith WO, Nelson DM (1986) Importance of ice edge phytoplankton production in the Southern Ocean. BioScience 36:251–257

Spindler (1990) A comparison of Arctic and Antarctic sea ice and the effects of different properties on sea ice biota. In: Bleil U, Thiede J (eds) Geological history of the polar oceans: Arctic versus Antarctic. Kluwer, Dordrecht, pp 173–186

Spindler M, Dieckmann GS (1986) Distribution and abundance of the planktic foraminifer *Neogloboquadrina pachyderma* in sea ice of the Weddell Sea (Antarctica). Polar Biol 5:185–191

Spindler M, Dieckmann GS, Lange MA (1990) Seasonal and geographic variations in sea ice community structure of the Weddell Sea, Antarctica. In: Kerry KR, Hempel G (eds) Antarctic ecosystems. Ecological change and conservation. Springer, Berlin Heidelberg New York, pp 129–135

Stonehouse B (1972) Animals of the Antarctic: the ecology of the far south. Peter Lowe, London, 171 pp

Stretch JJ, Hamner PP, Hamner WM, Michel WC, Cook J, Sullivan CW (1988) Foraging behavior of antarctic krill *Euphausia superba* on sea ice microalgae. Mar Ecol Prog Ser 44:131–139

Weeks WF, Ackley SF (1986) The growth, structure, and properties of sea ice. In: Untersteiner N (ed) The geophysics of sea ice. NATO ASI Series B: Physics vol 146. Plenum Press, New York, pp 9–164

Weissenberger J, Dieckmann G, Gradinger R, Spindler M (1992) Sea ice: a new technique to analyze and display the interstitial environment. Limnol Oceanogr 37:179–183

The Polar Ice Sheets: A Chronicle of Climate and Environment

Hans Oeschger[1]

1 Introduction

More and more interest is paid to the understanding of the complex Earth system processes. On the one hand, this is the basic goal of natural sciences and on the other hand, human activities have become a significant force affecting functions of the Earth system. Therefore, science is challenged to estimate the consequences of human interferences.

In the late 1970s the World Climate Research Programme (WCRP) and in the 1980s the International Geosphere-Biosphere Programme (IGBP) were established. Together, the Core Projects of these Programmes constitute the international framework of the quest for scientific understanding of climate and global change.

Great efforts are concentrated on the monitoring of processes in the atmosphere, in the ocean and on the continents. Faster computers are able to integrate the interactive equations and provide more and more accurate descriptions of Earth system processes. In addition to providing simulations of the evolution of global conditions beyond the current observational range.

For these model experiments it is indispensable to know the initial conditions and trends of the Earth system before significant human interference started. To disentangle the human induced component of the Earth system evolution from the one which had occurred without human interaction, it is necessary to understand the full spectrum of natural, externally and internally forced Earth system events and to model it. The better the models are capable of describing past Earth system events, the higher the fidelity of simulations in future evolution.

How can we provide the data base for the past Earth system evolution? A wealth of information is recorded in ocean and continental sediments, in geomorphological features, in peat bogs, tree-rings and in natural ice. In the following, we concentrate mainly on the polar ice which can be looked at as a chronicle of the history of climate and environment. Until the present only a small portion of the principally available information has been retrieved.

[1] University of Bern, Physics Institute, Sidlerstr. 5, CH-3012 Bern, Switzerland

Hempel (Ed.) Antarctic Science

Main advances in Global Change science have come from the analysis of deep ice cores drilled in Greenland and Antarctica.

In the following, a short overview on the present knowledge of anticipated climate change is given and questions are listed concerning information on Earth system history. Special emphasis will be given to results from ice core studies.

Finally, an outlook is presented basing on important recent results from studies on newly drilled ice cores.

2 Present Assessment of Global Climate Change

The greenhouse effect is a well-established phenomenon based on the absorption and re-emission of part of the infrared radiation emitted from the Earth's surface by the so-called greenhouse gases. The effect is to warm the surface and lower atmosphere. Emissions from human activities are substantially increasing the concentrations of the greenhouse gases CO_2, CH_4, N_2O and the chloroflurocarbons (CFCs). These increases will enhance the natural greenhouse effect (which keeps the Earth's surface 33 °C warmer than it would be if it were not present).

Based on the business as usual scenario (IPCC 1990) it is estimated that the global mean temperature during the next century will increase at a rate of 0.3 °C per decade which is greater than seen during the past 10 000 years.

During the past 100 years global mean temperature has increased by ca. 0.5 °C. The size of this warming corresponds to predictions at their lower end. The increase could be largely due to natural variability; alternatively, this variability and other man-made factors like emissions enhancing atmospheric turbidity and cloud properties could have offset a still larger man-made greenhouse warming (IPCC 1992).

Based on the observed increases of greenhouse gases and on a well-established greenhouse theory, there is little doubt, if at all, that the climate of the globe will change. However, large uncertainties exist in the timing and the magnitude of the anticipated changes. Global Change science is faced with the challenge of an added urgency to reduce the uncertainties:

As an example, the unequivocal detection of the enhanced greenhouse effect would significantly increase the readiness of governments for action that would slow down the rate of global change. In addition, it would enable the improvement of projections of climatic and environmental change on which environmental protection measures need to be based.

To separate the temperature history of the last 100 years and the anthropogenic signal, knowledge on the natural climate variability needs to be improved. It is composed of variations induced by external factors like changes in the albedo due to volcanic dust and possible changes in solar luminosity, internally forced variations, and a general system noise.

Information from natural archives, historical documents and for the past 100 to 200 years from instrument observation should be synthesized to a global record with high resolution in space and time of the evolution of climate and environment during the past 1000 to 2000 years.

Overlapping data from direct measurements and from proxy parameters help to calibrate proxy information which then serves to extend the records well back into the pre-instrument period.

From this record attempts should be made to extract the history of the albedo forcing of global climate due to volcanic eruptions and natural and anthropogenic dust (and possible effects on clouds). Special emphasis should be given to the synthesis of different types of information, e.g. how can one conclude from the acidity record in Greenland and Antarctic ice cores the global turbidity signal?

How can the recent satellite measurements of solar luminosity variations, the direct (Neutron Monitoring) and indirect (^{10}Be in ice and ^{14}C in trees) observations of solar modulation of cosmic radiation, the experimental and theoretical knowledge on the sun and other stars be combined to derive a quantitative history of intrinsic solar variations?

To what extent can the pre-industrial climate variations be explained by the albedo, solar and greenhouse forcing changes derived as suggested above?

What characteristics does the internal variability of the records show? Can one distinguish between general system noise and quasi- oscillatory phenomena, like the El Niño Southern Oscillation and the North Atlantic Oscillation?

These experimental and model studies would serve as a basis to project the natural Earth system evolution into the period increasingly influenced by human activities. What would have been the Earth system's evolution during the last 100 years without human interference?

From the actual system evolution, e.g. the mean global temperature trend, the estimated natural variability could be subtracted and the anthropogenic trend (including the overall uncertainty) assessed.

Such a detection study of the anthropogenic climate change would strongly draw on studies of the information recorded in natural archives, like on polar ice. In the following, the principal of global environmental system studies based on natural tracer analysis is discussed and the functioning of natural ice as an especially important archive for environmental system fingerprints is described.

3 The Global Environmental System (E.S.)

To illustrate the research in this field, a concept of the global environmental system is introduced (Fig. 1; Oeschger 1990). It includes the entirety of

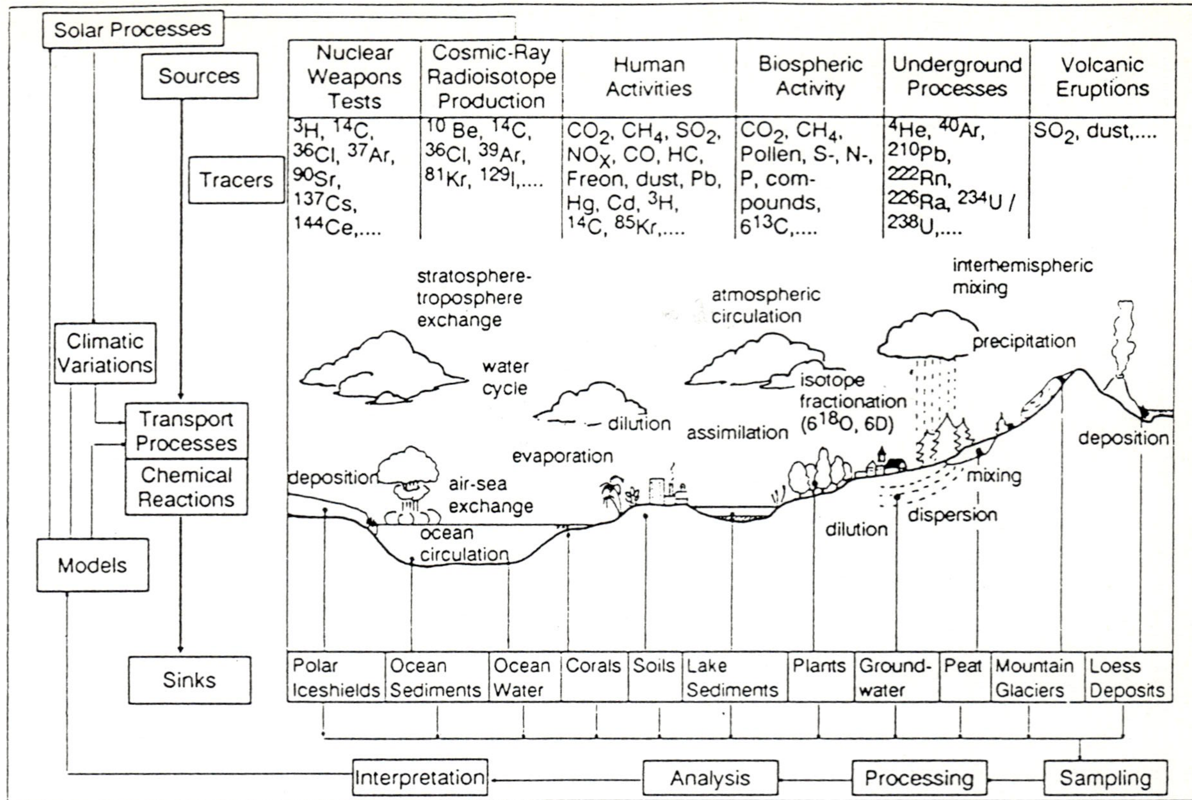

Fig. 1. Tracing physical, chemical and biological processes and interactions. Natural effects and human impact

physical, chemical and biological processes acting upon the earth's surface and in the atmosphere. The parts of the system interact in various dynamic sequences and are in contact with the planetary and galactic systems. The E.S., as defined here, agrees largely with the climate system as it is generally defined, but stronger emphasis is given to chemical and biological processes. Special attention is paid to those parameters which can be studied in natural archives and therefore enable the reconstruction of ancient system states.

The main components of the E.S. are the atmosphere, the hydrosphere, including the oceans and continental waters, the cryosphere, consisting of the polar ice sheets, sea ice and mountain glaciers, the biosphere, containing marine and continental living organisms, and the lithosphere, with bedrock and sediments which interface with the hydrosphere.

The energy of the sun drives the dynamic processes in the E.S. It causes atmospheric circulation and oceanic mixing and, due to evaporation and precipitation, the cycling of water. The energy balance determines the climatic conditions at individual locations on the Earth's surface. It is affected by scattering and reflection of short wavelength solar radiation in the atmosphere and on the Earth's surface, and the infrared radiation emission absorption and remission by the surface and by water vapour and gases (CO_2, O_3, . . .) in the atmosphere, together with the transfer of latent and sensible heat into the atmosphere and ocean.

The dynamic cycles of some elements, such as C, N and O, play an important role which is regulated by biospheric activity, chemical reactions, and physical exchange processes.

Dust particles and aerosols are injected into the atmosphere by wind action, volcanic eruptions, biospheric processes and human activities. They reappear on the earth's surface by dry fallout or wet deposition.

Of special interest for this approach to the understanding of the E.S. processes are the radioactive and stable isotopes which constitute ideal tracers for a variety of processes and their dynamics. Radioactive isotopes may have different origins:

- they are formed by the interaction of cosmic radiation with atoms in the upper atmosphere (^{19}B, ^{14}C, ^{36}Cl, ^{39}Ar, ^{81}Kr, . . .). This cosmic production is modulated by the changing shields of solar plasma and the Earth's magnetic field;
- they are also introduced to the E.S. as a result of society's use of nuclear fusion and fission (^{3}H, ^{37}Ar, ^{85}Kr, . . .); and finally
- they are released from the Earth's crust as products of the natural decay series of U and Th (^{222}Rn, ^{210}Pb, . . .).

Since half-lives vary from days (^{37}Ar, ^{133}Xe) to hundreds of thousands of years (^{10}Be, ^{36}Cl, ^{81}Kr), information about time constants of natural processes over a very wide range is attainable. Each isotope has its characteristic field of applications which may reach far beyond that of mere dating.

Stable isotopes (^2H, ^{13}C, ^{18}O) are other important sources of information. Phase transitions, chemical reactions and diffusion processes produce small changes in the natural isotope ratios. They reflect the conditions at which the processes occurred. Elements originating from different natural reservoirs can often be distinguished based on their different isotopic composition. Samples of air, water and ice, and organic materials and sediments, taken from many parts of the E.S., contain information on its static characteristics, like the partitioning of water between atmosphere, cryosphere and ocean, but also its dynamic characteristics, like mixing and circulation and exchange processes in and between the different system components. This and other information can be derived from the isotopic ratios, from the concentration levels of chemical elements and molecules, pollen and dust. A complete set of these parameters defines the state of the E.S. as "fingerprint parameters". They are continuously recorded in the natural archives, as in polar ice sheets, mountain glaciers, ocean and lake sediments, and organic materials like tree-rings or peat and coral deposits. Analyses of sequential samples allow the reconstruction of the historical evolution of the E.S.

The overall objective of investigating the E.S. is a consistent and systematic description of its complex processes and an understanding of the mechanisms controlling them.

4 Natural Ice as an Archive of the History of the Environmental System

Natural ice has unique properties as an archive of information on E.S. processes. During the past decades, ocean sediments had been the main source of information on the response of the climate system to perturbations like varying orbital parameters of the Earth. In the last few years new information, on short-term climatic variations, has been obtained by the study of information recorded in ice cores, obtained by drilling into the polar ice caps and alpine glaciers.

Particles of different origins suspended in the air serve as condensation nuclei for water vapour. At temperatures below freezing, snow flakes are formed. During their fall and in the surface layers of snow, additional particles and chemical compounds are absorbed. Under the load of subsequently deposited snowfall, the annual layers sink to greater depth, get compressed and thinner (Fig. 2). At depths of 50 to 70 m in Greenland, the sintering of the firn (neve) grains encloses atmospheric air in void spaces. Thus, the air bubbles in old glacier ice constitute physically occluded samples of the atmospheric gases at the time of pore close-off.

In the upper ice layers the time resolution is high, enabling the identification of single precipitation events or at least of seasonal variations. The thickness of annual layers thins with depth, and allows one to study the information recorded during longer time periods. So far, ice cores cover

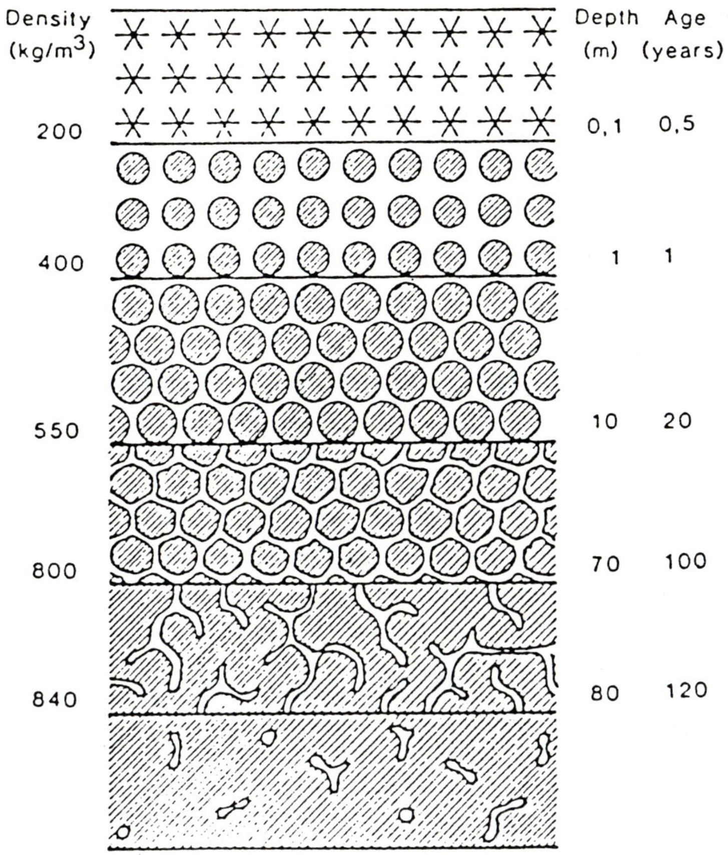

Fig. 2. Two-dimensional schematic of the metamorphosis of snow to firn and ice; depth-age relationships, typical for Greenland ice cores. (Oeschger et al. 1985)

up to approximately 150000 years, but at some locations, e.g. in Central Greenland or East Antarctica, much older ice should be recoverable at the bedrock. In summer 1992 in the frame of the European Greenland Ice Core Project (GRIP) on the summit of the Greenland ice sheet bedrock was reached at a depth of 3029 m. It is a great challenge to extract the information on the Earth system evolution covering with high resolution at least the last two glaciation cycles.

5 Examples of Environmental System Parameter Information in Natural Ice

(For an overview on the present state of art of studies of the environmental record in glaciers and ice sheets see Lorius 1989; Oeschger and Langway 1989.)

5.1 $^{18}O/^{16}O$ in Precipitation

$^{18}O/^{16}O$ ratios of precipitation water reflect fractionation of water molecules of different mass ($H_2{}^{18}O$ and $H_2{}^{16}O$) during evaporation and condensation. During evaporation of water the heavier water molecules get depleted compared to the lighter ones. Therefore, in newly formed water vapour the $^{18}O/^{16}O$ ratio is about 10% lower than in the water from which it escaped. During condensation the heavier isotopes get enriched in the liquid phase. In the water vapour of an air mass which is moving towards a colder region thus the $^{18}O/^{16}O$ ratio of the water molecules is decreasing due to the preferential loss of $H_2{}^{18}O$ (Fig. 3).

This, on the one hand, leads to seasonal variations of $^{18}O/^{16}O$ in precipitation, as shown by measurements on precipitation from the Jungfraujoch and the comparison with temperature (Fig. 4) and, on the other hand climatic changes lead to shifts which in a first approximation are proportional to the temperature change. Similar information can be obtained from studies of the $^2H/^1H$ ratio in the water molecules.

5.2 Dating of the Ice

The ice particles deposited on an ice sheet travel to greater depth and towards the margin where they melt or are extruded as part of an iceberg

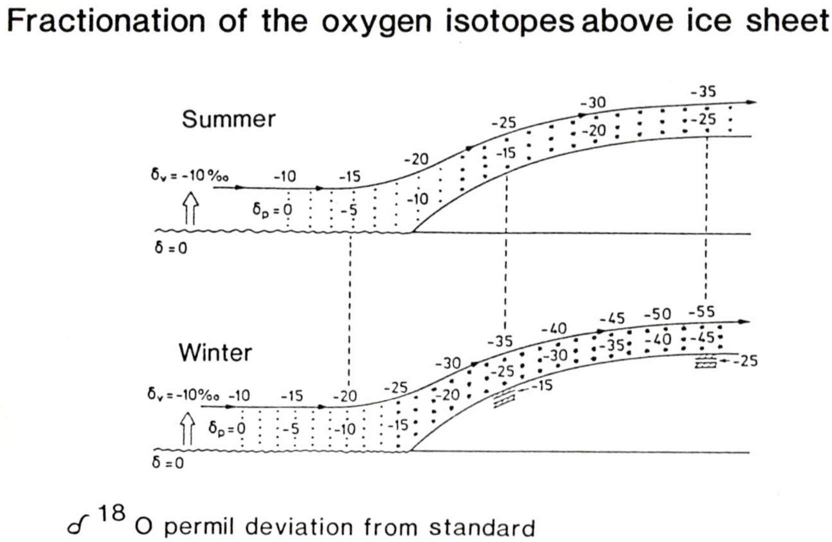

Fig. 3. Depletion of $H_2{}^{18}O$ in the water vapor of an air parcel travelling over an ice sheet and in the precipitation. δ-values give permille deviations from a standard. (Dansgaard et al. 1973)

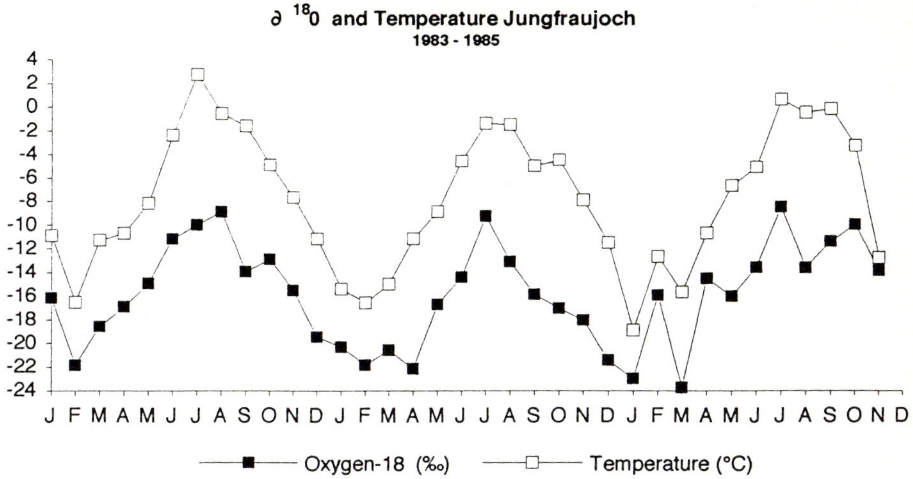

Fig. 4. $\delta^{18}O$ in monthly precipitation on the Jungfraujoch, Bernese Oberland, and mean monthly temperature. The good correlation, resulting in seasonal $\delta^{18}O$ variations, is clearly visible. (Physics Institute Berne 1985)

(Fig. 5). Due to the pressure of the overlaying ice the annual ice layers are gradually stretched and thinned and displaced to greater depth and by diffusion the seasonal variations of ^{18}O and ^{2}H get obliterated.

Studies of chemical constituents and microparticles provide additional information on seasonal variations. With high resolution the concentrations of H_2O_2, Ca^{2+}, HCHO and NH^4_+ can be studied in the field by flow injection analysis. Microparticle studies promise high resolution records far back in time, because the diffusion of particles is relatively small.

Dating by radioactive techniques is hampered by relatively large errors due to the statistical uncertainty of the count rates of the samples with generally very low radioactivity, but can be very helpful as a first orientation for the age of ice sampled e.g. in the marginal zones of ice sheets.

Common reference horizons may be found in ice cores obtained from different locations, enabling synchronization of the records. They conclude tephra, continental dust, extreme events expressed in different parameters, rapid climatic changes, rapid changes in CO_2 and CH_4 concentrations which are global, ^{10}Be spikes, etc.

Experimental and theoretical dating by ice flow modelling go hand in hand and should provide a consistent picture of the dynamic state of the ice sheet. Essential inputs for modelling experiments are the experimentally derived variations of ice accumulation rates (typically by a factor of 2 to 3 smaller during maximum glaciation than during an interglacial) and the apparent viscosity which also shows a strong dependence on the chemistry of the ice.

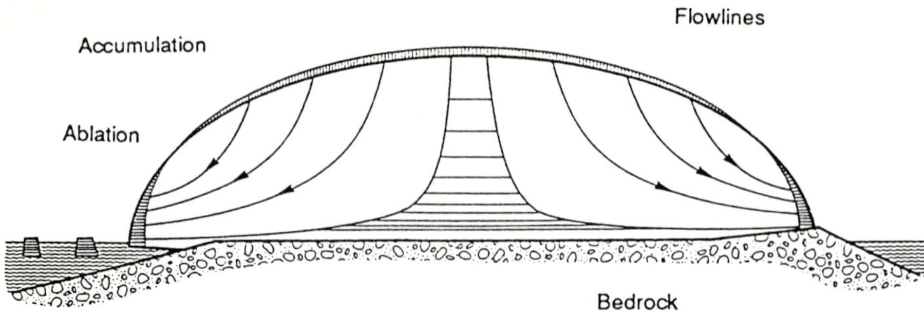

Fig. 5. Schematic ice flow pattern in a vertical cross-section of an ice sheet. (Stauffer 1988)

5.3 Climate History

The backbone information in a polar ice core is constituted by the stable isotope ratio profiles (either $^{18}O/^{16}O$ or $^{2}H/^{1}H$), reflecting the climatic history at the site where the precipitation fell. Figure 6 shows the lower parts of the ^{18}O – profiles of the ice cores from Dye 3, S.E. Greenland and Camp Century, N.W. Greenland (Daansgard et al. 1982). Clearly visible is the transition from Wisconsin to Holocene which occurred 13 000 to 10 000 radiocarbon years before present, a period with rapid changes between a cold and a mild climate state ca. 80 000 to 30 000 years before present and in the lowest part of the Camp Century core the transition to the Eemian interglacial with higher ^{18}O values than in the present interglacial. In the ice core from Vostok Station, Antarctica, the climate information even goes 160 000 years back in time and at the bottom shows the Eemian interglacial and the transition into the last part of the penultimate glaciation (Fig. 7).

5.4 Ocean Sediment – Ice Core Link

A link of special importance is provided by the $^{18}O/^{16}O$ ratio in the ice core and the ocean sediment records: ^{18}O changes in ocean water reflect changes in the continental ice volume. Compared to the ocean water, the continental ice of the last glaciation ^{18}O was by about 35% lower. An increase of the continental ice mass corresponds to a deposition of water, low in ^{18}O, stemming from the oceans and thus an increase of ^{18}O in the ocean water. ^{18}O of ocean water at a certain time is recorded in the $^{18}O/^{16}O$ ratio of the carbonate of foraminifera which lived at that time in the ocean water. Measurements of ^{18}O in the carbonate of foraminifera shells from ocean sediments enable reconstruction of the history of continental ice volume. In fact, the ^{18}O variations measured on deep ocean sediment samples are the

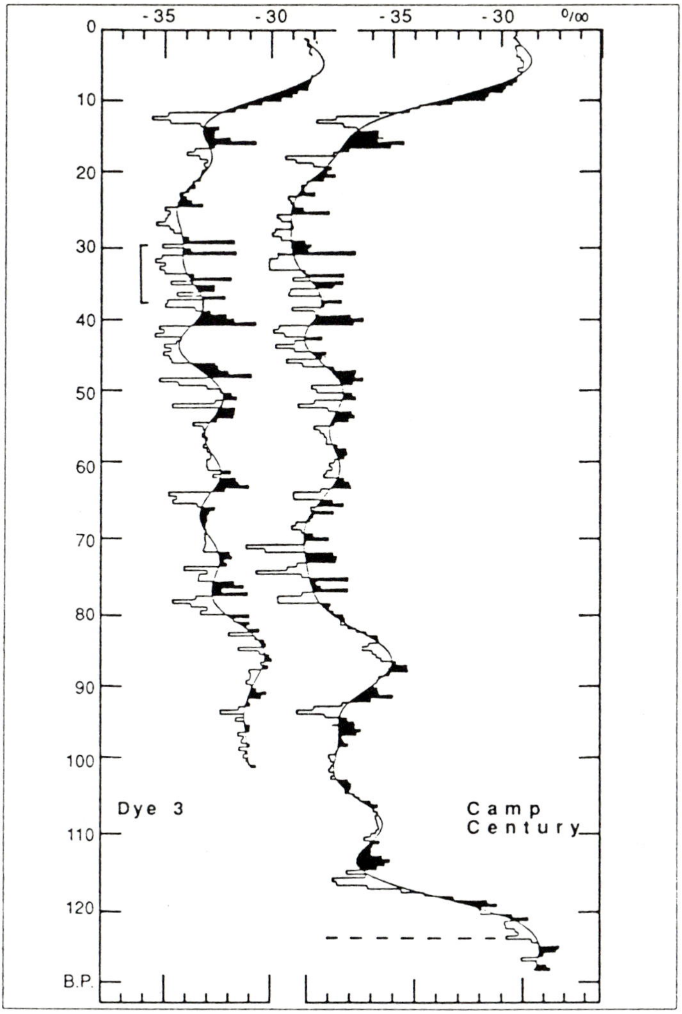

Fig. 6. $\delta^{18}O$ profiles for Greenland ice cores showing rapid bimodal fluctuations during the last glacial period. (Dansgaard et al. 1982)

master record of the ocean history covering the last million years. One of the most exciting new results in this field are the observations that the oceanic ^{18}O variations are directly reflected in the ice record: via photosynthesis and respiration in the ocean surface water atmospheric O_2 depicts ^{18}O changes in the surface water (Bender et al. 1985). ^{18}O measurements on the atmospheric oxygen extracted from ice cores do indeed reflect the changes expected from the observations on shells of planktonic foraminifera sampled in deep sea sediments.

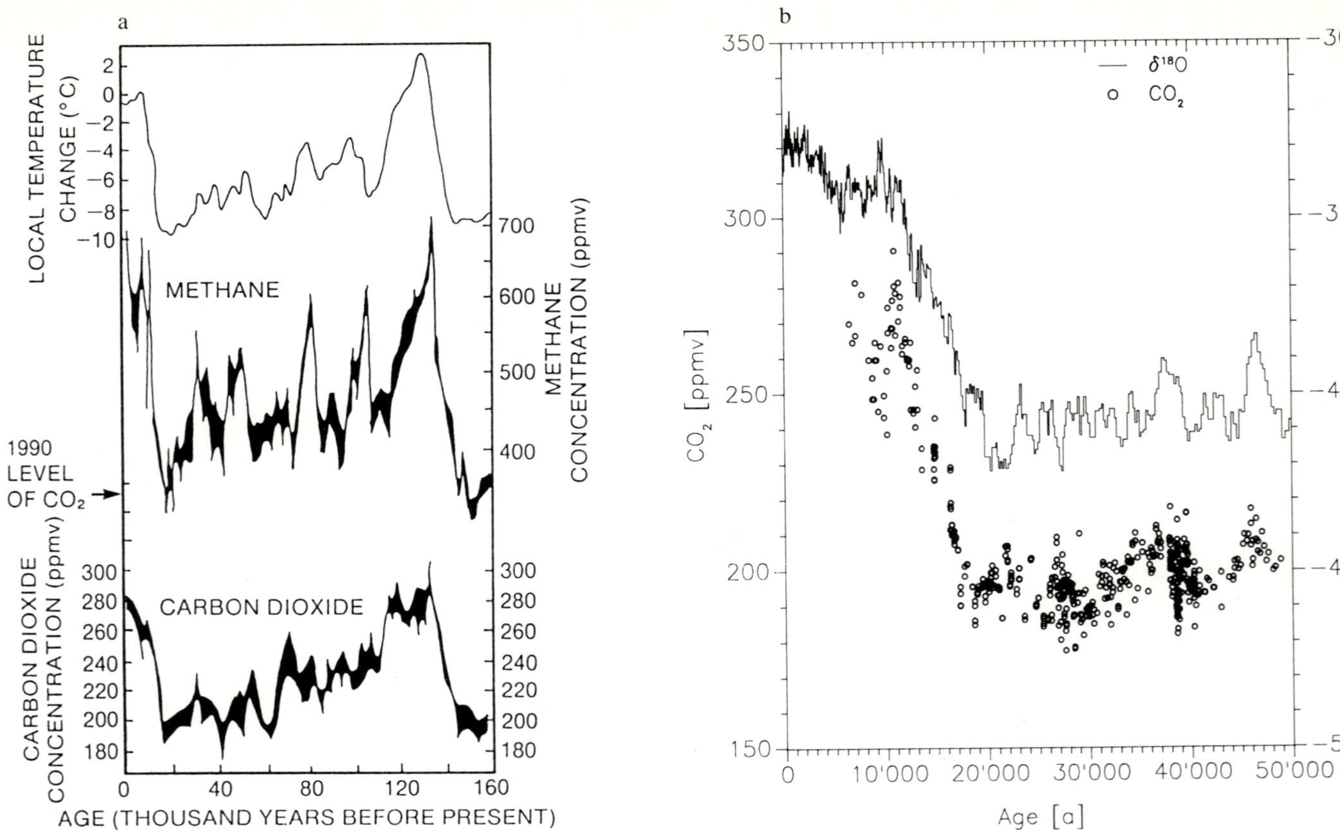

Fig. 7. **a** Vostok, Antarctica, ice core records for methane (*top curve*, Chappellaz et al. 1990), deviations in isotope temperature above the inversion layer (as a difference from the modern value of ca. −40 °C (*middle curve*, Jouzel et al. 1987) and carbon dioxide (Barnola et al. 1987). The CH_4 and CO_2 curves have been adjusted to fit the time-scale given (taking into consideration gas occlusion time). **b** Variations in CO_2 and $\delta^{18}O$ at Byrd Station, Antarctica, for the last 50 000 years (Neftel et al. 1988, and additional unpub. data). The 10% drop in CO_2 at the end of the last deglaciation (10 000 years B.P.) may have been caused by vegetative regrowth following the melting of continental ice sheets

5.5 ^{10}Be and ^{14}C Variations Give Information on Solar History

Interestingly enough, through changes of the concentration of radioisotopes produced by cosmic radiation in the atmosphere the sun writes its history into the atmosphere. As is known from cosmic ray produced neutron monitoring in the atmosphere, during periods of high solar activity the interior part of the solar system is more effectively shielded against the galactic cosmic radiation than during low solar activity. Similar to the neutron production the radioisotope production is modulated by the changing heliomagnetic shielding. ^{10}Be, a radioisotope with a half-life of 2 million years, shows these modulations almost directly and unattenuated. After its production it is attached to aerosol particles and deposited with precipitation on the ground a few month to a few years after production. Recent high resolution measurements of ice cores enabled the identification of the 11-year cycle of solar modulation and its production; during low sun spot activity the concentration in the ice is by about a factor of 1.5 higher than during high sun spot activity. It is known from direct boservation of the sun that the sun spot cycle is modulated. During periods of quiet sun, like the Maunder Minimum of solar activity, when essentially no sun spots were observed, the ^{10}Be concentration was increased by a factor of 1.6.

In the following, some pertinent results of ice core studies are shortly discussed. They comprise the reconstruction of the anthropogenic CO_2 increase and the last 150 000 years of history of climate and trace gases.

6 The Anthropogenic CO_2 Increase

6.1 The Control of the CO_2 Gas Concentration

During geologic times, the temperature of the Earth seems to have remained in the range of 0 °C to 100 °C. It is assumed that the control of the temperature is partly due to the chemical cycles in the Earth's crust, soils and ocean, regulating the atmospheric CO_2 content and thus the planet's greenhouse effect.

The atmospheric CO_2 concentration on time scales relevant to the human impact is controlled by the CO_2 exchange fluxes between the atmosphere and the terrestrial biota as well as between the atmosphere and the surface water of the ocean. By comparison, the inputs into the atmosphere from fossil fuel combustion and deforestation are much smaller. This implies the question whether the presently observed increase of the atmospheric CO_2 concentration is really due to human activities and not just the expression of an imbalance in the exchange fluxes between carbon reservoirs.

The most convincing answer comes from reconstruction's of the atmospheric CO_2 concentration history by the measurement of the CO_2 concentration of air in natural ice samples of known age. These measurement provide convincing evidence that the CO_2 increase of the last two centuries is an anthropogenic phenomenon, and also that, especially during periods of major climatic change, there were natural atmospheric CO_2 variations. In fact, the atmospheric CO_2 concentration turns out to be a most important paramenter reflecting the integrated physical. chemical and biological state of the Earth system.

6.2 The Anthropogenic CO_2 Increase and the CO_2 Production History

A consistent reconstruction of the atmospheric CO_2 concentration of the past two centuries has been obtained from an ice core from Siple Station, Antarctica (Fig. 8; Neftel et al. 1985; Friedli et al. 1986; Keeling et al. 1989). Since the atmospheric CO_2 is relatively well mixed, the Antarctica data are representative for the global CO_2 concentrations. The data obtained from the youngest ice samples overlap with the atmospheric data obtained at Mauna Loa, Hawaii. These results indicate that CO_2 started to rise around 1800 A.D. and showed a slight increase in the nineteenth century. In the twentieth century the growth rate further increased. The present CO_2 concentration has reached 353 ppm; the mean growth rate is about 1.8 ppm per year.

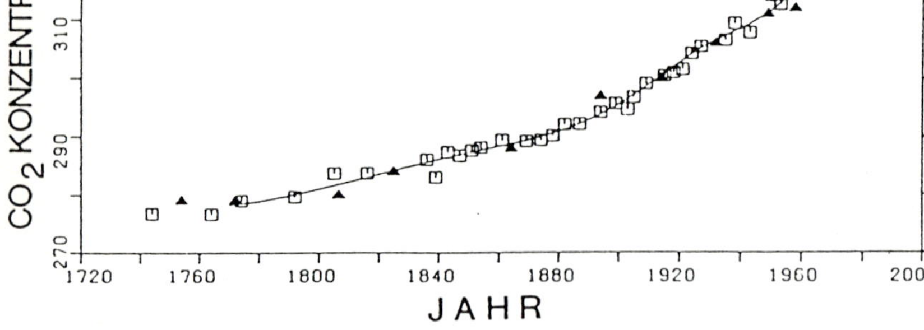

Fig. 8. Atmospheric CO_2 increase in the past 250 years, as indicated by measurements on air trapped in ice from Siple Station, Antarctica (*squares*, Neftel et al. 1985; Friedli et al. 1986) and by direct atmospheric measurements at Mauna Loa, Hawaii (*triangles*, Keeling et al. 1989)

How does this increase compare with the estimated CO_2 emissions and model predictions of the uptake of excess CO_2 by the carbon system? Part of the anthropogenic CO_2 can be taken up by the ocean and the terrestrial biomass. The net flux of CO_2 into the ocean is given by the product of a gas transfer coefficient and the CO_2 partial pressure difference between atmosphere and ocean. The partial pressure of CO_2 in the ocean surface depends on the rate at which anthropogenic CO_2 is transported from the surface to deeper ocean layers. From measurements of the radioactive isotope ^{14}C, it is known that in most ocean regions only the top several hundred metres have taken up significant amounts of anthropogenic CO_2. An exception is the North Atlantic where bomb-produced tritium has been observed even near the bottom of the ocean. The transport of biogenic detrital particles, which is important for the natural carbon cycle, does not significantly contribute to a sequestering of excess CO_2, since the activity of the marine biota is essentially controlled by other factors, such as light, temperature and limiting nutrients.

Carbon cycle models used to simulate the atmosphere-ocean system have been highly simplified, consisting only of a few well mixed or diffusive reservoirs. In these models, the oceanic transport mechanisms are parametrized. The parameters are derived from observations of (transient) tracers, like bomb-produced ^{14}C, that are analogues to the flux of anthropogenic CO_2 into the ocean.

A higher atmospheric CO_2 concentration leads to a fertilization effect, i.e. to a higher CO_2-flux into the land biosphereic reservoir. To estimate the global average fertilization effect is, however, very difficult, It is, therefore, preferable to combine the uptake of CO_2 due to fertilization and the CO_2 release due to deforestation and land managing changes into one quantity: the net biospheric CO_2 production, $P_{bio}(t)$. By deconvolution of the observed CO_2 increase O(t), using the response R(T) of the atmosphere-ocean system to an atmospheric CO_2 pulse input:

$$O(t) = \int_0^\infty P_{tot}(t - T)R(T)dT,$$

the total CO_2 production function $P_{tot}(t) = P_{bio}(t) + P_{fo}(t)$ can be calculated. The result is shown in Fig. 9 (Siegenthaler and Oeschger 1987). The total CO_2 production is relatively constant between 1780 and 1860. It is followed by a bump around 1910–1920 and steep increase after 1940, which seems to be essentially determined by the fossil CO_2 production $P_{fo}(t)$. The estimated net biospheric production rate $P_{bio}(t)$, $(P_{bio}(t) = P_{tot}(t) - P_{fo}(t))$, was roughly constant during the nineteenth century. During the twentieth century it decreased to about zero after 1940. It seems that during the past 40 years the release of CO_2 due to deforestation and changes in land managing and the uptake due to fertilization roughly cancelled. This observation is strongly supported by recent deconvolution calculations by Joos et al. (1992),

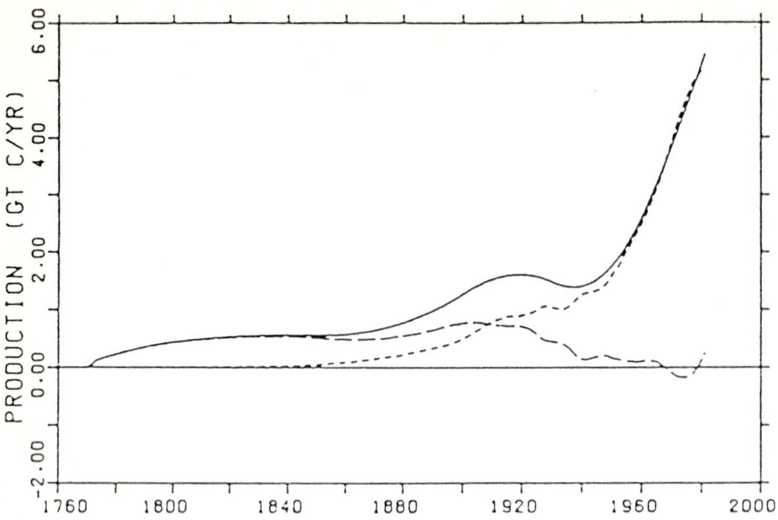

Fig. 9. CO_2 production rates obtained by deconvolving the measured CO_2 increase using a box-diffusion model for the CO_2 uptake by the oceans. (Siegenthaler and Oeschger 1987); *solid line* total production; *short-dashed line* estimated fossil CO_2 input; *long-dashed line* difference = CO_2 net biospheric input

concerning the atmospheric CO_2 increase from 1950–1990. In 1973 the growth of the fossil CO_2 production rate decreased from 4.4%/year to 2%/year. This decrease enabled a sensible carbon cycle model test: the deconvolved total CO_2 production during the entire time period agrees within +/− 1 GT C/y with the fossil production which in 1990 amounted to 6 GT C/y. If the net biospheric production had been comparable to the fossil production, then the change in the fossil CO_2 growth rate would not be so clearly visible in the observational date and the deconvolved production would have started to deviate.

In view of the complexity of the Earth system processes involved, it may be surprising that it was possible to

- determine, rather accurately the pre-industrial CO_2 concentration and its increase before the atmospheric measurements started;
- established trustworthy estimates of the CO_2 uptake by the ocean, and to arrive at a sensible carbon budget. In addition, it is not self-evident that the CO_2 system operated during the period of observation in the same manner as during the period of the past 30 years when it was calibrated. Also, the temperature increase of the past 100 years may have had an imprint on the system dynamics. Based on the ice core measurements of the past pre-industrial millennium, small natural CO_2 variations of the order of one to a few ppm/century cannot be excluded even during a period of minor climatic changes.

7 Climate and Trace Gas Variations of the Past 160 000 Years

7.1 Overview on the Climatic Variations of the Last Million Years

As a background to the following discussion of examples of ice core information in pre-industrial time a short overview on prominent climatic events in the recent Earth history is given below (Fig. 10; Folland et al. 1990).

This part of the Earth's history was dominated by the glacial-interglacial cycles with a pronounced period of 100 000 years. The global temperature varied typically by 5–7 °C, with large changes in ice volume and sea level.

The transition to the present interglacial took place 15 000 to 10 000 years ago. During the Holocene global temperature varied at most by 1 to 2 °C and 5000 to 6000 years ago worldwide temperatures were higher than at present.

There were relatively strong climatic variations during the past 1000 years. In the period from 1000 to 1300 A.D. climate in Europe was exceptionally warm; this period is called the Mediaeval warm period. But there were also cold, so-called neo-glacial episodes, the "Little Ice Age" 150 to 450 years ago deserving special interest.

7.2 The Glacial-Interglacial Cycles and the Atmospheric Trace Gases

The record of the ice ages was first seen in ocean sediment cores but in the last decade it was also clearly observed in the ice cores from Greenland and Antarctica. Figure 7 presents dT, CO_2 and CH_4 data spanning the past 160 000 years for a core drilled in Vostok Station, Antarctica. The previous and most recent deglaciations are marked by large, abrupt increases in temperature (ca. 9 °C), CO_2 (100 ppm) and CH_4 (300 ppb). The transition to glacial conditions started about 130 000 years ago but the glacial maximum occurred about 20 000 years ago. The Greenland ice cores show essentially the same trend, but much more variability during the glacial period, which will be discussed in the following section. In fact, a major contribution to our understanding of the Earth system has been the observation of a strong internal variability which till present can clearly be seen only in the high resolution ice core records.

7.3 Rapid Climatic Oscillation During the Previous Glacial Period

From Fig. 6 it can be seen that in Greenland between 80 000 and 30 000 years B.P. temperature (as reflected by ^{18}O) fluctuated back and forth between a cold and a mild climate state. The changes between these two states are reflected in all the parameters measured hitherto. They fall into

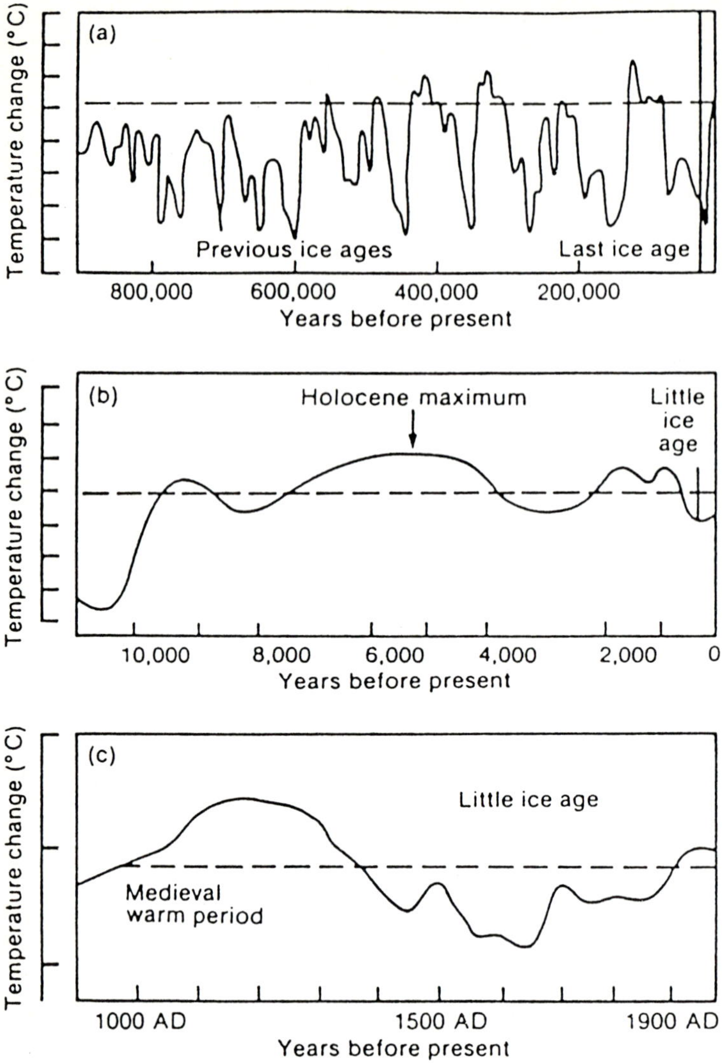

Fig. 10. Schematic diagrams of global temperature variations since Pleistocene: **a** the last million years; **b** the last ten thousand years; **c** the last thousand years. The *broken line* indicates conditions near the beginning of the twentieth century. (Folland et al. 1990)

distinct, narrow bands, suggesting that during the last glacial period the Earth system oscillated between two clearly defined, quasi-stable states (Oeschger et al. 1985).

It has been hypothesized that the main driver for the climatic oscillations is changing circulation in the North Atlantic Ocean (Broecker et al. 1985).

When water vapour is exported from one ocean basin drainage to another, salt is left behind by evaporation, increasing the salinity and density of the surface water mass, thus enhancing deep water formation. This, in turn, affects ocean circulation, sea surface temperature and the amount of heat available to warm the air currents that regulate the climate of the North Atlantic region, particularly the European subcontinent. The short duration of the bimodal North Atlantic oscillations precludes orbital forcing mechanisms and suggests the possibility of oscillations internal to the Earth system, resulting from complex Earth system feedback mechanisms (Broecker et al. 1985). Broecker and Denton (1989) point out that the ocean-atmosphere system is susceptible to mode switches, because of the non-linear coupling of these two reservoirs. Changes in water vapour transport, for example, can lead to changes in both the rate and pattern of ocean circulation (Broecker and Denton 1989).

Observed CO_2 variations deserve special attention. For the period 40000 to 30000 years, B.P. values between 180 to 200 ppm and 240 to 260 ppm have been observed for the cold and the mild climate states, respectively (Stauffer et al. 1984).

In contrast to the climatic signal recorded in the Greenland ice cores, the Antarctic cores lack a pronounced bimodal signal (Fig. 7a) (Neftel et al. 1988). One explanation may be that the switching in North Atlantic ocean dynamics does not produce large sharp climatic variations in the southern hemisphere but rather a smoother climate variability with milder periods reflecting clusters of North Atlantic flip-flops. Difficulties have hitherto been encountered regarding the detection of a bimodal CO_2 signal similar to that found in the Greenland cores. A series of detailed CO_2 measurements performed on the section of the Byrd station, Antarctica, ice core, which should have recorded rapid CO_2 variations, did not show such a characteristic signal, though indications of smaller variations were found. One explanation for the discrepancy may be a smoothing of the CO_2-variations due to the longer occlusion time of air in the Byrd core relative to the Greenland cores. It will be extremely important to check whether the rapid CO_2 variations will be detected in the ice core from Summit, Greenland, where bedrock was reached this summer. Drilling and analysis of a new, high accumulation ice core in Antarctica, in which the rapid CO_2 variations should be resolved, deserves high priority.

The atmospheric CO_2 concentration is an especially interesting, integrating parameter of the Earth system:

- Depending on the efficiency of the global oceans' biological pump which is steered by the oceans circulation the atmosphere CO_2 concentration can vary between 150 ppm (use of all the nutrients in the surface water) and 500 ppm (no biological activity in surface ocean). Thus a change in ocean circulation should be reflected in the atmospheric CO_2 level.

- Changing sea level and climate affect the amount of biomass stored on the continents and thus lead to changes in atmospheric CO_2.
- Changes in ocean chemistry (alkalinity) affect the partitioning between dissolved CO_2, bicarbonate and carbonate in the ocean surface effecting the atmospheric CO_2 concentration.

7.4 The Last Glacial Postglacial Transition

The transition from glacial to postglacial conditions in Greenland and Antarctica occurs at the same time. The timing of the transition can be inferred from changes in the CO_2 concentration. However, in Antarctica the transition is more monotonous than in Greenland, where strong climatic variations are superimposed on the general warming trend.

In Greenland (Dye 3 and Camp Century), initial warming occurred about 13000 years B.P. (Fig. 6). After a warm period of about 2000 years, a strong cooling followed which lasted about 1000 years. At about 10000 years B.P., within less than 100 years, the final transition to the Holocene followed (Daansgard et al. 1984). This climate sequence, even in the details, matches the $\delta^{18}O$ record in lake marl from Lake Gerzensee, Switzerland, and many other European records that have been radiocarbon-dated (Oeschger et al. 1984) (Fig. 11). The ^{14}C ages cited for the transition in the different archives differ somewhat from the absolute ages. Shortly before the initial warming (13000 years B.P.) the CO_2 concentration was still low (ca. 200 ppmv). During the Bolling-Allerod warm period, CO_2 was about 300 ppmv, and there are indications of lower values (250 ppmv) during the Younger Dryas cold period. However, the high values (>300 ppmv) at the beginning of the Holocene indicate that, during the postglacial, melt layers resulted in CO_2 values higher than the atmospheric concentrations.

In the Antarctic ice cores (see e.g. Fig. 7a,b), the increase in stable isotopes and CO_2 occurred parallel and, within the uncertainties, in phase. In all the Antarctic cores a slight cooling in the upper part of the increase is indicated. In the Vostok core it is of the order of 2 °C. Given the uncertainties in the time scales, this cooling probably corresponds to the Younger Dryas cold period observed in the Northern Hemisphere. This correspondence is supported by a significant decrease in the CH_4 concentrations occurring in the Greenland cores as well as in the Vostok core.

The problem of assessing the possible causes for the Younger Dryas is currently attracting considerable interest among the scientific community. The short duration of the cold event makes it difficult to tie it to cycles of orbital forcing; and it has been hypothesized that the oscillation is related to changes in the heat input into the North Atlantic caused, specifically, by the interruption of deep water formation in the Norwegian Sea. The Younger Dryas does not seem to be a unique event, but rather the last of a series of

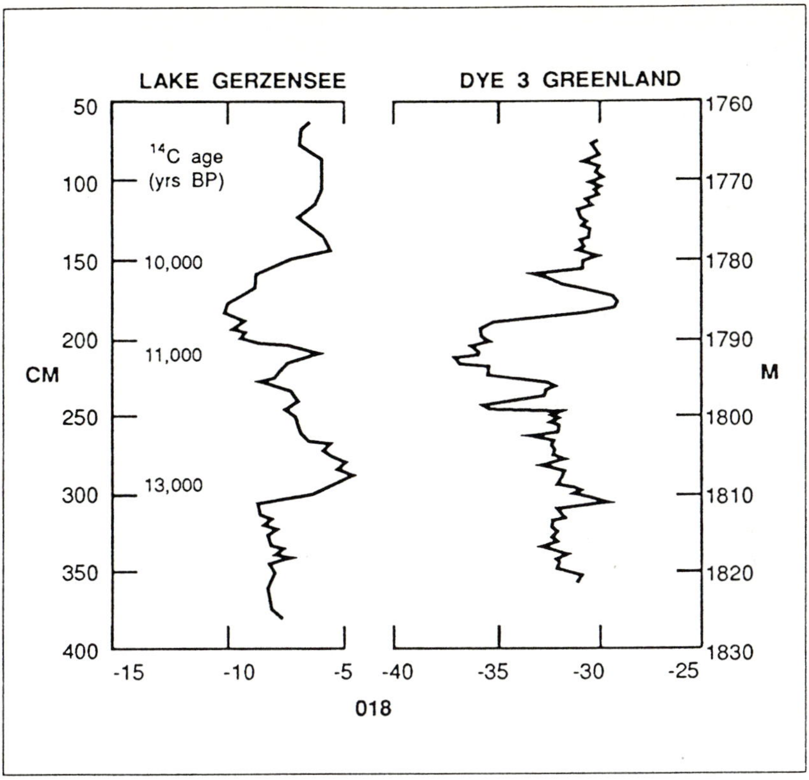

Fig. 11. Comparison of a section of the $\delta^{18}O$ profile from the Dye 3 ice core with the $\delta^{18}O$ record in lake carbonate from lake Gerzensee near Bern. The strong similarity suggests that both records represent the same sequence of climatic events

similar events during a major part of the last glaciation (see discussion of rapid oscillations in Greenland ice cores, above). The importance of ocean circulation in influencing the state of the climate system has been recognized and theoretically studied since the 1960s (e.g. Stommel 1961; Manabe and Stouffer 1988); and the possibility of more than one stable climatic state was shown by these researchers. New research will focus on two fundamental questions:

- Was the Younger Dryas a unique event related to the deglaciaton of North America, or rather the last of a series of similar events during a major part of the last glaciation?
- How was the signal of North Atlantic climate change spread over the globe? Candidates are via changes in ocean circulation and by reduction in the greenhouse effect (especially if not only CH_4, but also CO_2 showed a significant decrease).

7.5 The Holocene Climate

Was the pre-industrial Earth system in equilibrium? Records of the last 10 000 years are smoother and the climate appears to be more stable than during glacial times. These differences suggest that the dynamics of the climate system may be dependent on the state of the climate at any given time.

An interesting observation concerns the carbon balance: Measurements on the Byrd Station core, Antarctica, show a decrease in CO_2 from about 280 ppmv to 250 ppmv at the beginning of the Holocene (Fig. 7b), during a period of about 4000 years (Neftel et al. 1988). A probable explanation for this rapid decrease in CO_2 is the regrowth of soils and plants on the continents in areas which had been covered by ice during the preceding glacial period. Assuming that ocean chemistry was determined by processes similar to those operating in the present-day ocean, the 10% decrease in atmospheric CO_2 (requiring a 10% decrease in the partial pressure of CO_2 in the surface ocean) corresponds to an extraction of 400 to 500 Gt of carbon from the ocean-atmosphere system during this 4000-year period. In this long-lasting change in the CO_2 system, interaction with the ocean sediments did play a role.

8 Recent Results from Greenland Ice Cores; Answers and New Questions

In summer 1992, the European team conducting the Greenland Ice Core Project (GRIP) drilled to bedrock on the summit of the Greenland ice sheet and recovered a 3029 m long ice core which had already been partly studied in the field. In summer 1993, the US team conducting the Greenland Ice Sheet Programm 2 (GISP2) drilled to bedrock 28 km west of the summit. Though only some of the results of the analysis of the two cores are available at present, a wealth of new information evolved with, in part, enormous implications regarding Earth system science and the greenhouse warming issue. In the following, a short overview is given.

8.1 Confirmation of the Wisconsin Rapid Changes; Additional Evidence from North Atlantic Ocean Sediment Cores

The rapid glacial climatic variations discussed in Section 7.3 were confirmed in both new Greenland ice cores (GRIP Members 1993; Dansgaard et al. 1993; Grootes et al. 1993). The ^{18}O data in both cores suggest that the climate of the North Atlantic region during the Holocene had been remarkably stable, but showed large variations throughout the last glacial period, in agreement with those shown in Fig. 6.

The report of Bond et al. (1993) indicates correlations between the climatic records from North Atlantic sediments and Greenland ice cores. The North Atlantic sediments show rapid temperature oscillations closely matching those in the ice core records. They also show that the shifts in atmosphere-ocean temperature are bundled into cooling cycles lasting 10000 to 15000 years. Each cycle ended in a cold climatic state which showed ice-rafted debris, indicating an enormous discharge of icebergs into the North Atlantic. These cooling cycles may be driven by internal ice-sheet instabilities and the three to four rapid changes in a cycle by internal instability of the North Atlantic deep-water formation.

8.2 The Surprisingly Instable Eemian Climate

Isotope and chemical analyses of the GRIP ice core from Summit Greenland reveal that the last interglacial, the Eem/Sangoman interglacial, was interrupted by a series of cooling events which lasted from decades to centuries (GRIP Members 1993; Dansgaard et al. 1993). The cooling by several degrees centigrade occurred in years to decades. The Eemian interglacial seems to have been 2 to 3 °C warmer than the Holocene. This raises questions regarding the stability of the future, globally warmer climate due to the increasing greenhouse effect.

Studies of the GISP2 ice core drilled to bedrock in summer 1993 revealed a core section with similar climatic characteristics, but strong indications of rheological disturbances close to the bedrock (Taylor et al. 1993). At present, it is being considered, despite these indications, whether it will be possible to extract meaningful climatic information from the two cores. It seems, however, probable that the general instable character of the warm parts of the cores from the Eemian is not a rheological artifact.

8.3 The Greenhouse Gas Profiles; Open Questions

Methane measurements on the GRIP ice core (40000 to 8000 years old) show that concentrations varied parallel to the climatic variations, with regard to both the glacial variability and the warming to the present interglacial (Chappelaz et al. 1993). The authors propose that variations in the hydrological cycle at low latitudes may have been responsible for the variations in both methane and Greenland temperatures during the mild glacial periods.

Still open is the question whether the rapid changes during the glacial period were accompanied by significant CO_2 changes, as suggested from the data obtained on the older ice cores. CO_2 measurements in Greenland ice cores seem to be complicated because of interactions of CO_2 with chemical constituents in the ice.

These new discoveries have a dramatic impact on the science of global change and the awareness of the anticipated global warming (White 1993).

9 Recent Developments in Global Change Science; the Impact of Ice Core Results

9.1 Our Understanding of Climate

In the last decades climate was considered to be essentially driven by the orbital forcing. Power spectra show peaks for the periods 100k, 40k and 20k. The high resolution record obtained about 10 years ago in the Greenland ice cores strongly indicates that the variability of the internal Earth system plays an important climatic role. One might consider climate as an external (orbital) modulation of internal oscillators, primarily the North Atlantic deep-water formation flip-flops. The strong impact of the new Greenland results, which show drastic climatic changes in the Eemian interglacial, on the one side, and a relatively stable Holocene, on the other side, is a challenge for continental paleoscientists: they question whether the quiet Greenland Holocene is also representative for the lower latitudes where most of the world population lives. They are eager to recover paleorecords of the past few hundred thousand years from sites at lower latitudes which have a similarly high resolution and representative information as the ice cores. A high priority goal of the IGBP core project, Past Global Changes (PAGES), is to help to establish a continental network of paleorecords of hitherto unachieved quality and to study them with state-of-the-art-analytical techniques.

9.2 Climate Variability in the Center of Global Change Science

In the 1980s, when the WCRP and the IGBP were established, strong emphasis was placed on the modelling of the Earth system. Many key workers emphasized the need for faster computers and it is expected that modelling and predicting future climate will eventually become possible. Others emphasized the unexpected events which took place in the Earth system in the past, revealed by studies of information from natural chronicles like ice cores, ocean and lake sediments, peat bogs, tree rings, etc. Important examples are discussed in this chapter. Had these events been produced by model experiments they probably would have been considered as artifacts. During the last decade, the limits of Earth system modelling and the value of reconstructed past Earth system events have become evident.

The WCRP is faced with the task of improving the accuracy of seasonal to interannual climate prediction through programmes of coupled modelling of the global upper ocean, atmosphere, and land and ice systems. CLIVAR has been established to study climate variability. This project was initially intended to draw mainly from instrumental data, but information from high resolution paleodata, mainly from corals, must also be considered.

The concept of this project in the last year has changed. In addition to seasonal and interannual variability, interdecadal and centennial variability should also be addressed and the model experiments be interactive with coral data, including also information from ice cores, tree rings and other continental records. This should be accomplished within the framework of a CLIVAR/PAGES collaboration. Thus the importance of climate variability in the assessment of the human impact on the climate system has contributed to great advances in Gobal Change science. This new, interactive approach of modelers and paleoclimatologists is a promising step forward in this area of societal concern.

Acknowledgements. Thanks are due to Prof. G. Hempel for his continuous interest and instrumental support regarding the decision of the European Science Foundation to conduct the Greenland Ice Core Project in the frame of the Polar Network.

This paper summarized discussions in ice core studies with my colleagues. Ch. Langway, W. Dansgaard, C. Lorius and B. Stauffer. It has been also strongly influenced by the inspiring exchange of ideas with W. Broecker.

References

Barnola JM, Raynaud D, Korotkevitch YS, Lorius C (1987) Vostok ice core: a 160000 year record of atmospheric CO_2. Nature 329:408–414

Bender M, Labeyrie LD, Raynaud D, Lorius C (1985) Isotopic composition of atmospheric O_2 in ice linked with deglaciation and global primary productivity. Nature 318:349–352

Bond G, Broecker W, Johnson S, McManus J, Labeyrie L, Jouzel J, Bonani G (1993) Correlations between climate records from North Atlantic sediments and Greenland ice. Nature 365:143–147

Broecker WS, Denton GH (1989) The role of ocean-atmospheric reorganizations in glacial cycles. Geochim Cosmochim Acta 53:2465–2501

Broecker WS, Peteet D, Rind D (1985) Does the ocean-atmosphere system have more than one stable mode of operation? Nature 315:21–25

Chappellaz J, Barnola JM, Raynaud D, Korotkevich YS, Lorius C (1990) Ice-core record of atmospheric methane over the past 160000 years. Nature 345:127–131

Chappelaz J, Blunier T, Raynaud D, Barnola JM, Schwander J, Stauffer B (1993) Synchronous changes in atmospheric CH4 and Greenland climate between 40kyr and 8kyr BP. Nature 366:443–445

Daansgard W, Clausen HB, Gundestrup N, Gundestrup N (1973) Medd Gronl 197, 2

Daansgard W, Clausen HB, Gundestrup N, Hammer CU, Johnsen SF, Kristinsdottier PM, Reeh N (1982) A new Greenland deep ice core. Science 218:1273–1277

Daansgard W, Johnson SJ, Clausen HB, Dahl-Jensen D, Gundestrup N, Hammer CU, Oeschger H (1984) North Atlantic climatic oscillations revealed by deep Greenland ice cores. In: Climate processes and climate sensitivity, M Ewing Symp, vol 5. Geophys Monogr 29:288–298

Dansgaard W, Johnson SJ, Clausen HB, Dahl-Jensen D, Gundestrup NS, Hammer CU, Hvidberg CS, Steffensen JP, Sveinsbjoernsdottir AE, Jouzel J, Bond G (1993) Evidence for general instability of past climate from a 250-kyr ice-core record. Nature 364:218–220

Folland CK, Karl TR, Vinnikov KYA (1990) Observed climate variations and change. In: Houghton JT, Jenkins GJ, Ephraums JJ (eds) Intergovernmental Panel on Climate Change (IPCC) Climate change – the IPCC scientific assessment. Cambridge University Press, Cambridge

Friedli H, Loetscher H, Oeschger H, Siegenthaler U, Stauffer B (1986) Ice core record of the $^{13}C/^{12}C$ record of atmospheric CO_2 in the past two centuries. Nature 324:237–238
Greenland Ice Core Project (GRIP) Members (1993) Climate instability during the last interglacial period recorded in the GRIP ice core. Nature 364:203–207
Grootes PM, Stuiver M, White JWC, Johnson S, Jouzel J (1993) Comparison of oxygen isotope records from the GISP2 and GRIP Greenland ice cores. Nature 366:552–554
Intergovernmental Panel on Climate Change (1990) Climate change - the IPCC scientific assessment. In: Houghton JT, Jenkins GJ, Ephraums JJ (eds) Cambridge University Press, Cambridge
Intergovernmental Panel on Climate Change (IPCC) (1992) Climate change 1992 – the supplement report to the IPCC scientific assessment. In: Houghton JT, Callander BA, Varney SK (eds) Cambridge University Press, Cambridge
Joos F, Oeschger H, Siegenthaler U (1992) Control of fossil-fuel use and atmospheric carbon dioxide. Unpublished
Jouzel J, Lorius C, Petit JR, Genthon C, Barkov NI, Kotlyakov V, Petrov VN (1987) Vostok ice core: a continuous isotope temperature record over the last climatic cycle (160000 years). Nature 329:403–409
Keeling CD, Bacastow RB, Carter AF, Piper SC, Wohrf TP, Heimann M, Mook WG, Roeloffzen H (1989) A three dimenional model of atmospheric CO_2 transport based on observed winds. 1. Analysis of observational data in aspects of climate variability in the Pacific and the Western Americas. Geophys Monogr 55:165–236
Lorius CJ (1989) Polar ice cores and climate. In: Berger A et al. (eds) Climate and geosciences. Kluwer, Dordrecht, pp 77–103
Manabe S, Stouffer RJ (1988) Two stable equilibria of a coupled ocean- atmosphere model. J Climate 1:841–866
Neftel A, Moor E, Oeschger H, Stauffer B (1985) Evidence from polar ice cores for the increase in the atmospheric CO_2 in the past two centuries. Nature 315:45–47
Neftel A, Oeschger H, Staffelbach T, Stauffer B (1988) CO_2 record in the Byrd ice core 50000–5000 years BP. Nature 331:609–611
Oeschger H, Langway CC Jr (eds) (1989) The environmental record in glacier, and ice sheets. Dahlem Workshop Reports. John Wiley, New York
Oeschger H (1990) The ocean in human affairs. In: Singer S (ed) Paragon House, New York
Oeschger H, Beer J, Siegenthaler U, Stauffer B, Dansgaard B, Langway CC Jr (1984) Late glacial climate history from ice cores. In: Hansen JE, Takahashi T (eds) Climate processes and climate sensitivity, M Ewing Symp, vol 5. Geophys Monogr 29:299–306
Oeschger H, Stauffer B, Finkel R, Langway CC Jr (1985) Variations of the CO_2 concentration of occluded air and of anions and dust in polar ice cores. In: Sundquist ET, Broecker WS (eds) The carbon cycle and atmospheric CO_2: natural variations archean to present. Geophysical Monograph 32:132–142
Siegenthaler U, Oeschger H (1987) Biospheric CO_2 emissions during the past 200 years reconstructed by deconvolution of ice core data. Tellus 39B:140–154
Stauffer B (1988) Ziele und Ergebnisse klimatologisch-glaziologischer Forschungsprojekte in der Arktis und Antarktis in "Die Polarregion und die schweizerische Forschung", Publ Schweizerischen Naturforschenden Gesellschaft, Bd 2
Stauffer B, Hofer H, Oeschger H, Schwander J, Siegenthaler U (1984) Atmospheric CO_2 concentration during the last glaciation. Ann Glaciol 5:160–164
Stommel H (1961) Thermohaline convection with two stable regimes of flow. Tellus 13:224–230
Taylor KC, Hammer CU, Alley RB, Clausen HB, Dahl-Jensen D, Gow AJ, Gundestrup NS, Kipfstuhl J, Moore JC, Waddington ED (1993) Electrical conductivity measurements from the GISP2 and GRIP Greenland ice cores. Nature 366:549–552
White JWC (1993) Climate change: don't touch that dial. Nature 364:186

Biological Impacts of Seasonal Ozone Depletion

Harvey J. Marchant[1]

1 Introduction

Since the mid-1970s there has been a marked reduction in the concentration of Antarctic stratospheric ozone during springtime (Fig. 1). These observations were made originally by ground-based instruments but more recently satellite observations have been used to gain a broader picture of the so-called Antarctic ozone hole which develops in September and dissipates in November (Stolarski et al. 1986). The processes involved in the formation and breakup of the ozone hole are discussed by Rycroft (this Vol.). Following the breakdown of the polar vortex, ozone depleted stratospheric air is transported to midlatitudes (Atkinson et al. 1989). A consequence of a reduction in stratospheric ozone concentration is an increase in the amount of ultraviolet (UV) radiation reaching the surface of the Earth. The springtime UV irradiance at Antarctic coastal sites has been found to be as high or higher than at the summer solstice (Frederick and Snell 1988; Fig. 2). As well as an overall increase in the total amount of incident UV, reduced stratospheric ozone concentration leads to an increase in the amount of short wavelength UV (Fig. 3). The biological impact of UV irradiation is extremely wavelength dependent (Caldwell 1981; Smith and Baker 1989).

UV radiation is strongly absorbed by nucleic acids, proteins and other constituents of the cells of organisms and interferes in the processes of growth and reproduction. Any increase in the flux of UV radiation has deleterious effects on plant and animal life (Harm 1980; Tevini and Teramura 1989). Not surprisingly, the biological effects of enhanced UV exposure has attracted considerable attention. There is a substantial and fast growing literature on the biological effects of enhanced UV radiation, especially in aquatic ecosystems (see Calkins 1982; Worrest 1983; Hardy and Gucinski 1989; Häder and Worrest 1991 for reviews). Detailed reviews of the impacts of enhanced UV irradiation on Antarctic biota have been given by Bidigare (1989), Voytek (1990) and Karentz (1991). Here I give a brief overview of the potential and actual impacts of increased UV irradiance on Antarctic

[1] Australian Antarctic Division, Channel Highway, Kingston, Tasmania 7050, Australia

Hempel (Ed.) Antarctic Science

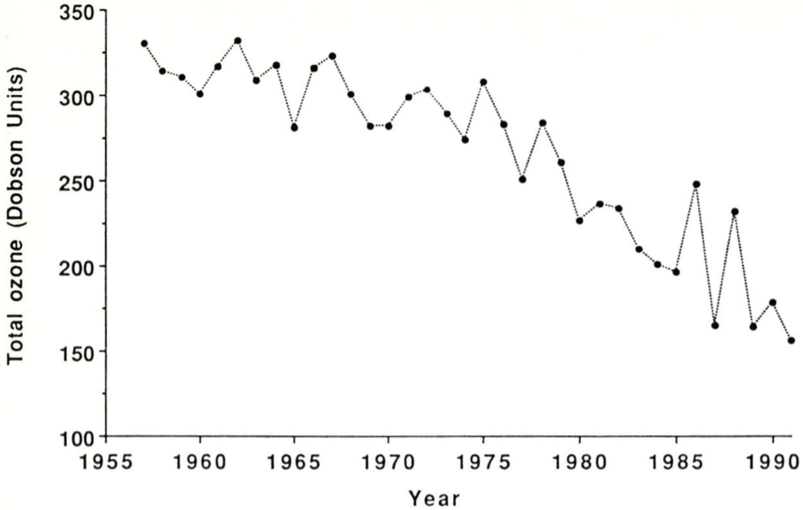

Fig. 1. Mean total ozone concentrations during October over Halley Bay, Antarctica. NB 1991 value is preliminary. (Redrawn from Farman et al. 1985 with recent data kindly provided by Brian Gardiner, British Antarctic Survey)

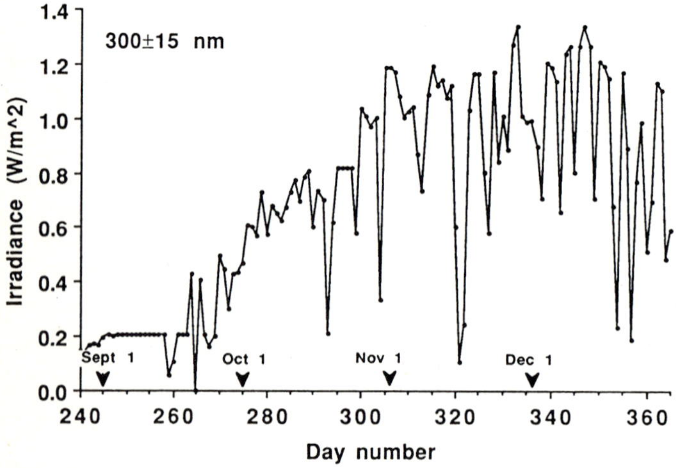

Fig. 2. UV-B irradiance (300 ± 15 nm) measured at solar noon from late August to 31 December 1990 at Casey Station, Antarctica. Data kindly provided by Colin Roy, Australian Radiation Laboratories

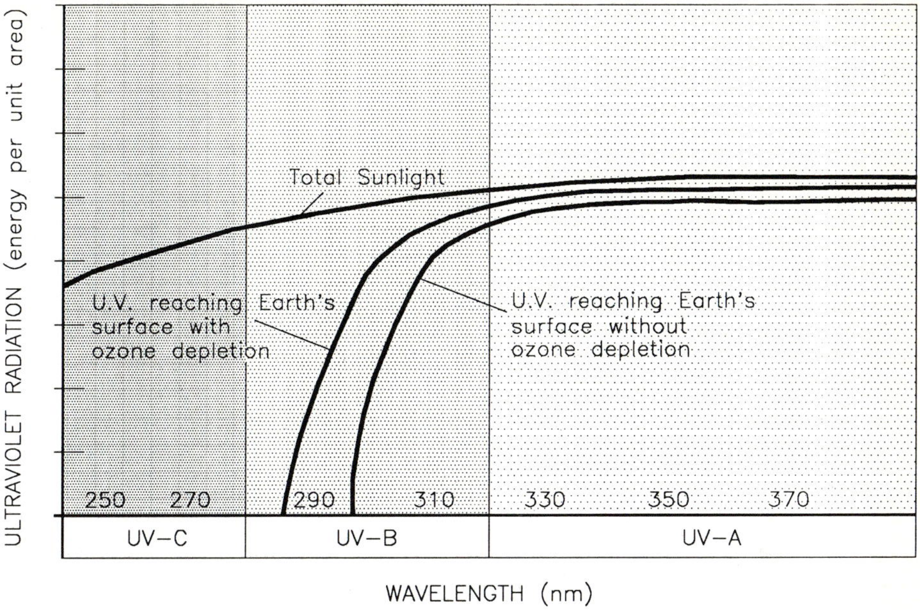

Fig. 3. UV-A is only slightly attenuated by the atmosphere, while UV-B is greatly attenuated. UV-C is completely blocked. Any decrease in ozone will lead to more and shorter wavelengths of UV-B reaching the Earth's surface

terrestrial and marine organisms and the mechanisms that they have evolved to minimize these impacts.

2 Impacts on Antarctic Terrestrial and Lacustrine Organisms

The principal components of the terrestrial flora of continental Antarctica are algae, lichens and mosses. When not protected by a covering of snow these organisms are directly exposed to solar UV radiation. The Antarctic mosses *Grimmia* and *Bryum* contain compounds which absorb radiation around the wavelengths of 323 and 262 nm. These compounds apparently function as effective UV filters as field experiments indicate that present ambient UV radiation does not contribute to photosynthetic stress in these bryophytes. The concentration of these UV absorbing pigments in both of the mosses was found to be high in early summer, when UV irradiance is also high, and declined over summer (Adamson and Adamson 1992). Markham et al. (1990) found that the concentration of flavonoids in herbarium samples of the moss *Bryum argenteum* collected from around the Ross Sea appeared inversely correlated with the ozone concentration measured in December at South Pole station. They suggested that the flavonoid levels of these organisms may provide an indicator of past stratospheric ozone concentrations.

Terrestrial Antarctic algae such as the chlorophyte *Prasiola* and the cyanobacterium *Nostoc* grow on the poorly developed Antarctic soil, in regions usually with a northerly aspect and enriched with nutrients derived from penguin rookeries or the nesting sites of other birds. *Prasiola* contains low concentrations of compounds which absorb at the same wavelengths as the mosses *Grimmia* and *Bryum*. Photosynthesis in this alga is depressed by ambient UV exposure. Preliminary investigations indicate that photosynthesis in the heavily pigmented lichen *Usnea* is not affected by ambient UV exposure (Adamson and Adamson 1992). Thus under increasing UV irradiation in Antarctica, the alga *Prasiola* would be expected to be more seriously effected than lichens or mosses. Scherer et al. (1988) reported the UV absorbing pigments in *Nostoc* were located in the sheath surrounding the cells in concentrations of up to 10% of the dry weight of this cyanobacterium. They also demonstrated that the concentration of these compounds increased with increasing exposure to UV irradiation and provided *Nostoc* with a high level of protection from UV radiation. Antarctic terrestrial plants support cryptic invertebrate communities of nematodes, tardigrades, rotifers and arthropods. Any diminution in the growth of the primary producers or change in their species composition would be expected to have an impact on the populations of these consumers.

Ekelund (1992) reports a preliminary study on the impact of UV-B on the phytoplankton of freshwater lakes and ponds on the sub-Antarctic island, South Georgia. He found that the motility of *Cryptomonas* sp. from ponds decreased on exposure to UV-B. However, growth rates of *Botryococcus*, *Lyngbya* sp. and *Staurastrum* sp., phytoplankton from lakes, apparently were unaffected by the UV-B treatments. Häder and Häder (1990) previously have reported reduction in motility and loss of phototactic orientation in *Cryptomonas*. The ponds, and to a lesser extent the lake, investigated on South Georgia contained humic material which absorb in the UV region of the spectrum. It is likely that the organisms in these water bodies receive substantially less UV radiation than they would if the humic material was absent. More investigations are needed to determine the impacts of UV on phytoplankton communities in Antarctic lakes.

3 Impacts on Antarctic Marine Organisms

Numerous Antarctic vertebrates which feed and spend most of their life in the sea breed or are raising young out of the water on the sea ice or coastal sites during the Antarctic spring and early summer. This is the time when incident UV is at its highest for the year. Emperor penguins hatch their chicks in July and feed them until they fledge in mid-December. Adelie penguins lay eggs in October which they incubate until December. The chicks fledge in late February. Crabeater and Weddell seals pup in October and November in the sea-ice zone and tend the young for only a few weeks. As UV is recognized to promote the incidence of cataract and various carcinoma of ocular and surrounding tissue (Charman 1990), concern has been expressed about the impact of increased UV on the eyes of these animals. Tears collected from southern elephant seals, which pup on sub-Antarctic islands in early November, show strong absorption at wavelengths shorter than 300 nm (Fig. 4). Seals lack naso-lachrymal ducts so that tears produced when these animals are on land run down their face. Thus some protection to the eyes and surrounding hairless tissue of these animals could be expected. Whether the tears of other seals similarly absorb UV remains to be ascertained.

Evidence from numerous studies indicates that UV exposure is harmful to fish larvae and juveniles as well as larval and adult crustaceans including shrimps, crabs and copepods (Hardy and Gucinski 1989). However, the impacts of UV on Antarctic invertebrates and fish are unknown. Many species of benthic Antarctic fish have a pelagic juvenile stage and Antarctic krill swarm close to the surface in the marginal ice edge zone and at the edge of the continental shelf (e.g. Higginbottom and Hosie 1989). They will

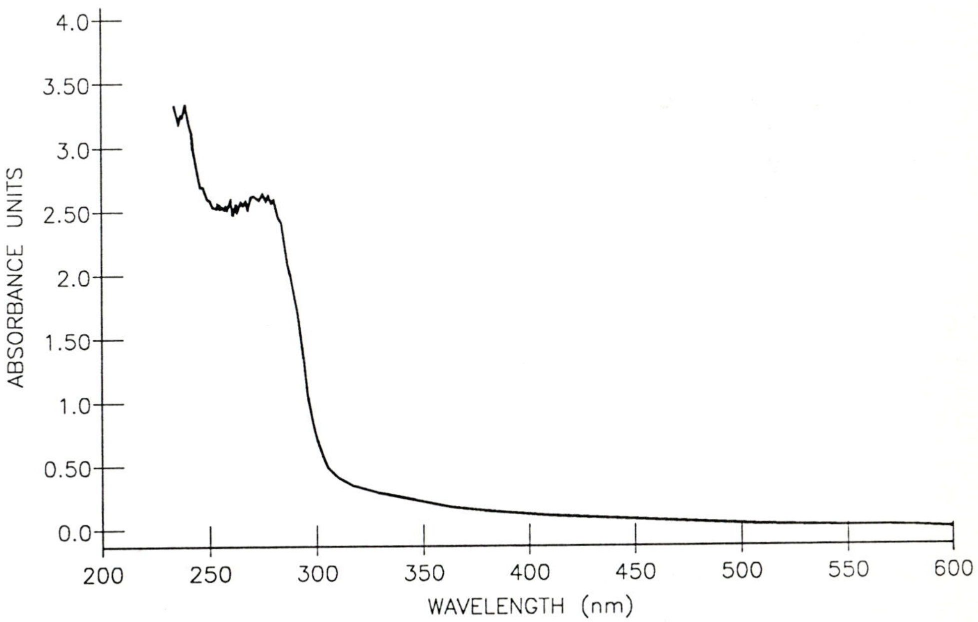

Fig. 4. Absorbtion spectrum of tears collected from southern elephant seal *Mirounga leonina*

be exposed to springtime UV irradiances. In addition, any reduction in primary productivity will indirectly limit the food availability for grazers.

In a survey of 57, mostly intertidal, Antarctic organisms comprising fish, algae and invertebrates, including the Antarctic krill (*Euphausia superba*) Karentz et al. (1991a) found that most contained mycosporine-like amino acids. They suggested that these compounds provided UV screening to the organisms possessing them.

By far the most attention to the impact of UV on Antarctic organisms has been given to the phytoplankton which constitute the base of the pelagic food web. Inhibition of primary production in phytoplantkton by UV irradiation has been demonstrated by numerous investigators (e.g. Jitts et al. 1976; Lorenzen 1979; Smith et al. 1980; Maske 1984). As well as inhibiting primary productivity, phytoplankton exposed to UV-B exhibit reduced amino acid, protein and pigment concentrations (Döhler 1984), nitrogenase activity (Döhler et al. 1987) and nitrogen metabolism (Döhler 1992). Motility and phototactic orientation are also impaired by UV-B irradiation (Häder and Worrest 1991). Phytoplankton differ greatly in their sensitivity to UV exposure (e.g. Calkins and Thordardottir 1980; Jokiel and York 1984; Ekelund 1991; Karentz et al. 1991b; Marchant et al. 1991). That organisms differ in their response to UV reflects differences in their concentrations of UV absorbing compounds (Yentsch and Yentsch 1982; Carreto et al. 1990; Marchant et al. 1991) and DNA repair capacity (Karentz et al. 1991b). Helbling et al. (1992) reported marked differences in resistance to UV radiation between tropical and Antarctic phytoplankton and that phytoplankton from below the pycnocline in tropical waters were very sensitive to solar radiation. On the basis of these observations they suggest that the resistance shown by phytoplankton is surface waters to UV exposure indicates a photoadaptive process. It is becoming clear that the response by phytoplankton to UV radiation is complex, and not simply a function of dose. Helbling et al. (1992) have investigated the effects of wavelength on phytoplankton from different latitudes and depths and Cullen and Lesser (1991) demonstrate that, at least for one diatom, for equal exposures of UV-B, a relatively short exposure to high irradiance inhibits photosynthesis more than a longer exposure of lower irradiance. These data indicate the importance of a knowledge of action spectra of organisms as well as time scales of vertical mixing of the water mass containing the phytoplankton under investigation. Becuase of the different responses of phytoplankton, any change in UV irradiance is likely to lead to a change in the species composition of phytoplankton populations (Worrest et al. 1981; Karentz 1991; Marchant and Davidson 1991).

That Antarctic marine phytoplankton are UV stressed was first shown by El-Sayed et al. (1990) at Palmer Station on the Antarctic Peninsular. In their experiments mixed phytoplankton assemblages were exposed to four treatments – enhanced UV-B, ambient UV, and two levels of reduced UV exposure. They found a marked reduction of primary productivity in the ambient and enhanced UV treatments compared with those treatments experiencing reduced UV exposure. In a study to establish the influence of the depth penetration of different wavelengths of UV on the productivity of Antarctic phytoplankton, Holm-Hansen et al. (1989) demonstrated that incident solar UV radiation significantly depressed photosynthesis to a depth of 10–15 m and that the radiation in the 305–350 nm spectral range accounted for some 75% of the inhibition of photosynthesis.

The depth penetration of UV in the ocean is strongly wavelength dependent with the shorter wavelengths being more readily absorbed (Smith and Baker 1979). In the transparent water of the Weddell Sea, 1% of surface UV-B irradiance reached 60 m depth (Gieskes and Kraay 1990) and in coastal waters off Davis we have found 2.5% of surface UV-B irradiance at 15 m depth. Using a DNA repair deficient strain of the bacterium *Escherichia coli* as a biological UV dosimeter, Karentz and Lutze (1990) elegantly demonstrated that significant amounts of UV-B penetrated Antarctic coastal waters, near the US Palmer station, to 10 m depth and found biological effects at depths of 20 and 30 m. Although the deleterious effect of UV radiation on biological systems is well known, relatively little is known of the ability of Antarctic marine organisms to cope with increase in this environmental stress.

4 Impacts in the Marginal Ice Edge Zone

Despite high concentrations of macronutrients, for most of the year the standing crop of phytoplankton and primary productivity of the Southern Ocean are low. Light limits productivity during the winter when the sun angle is low and the sea is ice and snow covered. In summer the high wind forcing over the Southern Ocean leads to deep mixing and low production. Overall the annual productivity of the Southern Ocean is similar to that of oligotrophic oceanic waters. At its maximum extent in September, the sea ice around Antarctica covers an area of some $20 \times 10^6\,km^2$ (Jacka 1983) The most productive part of the Southern Ocean is the marginal ice edge zone that is associated with, and trails the southward retreating sea ice in spring (Smith and Nelson 1986).

The release of freshwater from the melting ice produces a mixed zone some 10–20 m deep in which phytoplankton reach high numbers in the high light, nutrient rich environment (Smith 1987). Seeding of this ice edge zone by the sea ice community is likely to contribute to the developing planktonic population (Garrison et al. 1987) although the extent to which this occurs has recently been questioned (Riebesell et al. 1991). The major components of the phytoplankton in the marginal ice-edge zone are diatoms and the colonial stage in the life cycle of the prymnesiophyte *Phaeocystis pouchetii* (Fryxell and Kendrick 1988; Davidson and Marchant 1992). This southward moving region of high phytoplankton productivity is utilized by the higher trophic levels (Ainley et al. 1986) providing most of the carbon required to sustain the large populations of zooplankton, birds and mammals for which the Southern Ocean is noted (Sakshaug and Skjoldal 1989). The massive deposits of diatomaceous ooze in Southern Ocean sediments are thought to be due to reduced coupling of production and consumption in the marginal ice edge zone, compared with other oceanic areas, and a substantial amount of the biogenic production sinking rapidly from the euphotic zone. Thus grazing and sedimentation are apparently the principal fates of much of this ice edge bloom (Truesdale and Kellogg 1979; Smith and Nelson 1986; Riebesell et al. 1991). This breakup of the sea ice and the bloom in the marginal ice edge zone coincides with the ozone hole and the subsequent period of low ozone concentration over the Southern Ocean following the breakdown of the ozone hole.

The nutritional value of *Phaeocystis* is reported substantially lower than that of diatoms (Claustre et al. 1990), and euphausiids and copepods selectively graze diatoms (Meyer and El-Sayed 1983; Claustre et al. 1990). There is strong evidence from temperate waters that if diatoms are replaced by other species of phytoplankton, fish productivity is markedly reduced (Barber and Chavez 1983). While there is no doubt that diatoms contribute substantially to the vertical flux to the deep ocean both directly and indirectly as the faeces and moults of higher trophic levels, the contribution that *Phaeocystis* makes to the flux of organic material to deep water appears equivocal. Smith et al. (1991) report a massive bloom of *Phaeocystis* in the Greenland Sea associated with high flux rates of organic matter to water deeper than 1000 m. Wassmann et al. (1990) found that *Phaeocystis* contributes to vertical flux from the photic zone in the Barents Sea, but concluded that degradation of the mucilage by microbial activity and grazing at depths less than 100 m diminishes the carbon flux to the deep ocean.

Gibson et al. (1990) have calculated that up to 10% of the total global flux of dimethyl sulphide (DMS) to the atmosphere emanates from the Southern Ocean where *Phaeocystis* is reportedly the principal source of this gas. A major source of cloud condensation nuclei (CCN) over the ocean appears to be sulphate aerosol particles formed by the oxidation of DMS produced by marine algae (Ayers and Gras 1991). As the albedo of clouds, and as a consequence global climate, is determined by the concentration of CCN, a mechanism for the regulation of climate by marine biological activity has been proposed (Charlson et al. 1987). Thus any change in the distribution, abundance or activity of *Phaeocystis* may have climatic consequences.

The colonial stage of *Phaeocystis* produces UV-B absorbing compounds (Marchant et al. 1991), the concentrations of which are five to ten times higher in Antarctic strains of this alga than the strains examined from other parts of the world (Marchant and Davidson 1991). This stage in the life cycle of *Phaeocystis* from Antarctic waters is able to survive at higher irradiances of UV-B than the motile stage of this alga from Antarctica and the colonial stage of a strain from the East Australian Current which respectively produce either no, or less, of the UV-B absorbing compounds than Antarctic colonial strains (Marchant et al. 1991). Diatoms, the principal constituent of the phytoplankton, apparently have relatively low concentrations of UV-B absorbing compounds (H. Smith and S. W. Jeffrey unpubl.). All of the Antarctic diatoms so far examined have substantially lower concentrations of UV absorbing compounds than Antarctic colonial *Phaeocystis*. In addition, they show considerable variability in their ability to produce these compounds in response to exposure to UV (Fig. 5). Diatoms also differ substantially between species in their survival following exposure to UV-B (Calkins and Thordardottir 1980; Karentz et al. 1991b). On the basis of these findings we initially proposed that under increased UV irradiance it is likely that the ratio of the abundance of *Phaeocystis* to diatoms in the marginal ice edge zone would increase (Marchant and Davidson 1991). Our more recent investigations have focussed on the survival of diatoms exposed to various irradiances and different wavelength compositions of UV-B. Even though diatoms have relatively low concentrations of UV absorbing compounds, they vary greatly in their survival under UV exposure and this survival is not correlated with the possession of UV absorbing compounds. They therefore apparently use other mechanisms to mitigate UV damage. All species of Antarctic diatoms investigated so far are able to survive

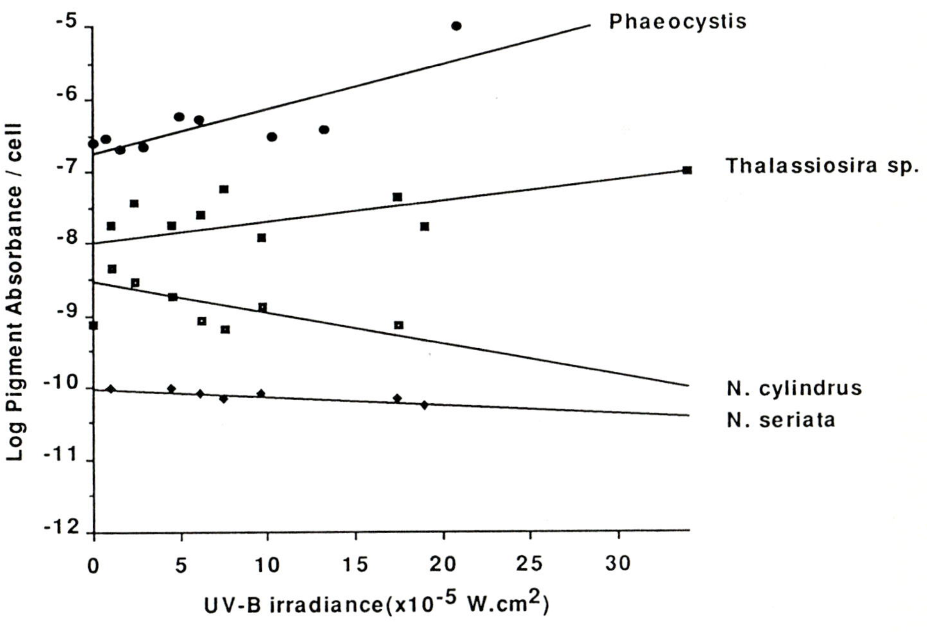

Fig. 5. Changes in the concentration of UV-B absorbing compounds per cell in Antarctic colonial *Phaeocystis*, *Thalassiosira* sp and two species of *Nitzschia* in response to different levels of UV-B irradiance

irradiances which are lethal to Antarctic *Phaeocystis* (Karentz et al. 1991b; Smith et al. 1992; Bramich et al. 1993). Karentz et al. (1991b) have related the impact of UV to the relative size of the organisms with the nuclear DNA as the principal cellular target. There is growing literature on the mechanisms employed by organisms to repair damage to DNA caused by UV irradiation. It is becoming apparent that diatoms and *Phaeocystis* solve the problem of mitigating UV damage in different ways. In the case of the colonial stage in the life cycle of *Phaeocystis* where it is possible to accumulate UV absorbing pigments in the mucilage surrounding the cells UV screening is the principal mechanism employed to minimize UV exposure. Diatoms, most of which lack a well-developed sheath of mucilage, may rely to some extent on absorption by the silica in their cell walls. In addition, they apparently have the ability to repair UV induced damage to their genetic material.

A detailed investigation of the penetration of UV into surface waters and the impact of UV exposure was conducted in the marginal ice zone of the Bellingshausen Sea in the spring of 1990 (Smith et al. 1992). They found that the ratio of UV-B to total irradiance increased at the surface of the sea and at depth with diminishing stratospheric ozone and that this change in spectral irradiance led to a number of physiological changes in the phytoplankton. Data obtained both under and outside the region of the "ozone hole" on this cruise enabled an estimate to be made of the impact of ozone depletion on primary productivity. A minimum of 6–12% reduction in primary productivity was calculated. During maximum ozone depletion up to 10% of surface incident UV radiation may reach the algal community growing on the underside of the sea ice (Trodahe and Buckley 1989; Ryan 1992). Ryan (1992) has reported a 5% reduction in primary productivity of sea ice algae. As discussed by Smith et al. (1992) reduction in primary productivity in the Antarctic ice zone should be viewed against a natural interannual variability of around 25% (Smith et al. 1988). While reduction in carbon fixation in Antarctic waters due to ozone depletion may be some three orders of magnitude less than global phytoplankton production, loss within the Antarctic ice edge zone is measurable and its ecological consequences remain unresolved.

Martin et al. (1990) have suggested that the growth of Antarctic marine phytoplankton is presently limited by the availability of iron. They have suggested that the addition of iron to Antarctic waters would promote phytoplankton growth, thereby increasing the removal of atmospheric CO_2. The idea of iron limitation of phytoplankton growth has prompted active debate (e.g. Banse 1990; de Baar et al. 1990; Joos et al. 1991; Sunda et al. 1991). Antarctic marine phytoplankton differ in their response to UV exposure. Therefore it appears essential to ascertain whether, if the growth of Antarctic phytoplankton is iron limited, the species being stimulated are effective in transferring carbon to deep water rather than those species more likely to fuel respirational losses by the organisms of the microbial loop in surface waters, thus exacerbating atmospheric CO_2 accumulation.

Other than phytoplankton, the impacts of enhanced UV exposure have been investigated in few other groups of Antarctic microorganisms. In an investigation on the vertical distribution on ATP in Halley Bay in the Weddell Sea and the impact of UV-B on the ATP content of microorganisms Vosjan et al. (1990) found the ATP to be concentrated in the upper 20 m of the water column. After 5 h of UV-B irradiance under their experimental conditions, ATP concentration of the microorganisms had declined by 75%. These data together with the finding by Herndl et al. (1993) of suppression of bacterial activity by about 40% in the top 5 m of the water column in temperate coastal waters signal the urgency to investigate the impact of UV on Antarctic marine bacteria and microheterotrophs.

Another related topic that has received little attention in Antarctic waters is the indirect effect of UV on phytoplankton due to photolysis of inorganic and organic chemical species in the water column. Any increase in UV radiation will increase the rate of transformation of macromolecules (Mopper et al. 1991). Thus enhanced UV flux could potentially lead to an increase in the products of photolytically cleaved macromolecules and a reduction in the activity of organisms that utilize these compounds. The significance of this with respect to the carbon budget of the Southern Ocean remains speculative. Photolysis of colloidal and particulate iron increases the proportion of the labile component thus enhancing the availability of iron to phytoplankton (Wells and Mayer 1991). It has yet to be ascertained whether increasing UV flux makes any contribution to the elevated levels of primary productivity in the marginal ice edge zone as a result of the photolysis of colloidal iron.

5 Conclusions

The concentration of stratospheric ozone in springtime over the Antarctic has been decreasing since the mid-1970s. Also, there have also been occasions in the past when stratospheric ozone concentrations have declined as a result of "natural" perturbations to the stratosphere by volcanic eruptions and massive meteor impacts such as the Tunguska impact on Siberia in 1908. It has been reported that the likely loss of stratospheric ozone in the Northern Hemisphere resulting from the Tunguska impact was 30 ± 15% which persisted for some years (Turco et al. 1981). Despite experimental evidence of Antarctic organisms (especially phytoplankton) being under UV-B stress, and that there has been stratospheric ozone depletion occurring for at least the last decade, and evidence of past incidents when there has been ozone depletion, there is no clear evidence of unequivocal changes in the marine ecosystem. This is hardly surprising considering that UV exposure is just one of a suite of environmental stresses to which these organisms are exposed. While there is a pattern of annual species succession, there is also interannual variability in the timing and abundance of individual species and it is not possible to ascribe this heterogeneity to a single stress. In addition, quantitative information on spatial and temporal Antarctic phytoplankton and populations of other organisms before the development of the ozone hole is sparse, making comparisons with present-day assemblages difficult.

References

Adamson H, Adamson E (1992) Possible effects of global climate change on Antarctic terrestrial vegetation. In: Quilty P (ed) Impacts of climate change on Antarctica. Department of the Arts, Sport, the Environment, Tourism and Territories. Australian Government Publishing Service, Canberra, pp 52–62

Ainley DG, Fraser WR, Sullivan CW, Torres JJ, Hopkins TL, Smith WO (1986) Antarctic mesopelagic micronekton: Evidence from seabirds that pack ice affects community structure. Science 232:847–849

Atkinson RJ, Matthews WA, Newman PA, Plumb RA (1989) Evidence of the mid-latitude impact of Antarctic ozone depletion. Nature 340:290–294

Ayers GP, Gras JL (1991) Seasonal relationship between cloud condensation nuclei and aerosol methanesulphonate in marine air. Nature 353:834–835

Banse K (1990) Does iron really limit phytoplankton production in the offshore subarctic Pacific? Limnol Oceanogr 35:772–775

Barber RT, Chavez FP (1983) Biological consequences of El Niño. Science 222:1203–1210

Bidigare RR (1989) Potential effects of UV-B radiation on marine organisms of the Southern Ocean: distributions of phytoplankton and krill during austral spring. Photochem Photobiol 50:469–477

Bramich D, Davidson AT, Marchant HJ, McMinn A (1993) Effects of UV-B irradiation on growth and survival of Antarctic marine diatoms. Mar Biol (in press)

Caldwell MM (1981) Plant responses to solar ultraviolet radiation. In: Lange OL, Nobel PS, Osmond CB, Ziegler H (eds) Encyclopedia of plant physiology, New Series, Physiological plant ecology, 1. Springer, Berlin Heidelberg New York, pp 169–197

Calkins J (1982) The role of solar ultraviolet radiation in marine ecosystems. Plenum Press, New York

Calkins J, Thordardottir T (1980) The ecological significance of solar UV radiation on aquatic organisms. Nature 283:563–566

Carreto JI, Carignan MO, Daleo G, De Marco S (1990) Occurrence of mycosporine-like amino acids in the red-tide dinoflagellate *Alexandrium excavatum*: UV-photoprotective compounds? J Plankton Res 12:909–921

Charlson RL, Lovelock JE, Andreae MO, Warren SG (1987) Oceanic phytoplankton, atmospheric sulfur, cloud albedo and climate. Nature 326:655–661

Charman WN (1990) Ocular hazards arising from depletion of the natural atmospheric ozone layer: a review. Opthalmic Physiol Opt 10:333–341

Claustre H, Poulet SA, Williams R, Marty J-C, Coombs S, Ben Mlih F, Hapette AM, Martin-Jezequel V (1990) A biochemical investigation of a *Phaeocystis* sp. bloom in the Irish Sea. J Mar Biol Assoc UK 70:197–207

Cullen JJ, Lesser MP (1991) Inhibition of photosynthesis by ultraviolet radiation as a function of dose and dosage rate: results for a marine diatom. Mar Biol 111:183–190

Davidson AT, Marchant HJ (1992) Protist interactions and carbon dynamics of a *Phaeocystis*-dominated bloom at an Antarctic coastal site. Polar Biol 12:387–395

de Baar HJW, Buma AGJ, Nolting RF, Cadée GC, Jacques G, Trégeur PJ (1990) On iron limitation of the Southern Ocean: experimental observations in the Weddell and Scotia Seas. Mar Ecol Prog Ser 65:105–122

Döhler G (1984) Effect of UV-B radiation on the marine diatoms *Lauderia annulata* and *Thalassiosira rotula* grown in different salinities. Mar Biol 83:247–253

Döhler G (1992) Impact of UV-B radiation on uptake of ^{15}N-ammonia and ^{15}N-nitrate by phytoplankton of the Wadden Sea. Mar Biol 112:485–489

Döhler G, Worrest RC, Biermann I, Zink J (1987) Photosynthetic $^{14}CO_2$ fixation and ^{15}N-ammonia assimilation during UV-B radiation of *Lithodesmium variabile*. Physiol Plant 70:511–515

Ekelund NGA (1991) The effects of UV-B radiation on dinoflagellates. J Plant Physiol 138:274–287

Ekelund NGA (1992) Studies on the effects of UV-B radiation on phytoplankton of Sub-Antarctic lakes and ponds. Polar Biol 12:533–537

El-Sayed SZ, Stephens FC, Bidigare RR, Ondrusek ME (1990) Effect of ultraviolet radiation on Antarctic marine phytoplankton. In: Kerry KR, Hempel G (eds) Antarctic ecosystems. Ecological change and conservation. Springer, Berlin Heidelberg New York, pp 379–385

Farman JC, Gardiner BG, Shanklin JD (1985) Large losses of total ozone an Antarctica reveal seasonal ClO_x/NO_x interaction. Nature 315:207–210

Frederick JE, Snell HE (1988) Ultraviolet radiation levels during the antarctic spring. Science 241:438–440

Fryxell GA, Kendrick GA (1988) Austral spring microalgae across the Weddell Sea ice edge: spatial relationships found along a northward transect during AMEREZ 83. Deep-Sea Res 35:1–20

Garrison DL, Buck KR, Fryxell GA (1987) Algal assemblages in the antarctic pack ice and in ice-edge plankton. J Phycol 23:564–572

Gibson JAE, Garrick RC, Burton HR, McTaggart AR (1990) Dimethylsulfide and the alga *Phaeocystis pouchetii* in antarctic coastal waters. Mar Biol 104:339–346

Gieskes WWC, Kraay GW (1990) Transmission of ultraviolet light in the Weddell Sea: report of the first measurements made in the Antarctic. BIOMASS Newsl 12:12–14

Häder D-P, Häder M (1990) Effects of UV radiation on motility, photo-orientation and pigmentation in a freshwater *Cryptomonas*. J Photochem Photobiol B 5:105–114

Häder D-P, Worrest RC (1991) Effects of enhanced solar ultraviolet radiation on aquatic ecosystems. Photochem Photobiol 53:717–725

Hardy J, Gucinski H (1989) Stratospheric ozone depletion: implications for marine ecosystems. Oceanography 2:18–21

Harm W (1980) Biological effects of ultraviolet radiation. Cambridge University Press, Cambridge

Helbling EW, Villifañe V, Ferrario M, Holm-Hansen O (1992) Impact of natural ultraviolet radiation on rates of photosynthesis and on specific marine phytoplankton species. Mar Ecol Prog Ser 80:89–100

Herndl GJ, Müller-Niklas G, Frick J (1993) Major role of ultraviolet-B in controlling bacterioplankton growth in the surface layer of the ocean. Nature 361:717–719

Higginbottom IR, Hosie GW (1989) Biomass and population structure of a large aggregation of krill near Prydz Bay, Antarctica. Mar Ecol Prog Ser 58:197–203

Holm-Hansen O, Mitchell BG, Vernet M (1989) Ultraviolet radiation in antarctic waters: effects on rates of primary production. Antarct J US 24:177–178

Jacka TH (1983) A computer data base for Antarctic sea ice extent. ANARE Research Notes, No 13 Australian Antarctic Division, Hobart

Jitts HR, Morel A, Saijo Y (1976) The relation of oceanic primary production to available photosynthetic irradiance. Aust J Mar Freshwater Res 27:441–454

Jokiel PL, York RH Jr (1984) Importance of ultraviolet radiation in photoinhibition of microalgal growth. Limnol Oceanogr 29:192–199

Joos F, Sarmiento JL, Siegenthaler U (1991) Estimates of the effect of Southern Ocean iron fertilization on atmospheric CO_2 Concentrations. Nature 349:772–775

Karentz D (1991) Ecological considerations of Antarctic ozone depletion. Antarct Sci 3:3–11

Karentz D, Lutze LH (1990) Evaluation of biologically harmful ultraviolet radiation in Antarctica with a biological dosimeter designed for aquatic environments. Limnol Oceanogr 35:548–561

Karentz D, McEuen FS, Land MC, Dunlap WC (1991a) Survey of mycosporine-like amino acid compounds in Antarctic marine organisms: potential protection from ultraviolet exposure. Mar Biol 108:157–166

Karentz D, Cleaver JE, Mitchell DL (1991b) Cell survival characteristics and molecular responses of Antarctic phytoplankton to ultraviolet-B radiation. J Phycol 27:326–341

Lorenzen CJ (1979) Ultraviolet radiation and phytoplankton photosynthesis. Limnol Oceanogr 24:1117–1120

Marchant HJ, Davidson AT (1991) Possible impacts of ozone depletion on trophic interactions and biogenic vertical carbon flux in the Southern Ocean. In: Weller G, Wilson CL, Severin BAB (eds) Proc Int Conf Role of Polar Regions in Global Change. Geophysical Institute, Fairbanks, pp 397–400

Marchant HJ, Davidson AT, Kelly GJ (1991) UV-B protecting pigments in the marine alga *Phaeocystis pouchetii* from Antarctica. Mar Biol 109:391–395

Markham KR, Franke A, Given DR, Brownsey P (1990) Historical Antarctic ozone level trends from herbarium specimen flavonoids. Bull Liasion Groupe Polyphenols 15:230–235

Martin JH, Gordon RM, Fitzwater SE (1990) Iron in Antarctic waters. Nature 345:156–158

Maske H (1984) Daylight ultraviolet radiation and the photoinhibition of phytoplankton carbon uptake. J Plankton Res 6:351–357

Meyer MA, El-Sayed SZ (1983) Grazing of *Euphausia superba* Dana on natural populations. Polar Biol 1:193–197

Mopper K, Zhou X, Kieber RJ, Kieber DJ, Sikorski RJ, Jones RD (1991) Photochemical degradation of dissolved organic carbon and its impact on the oceanic carbon cycle. Nature 353:60–62

Riebesell U, Schloss I, Smetacek V (1991) Aggregation of algae released from melting sea ice: implications for seeding and sedimentation. Polar Biol 11:239–248

Ryan KG (1992) UV radiation and photosynthetic production in Antarctic sea ice microalgae. J Photochem Photobiol 13:235–240

Sakshaug E, Skjoldal HR (1989) Life at the ice edge. Ambio 18:60–67

Scherer S, Chen TW, Böger P (1988) A new UV-A/B protecting pigment in the terrestrial cyanobacterium *Nostoc commune*. Plant Physiol 88:1055–1057

Smith RC, Baker KS (1979) Penetration of UV-B and biologically effective dose-rates in natural waters. Photochem Photobiol 29:311–323

Smith RC, Baker KS (1989) Stratospheric ozone, middle ultraviolet radiation and phytoplankton productivity. Oceanography 2:4–10

Smith RC, Baker KS, Holm-Hansen O, Olsen R (1980) Photoinhibition of photosynthesis in natural waters. Photochem Photobiol 31:585–592

Smith RC, Prezelin BB, Baker KS, Bidigare RR, Boucher NP, Coley T, Karentz D, MacIntyre S, Matlick HA, Menzies D, Ondrusek M, Wan Z, Waters KJ (1992) Ozone depletion: Ultraviolet radiation and phytoplankton biology in Antarctic waters. Science 255:952–959

Smith WO Jr (1987) Phytoplankton dynamics in marginal ice zones. Oceanogr Mar Biol Annu Rev 25:11–38

Smith WO Jr, Nelson DM (1986) Importance of ice edge phytoplankton production in the Southern Ocean. BioScience 36:251–257

Smith WO Jr, Keene NK, Comiso JC (1988) Interannual variability in estimated primary productivity of the Antarctic marginal ice zone. In: Sahrhage D (ed) Antarctic Ocean and Resources Variability. Springer, Berlin Heidelberg New York, pp 131–139

Smith WO Jr, Codispoti LA, Nelson DM, Manley T, Buskey EJ, Niebauer HJ, Cota GF (1991) Importance of *Phaeocystis* blooms in the high-latitude ocean carbon cycle. Nature 352:514–516

Stolarski RS, Krueger AJ, Schoeberl MR, McPeters RD, Newman PA, Alpert JC (1986) Nimbus 7 satellite measurements of the springtime Antarctic ozone decrease. Nature 322:808–811

Sunda WG, Swift DG, Huntsman SA (1991) Low iron requirement for growth in oceanic phytoplankton. Nature 351:55–57

Tevini M, Teramura AH (1989) UV-B effects on terrestrial plants. Photochem Photobiol 50:479–487

Trodahl HJ, Buckley RG (1989) Ultraviolet levels under sea ice during the antarctic spring. Science 245:194–195

Truesdale RS, Kellogg TB (1979) Ross Sea diatoms: modern assemblage distributions and their relationship to ecologic, oceanographic and sedimentary conditions. Mar Micropaleontol 4:13–31

Turco RP, Toon OB, Whitten RC, Pollack JB, Noerdling P (1981) Tunguska meteor fall of 1908: Effects on stratospheric ozone. Science 214:19–23

Vosjan JH, Döhler G, Nieuwland G (1990) Effect of UV-B irradiance on the ATP content of microorganisms of the Weddell sea (Antarctica). Neth J Sea Res 25:391–393

Voytek MA (1990) Addressing the biological effects of decreased ozone on the Antarctic environment. Ambio 19:52–61

Wassmann P, Vernet M, Mitchell BG, Rey F (1990) Mass sedimentation of *Phaeocystis pouchetii* in the Barents Sea. Mar Ecol Prog Ser 66:183–195

Wells ML, Mayer LM (1991) The photoconversion of colloidal iron oxyhydroxides in seawater. Deep-Sea Res 38:1379–1395

Worrest RC (1983) Impact of solar ultraviolet-B radiation (290–320 nm) upon marine microalgae. Physiol Plant 58:428–434

Worrest RC, Thomson BE, Dyke HV (1981) Impact of UV-B radiation upon estuarine microcosms. Phytochem Phytobiol 33:861–867

Yentsch CS, Yentsch CM (1982) The attenuation of light by marine phytoplankton with special reference to the absorption of near-UV radiation. In: Calkins J (ed) The role of solar ultraviolet radiation in marine ecosystems. Plenum Press, New York, pp 691–706

The Southern Ocean: Biogeochemical Cycles and Climate Changes

Paul Tréguer[1]

1 Introduction

For people concerned by global change due to enhanced greenhouse effect, the Southern Ocean is a very fascinating area. Its extensive cold waters (20% of the World Ocean surface) represent a priori potential sink for gases and could significantly contribute to control the excess atmospheric CO_2.

A significant part of the CO_2 exchanges between atmosphere and ocean is determined by biogeochemical processes (the so-called biological pump). Thus modelling of these CO_2 exchanges needs quantification of biogenic fluxes. Our present knowledge of these biogeochemical processes and our estimates for these biogenic fluxes are still very rudimentary. For example we do not have a complete explanation about the abnormally low rate of photosynthetic uptake of CO_2 so typical of Antarctic phytoplankters, i.e. we do not know the reasons why the exceptional nutrient richness of the Southern Ocean is almost not used.

In a global perspective it is worthwhile to note, because of this non-used nutrient richness, that the Southern Ocean is a huge source of nutrients for the surface waters of the rest of the World Ocean. Every year considerable amounts of nutrients are exported northwards, indirectly supporting the very large productivity of coastal upwellings in the southern hemisphere. Any modification in the yield of photosynthetic activity in the Southern Ocean indirectly entrained by climate change should have marked consequences for remote economically important marine ecosystems.

Thus to understand and to model the direct and/or indirect impact of climate change on biogeochemical processes and fluxes in the Southern Ocean is a major concern for IGBP related programmes (WOCE, SO-JGOFS, SO-GLOBEC and EASIZ). Before introducing the specific objectives of these programs, I will first recall the way this biogeosystem is

[1]URA CNRS 1513, European Institute for Marine Studies, University of Western Brittany, Brest, France

Hempel (Ed.) Antarctic Science

functioning today. Then I will deal with changes which both the World Ocean and the Southern Ocean experienced during the past (mainly during the last glacial age), because these changes could help understanding those expected for the future.

2 The Present

2.1 Physical Mechanisms Controlling CO_2 and Nutrient Transport Through the Antarctic Ecosystem

Three major processes (Fig. 1) govern the exchanges of CO_2 between the Southern Ocean and the atmosphere; each of these processes also governs the circulation of nutrients:

1. Thermohaline circulation transports poleward deep waters enriched with CO_2 and nutrients from remineralization of biogenic detritus. Upwelling of these Circumpolar Deep Water (CDW) at the Antarctic Divergence results first in a net flux of CO_2 towards the atmosphere at the sea-atmosphere interface, and second in fertilization of the upper layers of the Southern Ocean.
2. As these waters reach surface levels they are submitted to intense cooling during winter. These cold Antarctic surface waters have a high capacity to absorb gases from the atmosphere: they act as a sink for atmospheric CO_2. Entrained by eastward winds, part of these surface waters flows southward on the Antarctic shelves: there low temperature and increased salinity (brine rejection during freezing) make high density and nutrient rich Antarctic Bottom Water (AABW) that flows down to abyssal levels. Offshore, the other part of these surface waters are entrained northwards by westward winds. At the Polar Front Zone (PFZ) these waters meet warmer Sub-Antarctic waters and have no choice but to sink as cold and relatively less saline nutrient rich Antartic Intermediate Water (AAIW). The frontal zone is also a potential sink for CO_2. Both AABW and AAIW export huge quantities of nutrients towards the rest of the World Ocean.
3. As everywhere in the ocean, biological processes – the so-called biological pump – fix inorganic carbon and nutrients in surface waters. Most of this inorganic carbon and nutrients are recycled within the surface and subsurface layers. A small part of the organic matter generated in the surface layer escaped from recycling in this layer; biogenic matter is exported towards the deep and bottom layers, ultimately reaching the sea bottom. Only a few percent of the original carbon photosynthetic flux accumulates in sediments. Finally, the biological pump acts as a net sink for carbon (and thus for CO_2) and for related bio-elements.

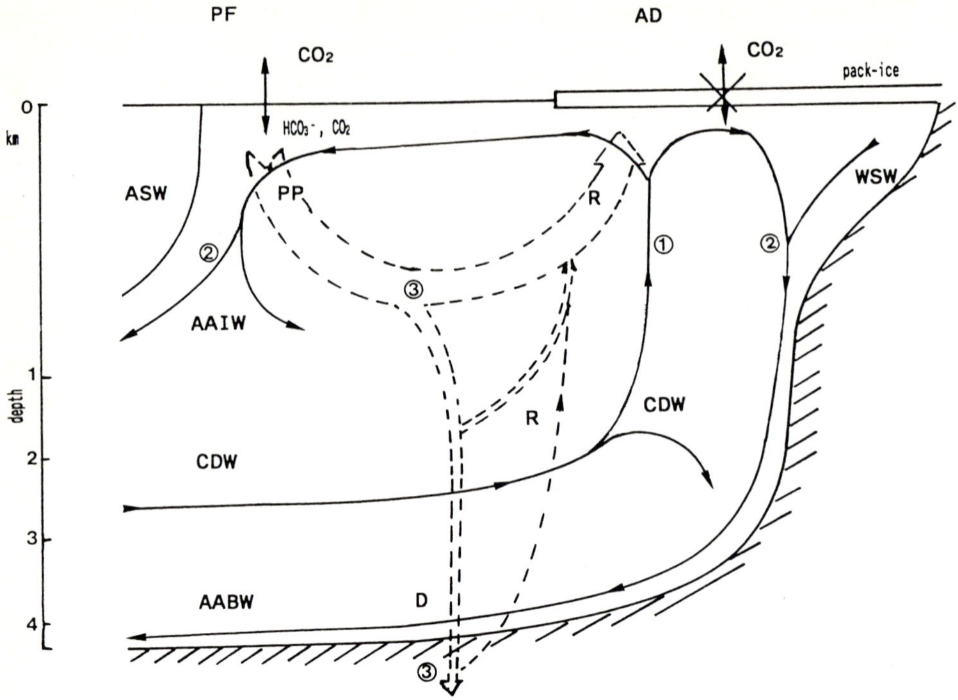

Fig. 1. Simplified diagram showing three major processes that govern the exchanges of CO_2 between the atmosphere and the Southern Ocean, and the transport of nutrients through the Southern Ocean: *1* upwelling of Circumpolar Deep Water at the Antarctic Divergence (*AD*): sources of CO_2 and of nutrients; *2* export of Antarctic Bottom Water (*AABW*) formed along the continental shelf by inputs of dense saline Western Shelf Water (*WSW* to *CDW*), and export of Antarctic Intermediate Water (*AAIW*) at the Polar Front: the northward flow of Antarctic Surface Water flows down beneath the SubAntarctic Water (*SAW*). Both processes involved cold waters (potential CO_2 sinks and export of nutrient rich waters); *3* uptake (*PP* Primary Production) and respiration (*R* recycling) of dissolved CO_2 (HCO_3^- predominates) via the biological pump result in a net sink of organic carbon buried in sediments. Biological fluxes are shown by *dashed lines*. The extension of the pack-ice depends on season, with dramatic consequences on exchanges of gases between atmosphere and ocean

2.2 CO_2 Balance at the Atmosphere-Water Interface

As far as the cycle of CO_2 is concerned, each of these three major processes results in very larges fluxes of carbon between the Southern Ocean and the atmosphere, but the overall net flux is very small. The balance between these processes may be susceptible to change, depending on the effects of climatic variables. For instance, the variable extent of the sea-ice coverage, which is very climate-dependent, will drastically modify the sea-atmosphere fluxes of gases (Fig. 1).

Our present knowledge about the CO_2 exchanges in the Southern Ocean is very rudimentary. Murphy et al. (1991), using a 150 years comprising climatological wind data set, showed the present CO_2 status of the South Pacific: a contour map of these data shows a band of near equilibrium with respect to the atmosphere that bisects the basin diagonally from the equator west of 170° W to approximately 55° S in the east. South and west of this band is a major sink region for CO_2. These results suggest the South Pacific may not be a major sink for CO_2, at least in the austral autumn. Metzl et al. (1991) demonstrated the large spatial variability of P_{co2} in the Indian sector of the Southern Ocean. He also showed a strong contrast between eastern and western portions. Between 1984 and 1990 partial pressure of CO_2 in surface water has been measured as well as in the air. The difference between water and air shows large zonal and latitudinal variabilities during austral summer that are related to frontal system, the circumpolar upwelling and biological conditions. On a mesoscale the 40° S–25° S band acts as a CO_2 sink over the whole study area (Metzl et al. 1991; Poisson et al. 1991).

Models of global CO_2 atmospheric distribution suggests that the Southern Hemisphere as a whole is not a net sink or a net source of CO_2, but this does not necessarily mean that there are no strong sources or sinks within the Southern Ocean.

2.3 Yield of the Biological Pump

As far as the yield of the biological pump is concerned, the Southern Ocean is very unique. In spite of the nutrient richness of surface waters, phytoplankton biomass and primary production are exceptionally low in Antarctica, being comparable to those of typical oligotrophic areas. This is the "Antarctic paradox". In fact, marginal ice zones and frontal areas are more efficient in using nutrients than the open ocean zone, but only in coastal and continental shelf zones have been reported severe depletion of nutrients (review in Tréguer and Jacques 1992).

Recent estimates for Antarctic annual net primary production (Smith et al. 1988; Lancelot et al. 1992) give about 120 tera mol-C (1.4–1.85 gigatons of carbon). This is not more than about 3–9% of the global marine production (ranging betwen 20 and 40 Gt-C, Longhurst 1991). Although the photosynthetically active radiation entering the surface water is largely sufficient from spring to summer, the Antarctic pelagic primary producers are unable to use up the huge amounts of nutrients available in surface waters. For instance, the yield of nitrate utilization in surface layers averages only about 20% on an annual basis (Tréguer and Jacques 1992), i.e. about 80% of the initial nitrate flux that enters the surface layer (at the Antarctic Divergence) leaves the Antartic system without being used by phytoplankters. Huge amounts of nutrients are exported northwards, indirectly contributing to the fertilization of the very productive coastal upwelling areas (such as those of

Benguela, Chile and Peru). Why is the Southern Ocean so unproductive at the first trophic level?

1. Physical Factors: The Irradiance-Mixing Regime Hypothesis. During spring and summer the photosynthetically active radiation (PAR) that reaches the sea surface is clearly non-limiting in Antarctica and primary producers can grow in surface layers, as the balance between photosynthesis and respiration gives a positive flux. Most of oceanographers here refer to the old Sverdrup's concept of "critical depth" (Z_c, Sverdrup 1953). This depth corresponds to the balance between vertically integrated rates of phytosynthesis by phytoplankton and respiration rates of the whole pelagic community. When the depth of the surface mixed layer becomes deeper than the critical depth no increase of phytoplankton biomass can occur. Obviously, the usual strong winds blowing around Antartica have the potential to create mixed-layer depth (Z_m) deeper than 100 m (except when ice melting produces stratification of the water column and counteracts wind mixing action). It makes sense to envisage Z_m as usually deeper than Z_c for the Antarctic Ocean. Nelson and Smith (1991) have recently revised Sverdrup's old equation, $Zc = 0.18\,I_o/kI_c$ (I_o and I_c being the values for the incident irradiance and for the critical irradiance, respectively) re-estimating I_o from more realistic field data and I_c. They arrive at the conclusion that, unlike what is usually held, both the MIZ and the ACC open ocean waters provide favourable irradiance/mixing regimes ($Z_c > Z_m$) for the initiation and early development of phytoplankton blooms, at least during spring and summer. These authors predict the highest levels of about $1\,\mu g\,l^{-1}$ that can be sustained in summer in open waters unstabilized by meltwater in the Weddell and Scotia Seas (may be less in areas experiencing stronger winds). This is in agreement with field observations.

2. Macronutrient Limitation: The Silicate Hypothesis. Unlike nitrate and phosphate distribution in surface waters, almost invariable south of the Polar Front (PF), that of silicate exhibits a large positive gradient southwards. In the vicinity of PF silicate concentrations as low as $5\,\mu M$ are frequent. From culture studies on Antarctic diatoms Jacques (1983) and Sommer (1991) have demonstrated the low affinity of these diatoms for silicate, giving support to the hypothesis that silicate could limit diatom growth in areas usually presenting silicate depletion (i.e. in the vicinity of the PF and in coastal zones such as the Ross Sea). Indeed, Nelson and Tréguer (1992) have detected weak but significant Si-limitation for natural populations of the Ross Sea during summer.

3. Micronutrients: The Iron Hypothesis. From data obtained in the North Pacific, Martin and Fitzwater (1988) have postulated that low aeolian inputs of iron in the Southern Ocean might explain low primary production in surface waters. Although it is not clear at present what the mechanism of this iron-enrichment response might be (e.g. Dugdale and Wilkerson 1990), long-term incubation experiments conducted using Ross Sea and Weddell Sea samples, enriched with unchelated iron (De Baar et al. 1990; Martin et al. 1990) give some support to this very much disputed hypothesis (Buma et al. 1991; Helbing et al. 1991; Martin et al. 1991; Morel et al. 1991).

4. Biomass: The Grazing Hypothesis. Active grazing by herbivores can maintain phytoplankton stocks at low levels and this has been proved for the SIZ of the Weddell Sea. Smetacek et al. (1990) and Lancelot et al. (1991) have illustrated the importance of episodic passages of krill swarms in depleting diatom blooms in ice-free planktonic communities. The major role played by protozoans in regulating phytoplankton development in the MIZ has also been highlighted by Lancelot et al. (1991).

2.4 A Huge Nutrient Import-Export System

The relative non-use of nutrients by phytoplankters gives the Antarctic Ocean a unique status as compared with the rest of the World Ocean: basically the Antarctic Ocean is a huge import-export system. This is illustrated in Fig. 2, illustrating our present view of the conservative and non-conservative fluxes of silicon running through a simplified two-box model for the Antarctic Ocean (south of the Polar Front). It is worthwhile to note that numbers in Fig. 2 are very tentative, although consistent with our present knowledge of the silicon biogeochemistry in this ocean. These numbers are only rough estimates and of course have to be determined to get a correct Si balance (conservative fluxes are probably known with an accuracy of about 20–40%, non-conservative fluxes are estimated at $\pm 100\%$).

The conservative Si fluxes are based on Gordon and Taylor's (1975) estimates for circulation and meridional fluxes of water masses and on average values for silicic acid concentrations in the different water masses. Circumpolar Deep Water (around $100\,\mu$M Si) enters the Antarctic system with a rate of about $50\,10^6\,m^3\,s^{-1}$ (=50 Sverdrup, Sv). At the Antarctic Divergence, CDW (reinforced by recycling of AAIW) enters the surface layer at a rate of 60 Sv. Then it splits in two parts under the infuence of eastward winds along the continent and westward winds offshore, leading to the formation of Antarctic Bottom Water (AABW, 39 Sv), on the one hand, and of Antarctic Winter Water (WW, 21 Sv), on the other. At the Polar Front the Winter Water feeds the formation of Antarctic Intermediate Water (11 Sv, $40\,\mu$M-Si). The AABW ($113\,\mu$M-Si) leaves the Antarctic system at the rate of 39 Sv (Gordon and Taylor 1975).

At the Polar Front Zone the Antarctic Ocean imports about $158\,10^{12}$ mol-Si (=158 tera mol-Si) per year, i.e. over 100 times as much as the Amazon river input to the Atlantic Ocean. It exports annualy about 14 tera mol-

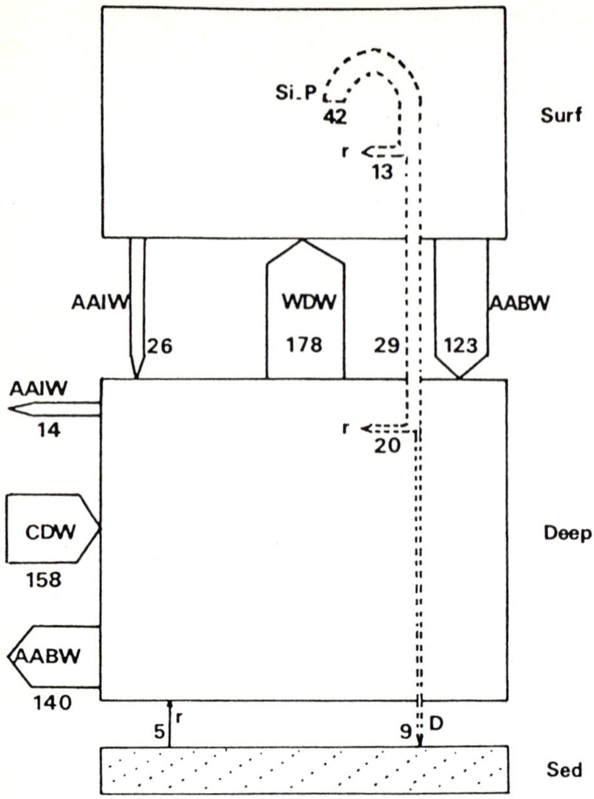

Fig. 2. Silica cycle in the Antarctic Ocean. All fluxes are in tera (10^{12}) mol-Si per year. *Dashed lines* indicate biological fluxes, *solid lines* physical fluxes; *CDW* circumpolar deep water; *WDW* warm deep water; *AABW* Antarctic bottom water; *AAIW* Antarctic intermediater water; *Si-P* production of biogenic silica; *r* dissolution of biogenic silica; *D* opal deposit

Si through the AAIW (the most important nutrient resource for South American and South African coastal upwellings) and about 140 tera mol-Si through the northwards AABW circulation. In this respect the Antarctic Ocean appears as a *net silicon sink of about 4 tera mol-Si per year*.

Biological processes are responsible for this Si gap. According to Tréguer and Van Bennekom (1991) the gross annual production of biogenic silica in the surface layer averages about 42 tera mol-Si per year. About 30% (Nelson and Gordon 1982) is recycled within the surface layers. From the 29 tera mol of particulate biogenic silica annually entering the deep layers, only 9 reach the sediments, where a dissolution process goes on. A flux of about 4 tera mol-Si per year ultimately accumulates as opal (Ledfford-Hoffman et al. 1986); this Antarctic deposit of silica is about two-thirds that of the World Ocean.

3 The Past (The Last Glacial Age, 18 000 Years B.P.)

Between the present climate conditions and the last glacial paroxysm lie about 18 000 years. The mean global temperature 18 000 years ago was about 4 °C to 5 °C lower than at present. Of course, during this last glacial age, ice sheets were more extended than today: for example the Drake Passage (between South America and the Antarctic Peninsula) was almost completely ice-covered.

3.1 Atmosphere/Oceanic Interactions and Modifications During the Glacial Age

Antarctic ice cores (Delmas et al. 1980; Neftel et al. 1982; Barnola et al. 1987) showed that, during the glacial age, atmospheric CO_2 dropped to about 190 vpm (Table 1). The mean surface temperature was about 4 to 5 °C below that of present but large latitudinal variations were observed: temperatures in the equatorial and tropical zones did not sink more than by 1–3 °C, although they sank more at higher latitudes. In parallel with the strong temperature gradient between equator and polar zones, intense winds swept oceans and continents at intermediate latitudes. Along with the drop of the sea level down to 120–150 m below that of present, consequences of these climate change were:

1. increased land erosion in intermediate latitudes, which means increased transport of aeolian lithogenic material (including iron) to oceanic waters;
2. additional inputs of nutrients in coastal waters of temperate latitudes due to increased erosion of rewashed sediments in estuaries. According to Broecker (1982), phosphate concentrations increased considerably during the early parts of the glaciation periods (as compared with the modern discharge). Unlike nitrate and phosphate during this period, the net global river input of orthosilicic acid (= silicate) to the oceans did not significantly change (Table 1), as compared with the input of today (Lisitzin 1985): two-thirds of silicate river inputs were (and still are) indeed originating from the equatorial zone, where climate changes were less important;
3. the activity of subtropical coastal upwellings (created by trade winds and more generally by winds blowing equatorward parallel to the coastline) increased to a great extent (intensification of subsurface deep water upwelling) and tended to migrate to the equator (Lisitzin 1985);
4. generally, the vertical oceanic circulation was reinforced for the upper levels ($z < 2000$ m) of the World Ocean at least in non-polar regions, but the deep layers were considerably less ventilated (e.g. Boyle and Keigwin 1982; Streeter et al. 1982), which made possible CO_2 storage in bottom waters, thus possibly contributing to low atm CO_2 levels;

Table 1. Changes in average composition of atmosphere and of deep water of the World Ocean (Broecker and Peng 1982; Lisitzin 1985; Sarmiento and Toggweiler 1984; Boyle 1988), and global marine primary production (in tera = 10^{12} mol of carbon per year). *Arrows* indicate known or possible variations relative to present (reference)

Time	Atm	Earth	Average deep water (W.O.)									World Ocean
(y)	CO_2 vpm	Surf ΔT °C	Θ °C	S psu	ALK	ΣCO_2 $\mu g\ kg^{-1}$	$CO_3{=}$	NO_3^-	$HPO_4{=}$ $\mu mol\ kg^{-1}$	$Si(OH)_4$	O_2	Prim. prod.- Tera mol C y^{-1}
−18 000	190	−4–5	0	35.9	2557	2460	89	48	3.2	120	90–110	↑
Variation	↓	↓	↓	↑	↑	↑	↑	↑	↑	↗	↓	
Present reference	350	0	1	34.7	2365	2250	86	33	2.2	100	190	3300

5. reduction of oceanic convection or of vertical exchanges between surface and deep levels in high latitudes (Siegenthaler and Wenk 1984; Ennever and McElroy 1985; Toggweiler and Sarmentio 1985; Wenk and Siegenthaler 1985).

3.2 Palaeo-Composition of the Deep World Ocean (Table 1)

As mentioned above, the mechanism leading to increased nutrient concentrations in the palaeo-ocean was erosion of nutrient-rich sediments on the continental shelves. These margins received rich deposits of organic matter during the last interglacial high-sea-level period (Broecker 1982). Deposition and erosion are slow processes, depending on major changes in sea levels, which cannot be responsible for rapid changes in CO_2, such as were detected to have happened during the Ice Age (Siegenthaler and Wenk 1984).

3.3 Palaeo-Productivity, Carbon and Nutrient Cycles (with Emphasis on the Cycle of Silicon) in the Southern Ocean

Low atmospheric CO_2 during the last glacial age has been interpreted by a number of models (Knox and McElroy 1984; Toggweiler and Sarmiento 1985; Wenk and Siegenthaler 1985; Broecker and Peng 1986; Boyle 1988; Keir 1988). They help understanding the intriguing relationships between atmospheric CO_2, ocean chemistry and ocean circulation during climate change events.

To explain reduced total amount of CO_2 in the atmosphere, three main hypotheses have been considered: enhanced global marine primary production linked to increased availability of nutrients; decrease in the production of calcium carbonate; decrease in vertical exchanges between surface and deep waters at high latitudes. The decrease in atmospheric CO_2 has also been described in direct response to increased oceanic alkalinity, along with vertical oceanic nutrient fractionation (Boyle 1988).

A decrease in atmospheric CO_2 during this glacial age is indeed consistent with the hypothesis of enhanced global marine primary production, especially in areas where nutrients previously limit photosynthetic activity. Variations in nutrient inputs to surface layers were due to: changes in the pattern of ocean mixing; extra amounts of macronutrients in coastal areas from rewashing of exposed continental shelves; increased aeolians inputs of micronutrients (for instance trace metals).

As regards subtropical coastal upwelling zones, since 1952 Arrhenius has postulated that during glacial ages increase in the planetary temperature gradients brought increased wind velocities. This stimulates both the upwelling of nutrient-rich thermocline oceanic waters and coastal upwellings. Higher primary production in coastal upwelling zones during this epoch can be traced to increased biogenic silica accumulation (Lisitzin 1985).

Enhanced primary production in the Southern Ocean, for the glacial age, has also been postulated, e.g. by Knox and McElroy (1984), Sarmiento and Toggweiler (1984), Broecker and Peng (1987). Knox and McElroy (1984), Sarmiento and Toggweiler (1984) discussed the idea that net high-latitude productivity was coupled with a decrease in the thermohaline overturning. The hypothesis of higher primary production in the last glacial age has been tested using opal deposition as a tracer of primary production in surface layers (diatoms play a major role in Antarctic phytoplankton). The Southern Ocean is characterized by a huge belt, about 1500–2000 km wide, located south of the Polar Front, where very large amounts of biogenic silica accumulate (DeMaster 1981; Lisitzin 1985). During the last glaciation this belt shifted about 500 km northwards (review in Bareille 1991). A controversy is arising about estimates of biogenic silica production/primary production for the last glacial period. Since it has been proven that diatoms are able to fractionate elements in surface waters (e.g. Murnane and Stallard 1988; Shemesh et al. 1989), Cd/Ca or Ge/Si ratios in plankton can be used as indices for nutrient palaeo-distributions and palaeo-productivity (Boyle 1988; Froelich et al. 1989). For the Atlantic sector of the Southern Ocean Cd/Ca variations indicate a primary production a little lower for the last glacial period compared to the Holocene, in parallel with low variations in nutrient distributions. Ge/Si variations also indicate a silicate increase, attributed by Mortlock et al. (1991) to a palaeo-productivity decrease. Lower production in the Atlantic sector is explained by the northwards extension of the ice coverage.

Palaeo-accumulation rates of biogenic silica have been determined by Bareille (1991) for the Indian sector of the Southern Ocean. The rich 1000-km-wide biogenic silica belt ($>2\,g\,cm^{-2}\,ka^{-1}$) did not exhibit higher accumulation rate during the last glacial period but only a northward shift of about 5° in latitude. From transfer functions Bareille has calculated Southern Ocean palaeo-productivy rates very comparable to those of present (Tréguer and van Bennekom 1991). It is worthwhile to note that Bareille's conclusion for the Indian sector might differ from other estimates for other sectors of the Southern Ocean. Palaeo-production in the south of Australia for this epoch is around 100% higher than for present (L. Labeyrie, pers. comm.). No estimate is available for the Pacific.

Total enhanced austral palaeo-productivity could be only about 50% over that for present (L. Labeyrie, pers. comm.), i.e. the contribution of the Southern Ocean to an atmospheric CO_2 decrease due to biological mechanisms should have been more limited than scientists usually estimated a few years ago. Other mechanisms bound to be envisaged to explain an Antarctic contribution to the CO_2 decrease during the glacial age are e.g.:

1. fluctuations in the ratio of the production of organic carbon to the production of inorganic carbon. Actually, during the glacial ages the Southern Ocean was characterized by rates of accumulation of calcium carbonate

considerably lower than during interglacial periods (for Antarctic and sub-Antarctic domains, Bareille 1991);
2. decrease in the convection at high latitudes;
3. increase in the production of AABW. This last hypothesis was suggested by Gordon (1971) who indicated that a more active ACC would lead to higher salinity near the Antarctic. This will favour thermohaline convection and larger AABW production. Bareille (1991) gives experimental support to this idea based on variations in fluxes of detritic material measured in the Indian sector of the Southern Ocean.

4 Questions and Objectives for the Future

Dramatic climatic changes are expected in response to global warming of the surface temperature owing not only to an increase in the atmospheric CO_2 content (expected to rise by 2050 twice as high as the present concentration) but also to increases in CH_4 and N_2O rejections in the atmosphere.

The tight coupling between both atmospheric and oceanic systems means we may expect drastic consequences for the ocean and specifically for the biogeochemical cycles. Nonetheless, the response to perturbations of the two components of the system (i.e. atmosphere and continents, on the one hand, and ocean, on the other) are not to be expected on the same time scale. Present models are relatively realistic to illustrate changes in the rapid component (i.e. on a time scale of years and decades for atmosphere and surface of continents), but not for changes eventually occurring in the slow component (i.e. on a time scale of decades for surface oceanic waters and up to 100 years for deep oceans and to 10000 to 100000 years for polar ice sheets).

In spite of some inadequacies, models are necessarily to be used in order to know about consequences of climate changes in the ocean biogeochemistry, the past being in fact more helpful in identifying major processes to be involved in scenarios than in quantitatively predicting the future. For instance, 100 million years ago atmospheric CO_2 content was probably as high as 1000 ppm and this period could be considered as an analogue for our near future; however, tropical ambient conditions that prevailed during the Cretacean are not transposable to the present, because the locations of the continents have significantly changed.

4.1 Possible Impacts of Climate Change on the Circulation of the World Ocean and on Its Biogeochemistry

1. Doubling the atmospheric CO_2 content means increasing the surface temperature of the Earth by about 4 °C (Table 1). It is worthwhile to note

that this enhancement will be unequally distributed (see e.g. Mikolajewicz et al. 1990) between polar regions (5–7 °C) and low latitudes (only about 3 °C). As a consequence of the warming of the low atmosphere, larger amounts of water vapour in the atmosphere are expected, which will result in abundant rains (in contrast, enhanced evaporation is expected for instance in the Mediterranean region). As already mentioned, the buffer effect of the oceans and of the polar ice sheets will make the occurrence of these consequences of the general warming up delay by about 50 to 100 years (Duplessy and Morel 1990).

Because the mean ocean temperature will rise, less CO_2 will be dissolved: Duplessy and Morel (1990) do not exclude the possibility that oceans could become saturated by atmospheric CO_2!

2. The consequences of such a strong and rapid climate change in the oceanic circulation are still under discussion. Mikolajewicz et al. (1990), using the Hamburg large-scale geostrophic OGC model, which exhibits a realistic thermohaline circulation, predict that global deep-water production could decrease to a minimum by year 40 (after today), the fractional reduction being stronger in the North Atlantic than in the Southern Ocean. After year 40 the global deep-water formation should increase again due to increased formation of AABW. This decrease in deep-water formation is expected to have a strong impact on atmospheric CO_2, with less CO_2 transported to the deep ocean by the newly formed deep and bottom water, which in turn will lead to amplification of the CO_2 increase.
3. Models show that doubling of the atmospheric CO_2 content yields a global mean sea level rise (from thermal expansion alone) of about 20 cm in 50 years. Nonetheless, very large discrepancies in this rise are to be expected according to Mikolajewicz et al. (1990): for example a rise of 40 cm is projected for the North Atlantic and only 10 cm in the Ross Sea.
4. A direct biological response to raised levels of atmospheric CO_2 is generally not expected for the Southern Ocean (nor for other areas of the World Ocean), since primary production is not CO_2-limited. However, the well-known sensitivity of marine production to hydrodynamic conditions and structures means that considerable indirect effects should be expected in the event of temperature-induced changes.

4.2 Objectives of IGBP Related Programmes

Three major objectives have been established by IGBP related programmes for understanding and modelling the interactions of global change with the biogeochemical cycles in the Antarctic:

1. To identify the sources, sinks and transport of gases and trace chemicals in the Antarctic troposphere and stratosphere. This is especially the task of the Global Atmospheric Chemistry (IGAC) programme.

2. To determine the exact role of the oceans, sea ice and biosphere in the control of today's biogeochemical cycles (C, N, P, Si, S). The Southern Ocean – Joint Global Ocean Flux Study (SO – JGOFS) focusses on modelling these cycles in the whole water column. In particular, this programme addresses the question of the key factors that control primary production in the different subsystems of this ocean. The Southern Ocean – Global Ecosystem Dynamics (SO – GLOBEC) programme pays closer attention to secondary production processes.
3. Finally, the Ecology of the Sea-Ice Zone (EASIZ) programme seeks to determine what will happen at the ice-edge zone especially sensitive to climate change.

References

Bareille G (1991) Flux sédimentaires: paléoproductivité et paléocirculation de l'Océan Austral au cours des 150 000 dernières années. Thèse Université de Bordeaux 1, 259 pp

Barnola J, Raynaud D, Korotkevitch Y, Lorius C (1987) Vostok ice core provides 160 000-year record of atmospheric CO_2. Nature 329:408–414

Boyle EA (1988) Vertical oceanic nutrient fractionation and glacial/interglacial CO_2 cycles. Nature 331:55–56

Boyle EA, Keigwin LD (1982) Deep circulation in the North Atlantic during the 200 000 years: geochemical exidence. Science 218:784–787

Broecker WS (1982) Glacial to interglacial changes in ocean chemistry. Prog Oceanogr 11:151–197

Broecker WS, Peng TH (1982) Tracers in the sea. LDGO, Columbia Univ, Palisades, New York, 10964, 1–690

Broecker WS, Peng TH (1986) Carbon cycle: 1985 glacial to interglacial changes in the operation of the global carbon cycle. Radiocarbon 28:309–327

Broecker WS, Peng TH (1987) The role of $CaCO_3$ compensation in the glacial to interglacial atmospheric CO_2 change. Global Biogeochem Cycles 1:15–29

Buma AG, De Baar HJW, Nolting RF, Van Bennekom AJ (1991) Metal enrichment experiments in the Weddell-Scotia Seas: effects of iron and manganese on various plankton communities. Limnol Oceanogr 36:1865–1878

De Baar HJW, Buma AGJ, Nolting RF, Cadée GC, Jacques GJ, Tréguer P (1990) On iron limitation of the Southern Ocean: Experimental observations in the Weddell and Scotia Seas. Mar Ecol Prog Ser 65:105–122

Delmas RD, Asciencio JM, Legrand M (1980) Polar ice evidence that atmospheric CO_2 20 000 yr BP was 50% lower of present. Nature 284:155–157

DeMaster DJ (1981) The supply and accumulation of silica in the marine environment. Geochim Cosmochim Acta 45:1715–1732

Dugdale RC, Wilkerson FP (1990) Iron addition experiments in the Antarctic: a reanalysis. Global Biogeochem Cycles 4:13–19

Duplessy JC, Morel P (1990) Gros temps sur la planète. Odile Jacobs, Paris, 296 pp

Ennever F, McElroy M (1985) Changes in atmospheric CO_2: factors regulating the glacial to interglacial transition. In: Sundquist E, Broecker WS (eds) Natural variations archean to present. Geophys Monogr 32:154–162

Froelich PN, Mortlock RA, Shemesh A (1989) Inorganic germanium and silica in the Indian Ocean: biological fractionation during $(Ge/Si)_{opal}$ formation. Global Biogeochem Cycles 3:79–88

Gordon AL (1971) Oceanography of the Antarctic waters. In: Reid J (ed) Antarctic Oceanology. Antarct Res Ser 15:169–203

Gordon AL, Taylor HW (1975) Heat and salt balance within the cold waters of the World Ocean. Numerical models of circulation. National Academy of Sciences, Washington, DC, pp 54–56

Helbing WE, Villafane V, Holm-Hansen O (1991) Effect of iron on productivity and size distribution of Antarctic phytoplankton. Limnol Oceanogr 36:1886–1898

Jacques G (1983) Some ecophysiological aspects of the Antarctic phytoplankton. Polar Biol 2:27–33

Keir RS (1988) Paleoproduction and atmospheric CO_2 based on ocean modeling. In: Berger WH, Smetacek VS, Wefer G (eds) Productivity of the ocean: present and past. John Wiley, New York, pp 395–406

Knox F, McElroy M (1984) Changes in atmospheric CO_2: influence of the marine biota at high latitudes. J Geophys Res 89:4269–4637

Lancelot C, Veth C, Mathot S (1991) Modelling ice-edge phytoplankton bloom in the Scotia-Weddell Sea sector of the Southern Ocean. Mar Chem 35(1-4):305–324

Lancelot C, Mathot S, Becquevort S, Dandois JM, Billen G (1992) Carbon and nitrogen cyling through the microbial network of the marginal ice zone of the Southern Ocean with particular emphasis to the NW Weddell Sea. Report Belgian Program Scientific Research on the Antarctic, Bruxelles, Université Libre de Bruxelles Publ, p 106

Ledfford-Hoffman PA, DeMaster DJ, Nittrouer CA (1986) Biogenic silica in the Ross Sea and the importance of Antartic continental-shelf deposits in the marine silica budget. Geochim Cosmochim Acta 50:2099–2110

Lisitzin AP (1985) The silica cycle during the last ice age. Paleooceanogr Paleoclimatol Paleoecol 50:241–270

Longhurst AR (1991) Role of the marine biosphere in the global carbon cycle. Limnol Oceanogr 36:1507–1526

Martin JH, Fitzwater SE (1988) Iron deficiency limits phytoplankton growth in the NE Pacific subarctic. Nature 331:341–343

Martin JH, Gordon RM, Fitzwater SE (1990) Iron in Antarctic waters. Nature 345:156–158

Martin JH, Gordon RM, Fitzwater SE (1991) The case for iron. Limnol Oceanogr 36:1783–1792

Metzl N, Beauverger C, Brunet C, Goyet C, Poisson A (1991) Surface water CO_2 in the SW Indian sector of the Southern Ocean: a highly variable CO_2 source/sink region in summer. Mar Chem 1-4:85–96

Mikolajewicz U, Santer BD, Maier-Reimer E (1990) Ocean response to greenhouse warming. Nature 345:589–593

Morel FMM, Hudson RJM, Price NM (1991) Limitation of productivity by trace metals in the sea. Limnol Oceanogr 36:1742–1755

Mortlock RA, Charles CD, Froehlich PN, Zibello MA, Saltzman J, Hays JD, Burckle LH (1991) Evidence for lower productivity in the Antarctic Ocean during the last glaciation. Nature 351:220–223

Murnane RG, Stallard RF (1988) Ge/Si fractionation during biogenic opal formation (paper 890408). Paleooceanography 3:461–469

Murphy PP, Feely RA, Gammon RH, Kelly KC, Waterman LS (1991) Autumn air-sea disequilibrium of CO_2 in the South Pacific Ocean. Mar Chem 1-4:77–84

Neftel A, Oeschger H, Schwander J, Stauffer B, Zimbrunn R (1982) Ice core sample measurements give CO_2 content during past 40000 years. Nature 295:220–223

Nelson DM, Gordon LI (1982) Production and pelagic dissolution of biogenic silica in the Southern Ocean. Geochim Cosmochim Acta 46:492–501

Nelson DM, Smith WO (1991) Sverdrup revisited: critical depths, maximum chlorophyll levels, and the control of the Southern Ocean productivity by irradiance-mixing regime. Limnol Oceanogr 36:1650–1661

Nelson DM, Tréguer P (1992) Role of silicon as a limiting factor to Antarctic diatom: evidence from kinetic studies in the Ross Sea ice edge. Mar Ecol Prog Ser 80:255–264

Poisson A, Metzl N, Brunet C, Schauer B, Bres B (1991) CO_2 sources and sinks in the Indian sector of the Southern Ocean. Abstr Int Symp Roma, September 1991

Sarmiento JL, Toggweiler JR (1984) A new model for the role of the oceans in determining atmospheric P_{CO_2}. Nature 308:621–624

Shemesh A, Burckle L, Froelich PN (1989) Dissolution and preservation of Antarctic diatoms and the effect on sediment thanatocoenoses. Quat Sci 31:288–308

Siegenthaler U, Wenk T (1984) Rapid atmospheric CO_2 variations and ocean circulation. Nature 624:626

Smetacek VS, Scharek R, Nöthig EM (1990) Seasonal and regional variations in the pelagial and its relationship to the life history cycle of krill. In: Kerny R, Hempel G (eds) Antarctic ecosystems: ecological change and conservations. Springer, Berlin Heidelberg New York, pp 103–114

Smith WO, Keene NK, Comiso JC (1988) Interannual variability in estimated primary productivity in the Antarctic marginal ice zone. In: Sahrage D (ed) Antarctic Ocean and resources variability. Springer, Berlin Heidelberg New York, pp 131–139

Sommer U (1991) Comparative nutrient status and competitive interactions of the two Antarctic diatoms *Corethron criophilum* and *Thalassiosira antarctica*. J Plankton Res 13:61–75

Streeter SS, Belanger PE, Kellog TB, Duplessy JC (1982) Late Pleistocene paleo-oceanography of the Norwegian-Greenland Sea: benthic foraminiferae evidence. Quat Res 18:77–90

Sverdrup HV (1953) On conditions for the vernal blooming of phytoplankton. J Cons Perm Int Explor Mer 18:287–293

Toggweiler R, Sarmentio J (1985) Glacial to interglacial changes in atmospheric CO_2: the critical role of ocean surface water in high latitudes. In: Sundquist E, Broecker WS (eds) Natural variations archean to present. Geophys Monogr 32:99–110

Tréguer P, Jacques G (1992) Dynamics of nutrient and phytoplankton in the Southern Ocean: a review. Polar Biol 12:149–162

Tréguer P, Van Bennekom AJ (1991) The annual production of biogenic silica in the Southern Ocean. Mar Chem 1-4:477–488

Wenk T, Siegenthaler U (1985) The high latitude oceans as a control of atmospheric CO_2. In: Sundquist E, Broecker WS (eds) Natural variations archean to present. Geophys Monogr 32:185–194

Detection of Change in Antarctica

H. Jay Zwally[1]

1 Introduction

Detection of change in Antarctica is of global interest because potential changes in the Antarctic in environment are coupled to the Earth system. An interdisciplinary research theme identified by the SCAR Steering Committee (Weller et al. 1989) for the International Geosphere-Biosphere Program (IGPB) is: "*detection of changes of global importance, best observed in Antarctica*, which are fundamental to establishing the current nature of trends of change and thereby providing a foundation for understanding the underlying processes." The goals of Antarctic detection-of-change research should be to determine what is happening now throughout Antarctica and the surrounding seas, attempt to relate the current situation to the past, and understand why changes are occurring.

Detection of change requires, first of all, critical measurements to provide reliable data and improve the understanding of the physical processes involved. Evaluation of the long-term nature and the large-scale characteristics of any observed change is essential to assess its significance, particularly within the context of the IGBP. Typically, a variety of measurements, studies of the physical processes, and numerical models are needed to understand the role of changes of the Antarctic environment in the Earth system as a whole. Some important measurements have begun, and some specific examples of ongoing studies of changes are described in the following.

2 General Strategy

Research on detection of change is based on the concept that certain key variables may sufficiently describe the environment so that fundamental changes may be detected by monitoring those variables. Common examples are atmospheric temperature and precipitation. Implementation of this

[1]NASA Goddard Space Flight Center, Oceans and Ice Branch, Code 971, Greenbelt, Maryland 20771, USA

Hempel (Ed.) Antarctic Science

concept, however, is not so simple. The first step is selection of key variables for measurement, based on current understandings of the physical, chemical, and biological processes. Then, the available measurements and data analysis may provide insights and documentation on the occurrence of significant changes. The resulting improved understanding of the processes of change, in turn, leads to better selection of the variables and definition of the processes to be studied further. In practice, detection-of-change research has been evolving in such an iterative manner.

The SCAR report (Weller et al. 1989) identified three principal areas for detection-of-change research in Antarctica: atmospheric composition, indicators of climate change, and Antarctic ice and sea level. The category of atmospheric composition includes three topics: changes in Antarctic ozone, temporal variability in atmospheric carbon dioxide, and potential changes in other trace gases. A minimum list of potential indicators of climate change that should be monitored includes: sea ice concentration, extent, and thickness; temperatures (surface, tropospheric, and stratospheric); upward terrestrial and reflected solar radiation at the top of the atmosphere; and the distribution of surface melting on the ice sheet and shelves. The category of Antarctic ice and sea level includes detection of changes in: surface elevation over the complete ice sheet, ice-sheet extent and surface characteristics, and ice velocity fields.

Principal considerations in evaluating observed changes are confirmation of their persistence on decadal time scales, assessment of their likelihood for continuance in the future, and determination of their large-scale characteristics and global interactions. Methods for evaluation of changes are principally temporal and spatial analysis of the data, examination of quantitative relationships and possible correlation among variables, and numerical modeling of the interactive processes.

Detection of change requires careful analysis with respect to both the spatial and temporal variabilities of the possible signal of change. The short temporal length of many important data records is an often-stated limitation in analysis of environmental data. However, in many instances, lack of spatial coverage is even more serious than shortness of the record. For example, a lengthy time series of a variable at only a few locations may describe only a localized phenomenon or represent a spatial aliasing of the data. In some cases, measurements over large areas for even short periods may be more informative than long-period observations at a few locations. Reversals of apparent trends in existing data records, or the lack of consistency from region to region, have clearly demonstrated the need for better spatial and temporal sampling strategies, as well as long-term measurements and wider spatial coverage.

It is important to distinguish between measurement of a definitive change and detection of a trend in the data. An implicit question concerning detection of change is whether the variable is tending to increase or decrease. What is the future value? Attention to basic principles of measurement sampling and data analysis is especially important for evaluation of so-called trends. For illustration, consider a measurement of some value at time T1 and location P1 and again at time T2 and location P2 (Fig. 1). If the difference between V(T2, P2) and V(T1, P1) is large compared to the measurement error, then the change in V is definitive. In this case, a clear change would be unambiguously detected.

Of course, two measurements of a variable provide very little information about the past, future, or intervening values. A third measurement may show that the variable is not changing linearly or monotonically, but provides little information on past or future values, as illustrated in Fig. 1. Other pitfalls of inadequate sampling, which can be interpreted either in terms of time or space or both, are illustrated in Fig. 2. In Fig. 2a, the ten data points at nearly equal intervals (connected by a dashed line) clearly suggest a downward trend. However, the addition of ten more interspersed samples suggests a very noisy signal, with a small downward trend indicated by the linear fit. Adding 25 more samples (Fig. 2b) suggests a cyclical variation, still with a small downward trend. However, these data samples are taken from a perfect sine wave, which has zero long-term trend. Clearly, more data usually add information on the variability of a variable until adequate sampling is achieved. Unfortunately, some studies of data records have been fraught with false conclusions based on insufficient data or improper interpretations regarding trends.

In general, numerical models have been useful for understanding changes and their significance, and for making predictions. However, models are too often seriously limited by inaccurate or insufficient quantitative descriptions of the interactive processes. If, for example, the quantitative effects of specific changes in cloud cover of various types and heights are not adequately known, the climatic effects of such changes cannot be correctly modeled. Similarly, if the potential causes of changes in cloud cover, such as

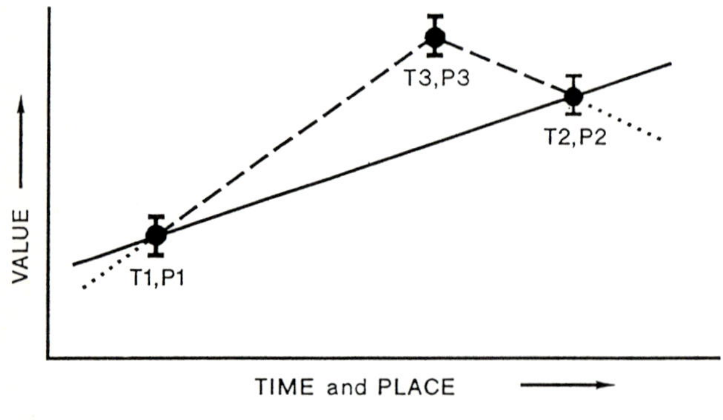

Fig. 1. The measured difference in value between 1 and 2 is large compared to the measurement error, so the observed change is definitive. However, evaluation of the cyclical characteristics of the variation and possible trends clearly requires more data

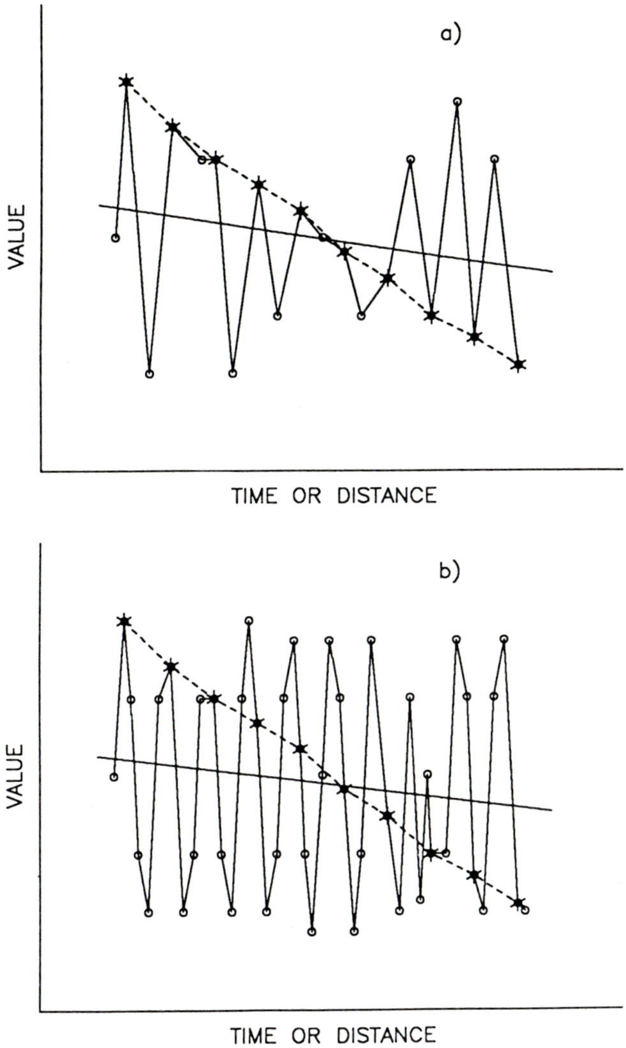

Fig. 2. Inadequate sampling (spatial or temporal) may falsely suggest a trend, as shown by the linear character of the 10 data points (at near equal spacings) in **a** and **b**, in contrast to the noisy character of 20 points in **a** and the cyclic character of 45 points in **b**. All data samples are from a perfect sine wave with a zero long-term trend

changes in sea ice concentration, are not quantitatively described, then the models must be incomplete or inaccurate. These and other examples demonstrate the strong need for more and better quantitative measurements. Therefore, an improved measurement program must have a high priority for detection of change.

3 Sample Studies of Change

Some important measurements have begun. Temperatures are monitored at stations and sea ice extent is observed by satellite, for example. Recent results include observation of ozone depletion, changes in ice stream flow, and ice-shelf disintegration. New results include satellite monitoring of surface melting. Techniques are being developed for measuring changes in ice sheet volume, equilibrium-line position, surface temperature, and phytoplankton distributions by satellite sensing. In this section, a few specific examples are given to illustrate the state of progress, as well as the principles discussed in the previous section.

3.1 Antarctic Ozone Hole

The discovery of large decreases in stratospheric ozone over Antarctica during spring has attracted worldwide attention. The discovery was based on routine monitoring of ozone at Halley Bay station since 1957 (Farman et al. 1985), followed by analysis of satellite data acquired since 1978 (Stolarski et al. 1986; Newman et al. 1991). The record of the October mean total ozone (Fig. 3) shows the marked decreases occurring each year since about 1980.

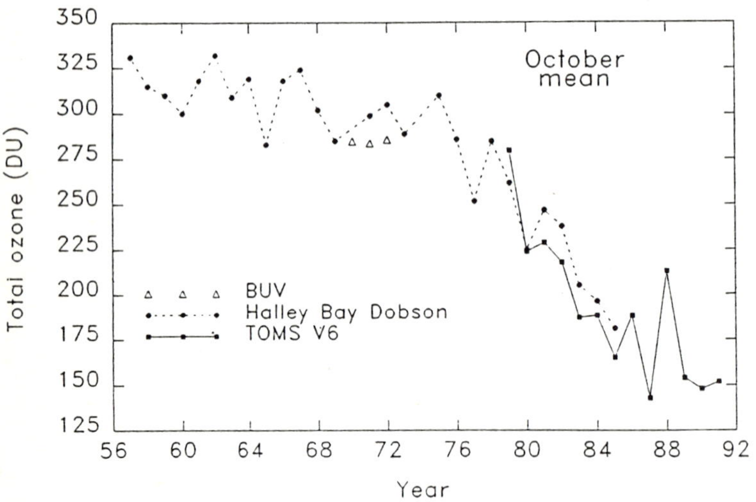

Fig. 3. Satellite and ground-based measurements of the mean total-ozone column in October, showing springtime development of the Antarctic ozone hole in recent years. Data with *dashed line* are from Dobson instrument at Halley Bay (Farman et al. 1985). Data with *solid line* are from Nimbus 7 TOMS minimum values on 2° latitude by 5° longitude grid. *Triangles* are data from first 3 years of Nimbus 4 BUV. (Newman et al. 1991)

The cause of the decreases is now well established, as a result of aircraft, surface, and satellite meaurements and theoretical studies in recent years (Anderson et al. 1989). The changes are caused by photochemical processes in the atmosphere involving ozone, ultraviolet radiation from the sun, C10 from chlorofluorocarbons (CFCs), and Polar Stratospheric Clouds (PSCs). CFCs introduced into the atmosphere worldwide by human activities interact with the atmosphere producing C10. During the polar night, the C10 concentration builds in the vicinity of the PSCs. In spring, exposure to the solar radiation causes the constituents to interact, decreasing the ozone markedly in a period of only several weeks.

Decreases in ozone are of global concern, because ozone naturally filters ultraviolet solar radiation that is destructive to marine and land organisms. Although the implications of the Antarctic findings for the global distribution of atmospheric ozone are not yet fully understood, the linkages established between ozone depletion and human activities have already led to limitations on the worldwide production of CFCs and the search for alternatives. The activities progressed quickly from routine monitoring of Antarctic ozone levels to intensive research of the interactive processes, to understanding the cause of changes, to modification of industrial practices and governmental policies. The results dramatically illustrate the importance of making accurate and systematic measurements of key variables, even though their specific value may not be fully understood or anticipated.

3.2 Sea Ice and Temperature

Sea ice is generally more extensive during colder periods. Therefore, variations in the extent of sea ice are often considered to be an indicator of climate change. Modeling of variations in the Antarctic sea ice extent as a function of temperature suggests a change in ice extent of 1.4° latitude/Kelvin air temperature (e.g. Parkinson and Bindschadler 1984). Sea ice variations are also of particular interest, because climate models of greenhouse warming have suggested that the warming would be greatest in polar regions, mainly as a result of the positive feedbacks from reduced sea ice cover (e.g. Dickinson et al. 1987).

Attempts have been made to link observed changes in Antarctic sea ice to possible ongoing climate change (e.g. Kukla and Gavin 1981), but others have noted the large amount of regional and interannual variability that limits the ability to deduce trends from the data (e.g. Zwally et al. 1983). The yearly average Antarctic sea ice extent from 1973 to 1976 and 1979 to 1986 is shown by region in Fig. 4 from Nimbus-5 and Nimbus-7 satellite passive microwave data. A 10% decrease in overall area occurred into 4 years from 1973 to 1976, but from 1979 to 1986 essentially no overall change occurred.

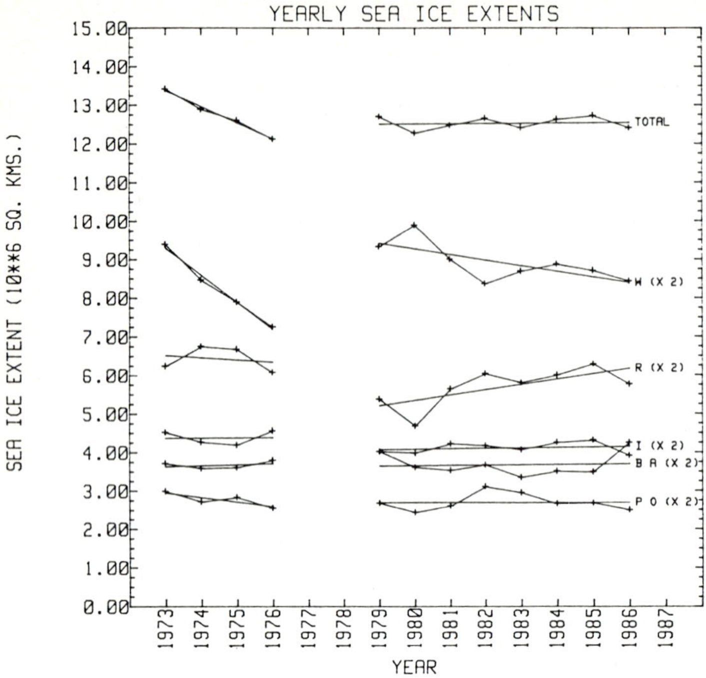

Fig. 4. Antarctic sea ice extent (yearly averages) for the entire Southern Ocean and the Weddell Sea (60° W to 20° E) (*W*), Indian Ocean (20° E to 90° E) (*I*), Pacific Ocean (90° E to 160° E) (*PO*), Ross Sea (160° E to 130° W) (*R*), and Bellingshausen-Amundsen Sea (130° W to 60° W) (*BA*) regions, showing a sharp decline from 1973 to 1976 and no overall change in recent years. From 1979 to 1986, the ice extent increased in the Ross Sea region, but decreased in the Weddell Sea region. Data are from the passive microwave sensor on Nimbus-7

On a regional basis, correlations between warmer temperatures and decreased sea ice extent were shown to be significant (Weatherly et al. 1991). Jacka and Budd (1991) also found a change in ice extent of about 1.5° latitude/Kelvin to hold regionally, which is consistent with the modeling result. However, attempts to deduce long-term trends in the overall ice extent or temperature have been ambiguous. From analysis of temperature data through 1982, Raper et al. (1984) found evidence of a warming trend, but little relationship between linear trends in hemispheric sea ice extent and areally averaged temperatures. The analysis of sea ice data fròm 1978 to 1987 by Gloersen and Campbell (1991) found no significant trend in the Antarctic, but a small significant negative trend in the Arctic. On the other hand, Jacka and Budd (1991) deduced "small significant trends of temperature increase and sea ice decrease" in the Antarctic for the period 1973 to 1989. However, 1973 had the largest Antarctic sea ice extent during this period, followed by a marked decline in the mid-1970s associated with the

Weddell polynya, and recovery by 1981 (Zwally et al. 1983). Prior to 1973, the sea ice data available from 1966 showed an increasing ice cover until 1973 (Zwally et al. 1983), during a period when temperatures were increasing. Therefore, the deductions about sea ice trends in recent decades are significantly influenced by the unusually large ice cover around 1973 and the subsequent anomalous decline in the mid-1970s.

The large variability of sea ice probably occurs because factors other than atmospheric temperature are significantly forcing the sea ice anomalies. For example, short-term regional variations in the extent of the Antarctic ice pack have been shown to be related to changes in the atmospheric circulation (Cavalieri and Parkinson, 1981). The map of the difference in winter ice cover between September 1980 and September 1985 illustrates the large regional variability and the tendency for opposing trends in different regions (Fig. 5). During this period, the sea ice cover decreased significantly in the Weddell sea region and increased significantly in the Ross Sea region, while the overall ice area did not change. Analysis of the leads and lags between sea ice changes and temperature in different seasons provides some insights as to whether sea ice anomalies cause temperature changes or vice versa and to the role of the ocean in the persistence of anomalies (Weatherly et al. 1991). In general, a better understanding of the processes controlling the variations in the sea ice cover is needed to determine their causes and their relation to climate change.

3.3 Ice-Sheet and Ice-Shelf Surface Melting

Measurement of surface melting is of interest for studies of glacier dynamics and for monitoring climatic changes in polar regions. Under current climatic conditions, surface temperatures remain below freezing over most of the Antarctic ice sheet throughout the year. Some surface melting extends inland from the Ross ice shelf onto the ice sheet of Marie Byrd Land (Giovinetto 1964; Alley and Bentley 1988). However, in most places on the ice sheet, the dry snow line lies only a few tens of kilometers inland from the coast or inland from an ice-shelf grounding line (Giovinetto 1964). Extensive surface melting usually occurs each year on many of the ice shelves, specifically the shelves in the Antarctic Peninsula, those adjacent to the Amundsen and Bellingshausen Seas, and those along the East Antarctic coast, including the Amery ice shelf. Of the major ice shelves, the Ross ice shelf occasionally has surface melting, but the Ronne Filcher ice-shelf lies almost entirely within the dry snow line (Giovinetto 1964).

Where melting occurs, downward percolation of meltwater from the firn surface, followed by refreezing at colder depths, transports heat downward at a much greater rate than thermal conduction and diffusion. Enhancement of this melt – freeze process during a greenhouse warming might raise the ice shelves to the pressure melting point throughout and cause their dis-

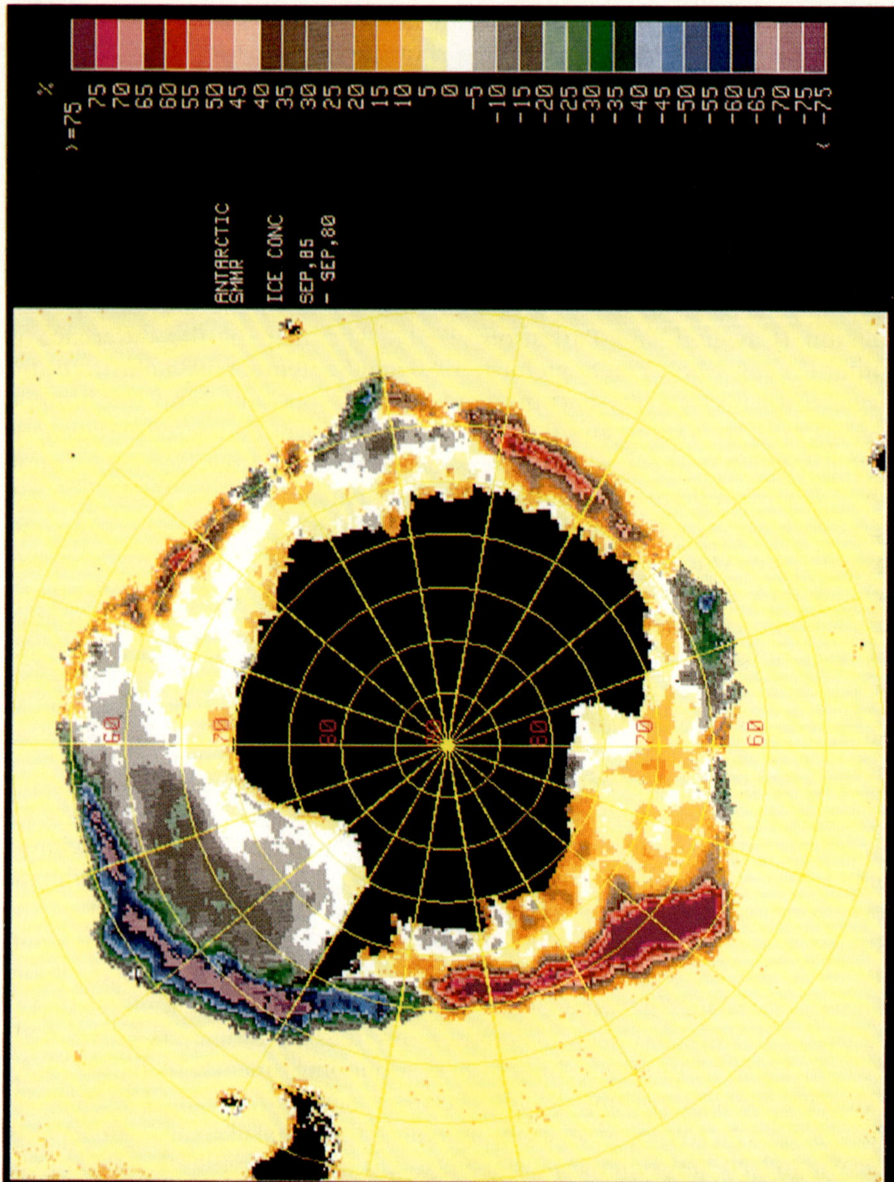
ANTARCTIC
SMMR
ICE CONC
SEP,85
- SEP,80
%
>=75
75
70
65
60
55
50
45
40
35
30
25
20
15
10
5
0
-5
-10
-15
-20
-25
-30
-35
-40
-45
-50
-55
-60
-65
-70
-75
< -75
60
70
80
80
70
60

Fig. 5. Difference in sea ice concentration (0 to 100% areal coverage) between September 1980 and September 1985, showing typical regional characteristics of interannual sea ice anomalies. In the Weddell Sea region, the sea ice concentration in 1985 is more than 50% greater in the outer 400 km of the pack and as much as 20% greater in the inner part of the pack. In the Ross Sea and Bellingshausen-Amundsen Sea regions, similar differences of opposite sign are shown

integration according to Mercer (1978) and Stuiver et al. (1981). However, Paterson (1984) concluded that the projected greenhouse warming would not be sufficient to induce disintegration, but noted the need for more quantitative data.

Passive microwave images of the polar regions, from the Nimbus-5 satellite, showed large increases in microwave brightness temperature (T_b) during summer on the percolation zone of Greenland and on the Antarctic ice-shelves where surface melting occurs (Zwally and Gloersen 1977). The extent and duration of surface melting on the Antarctic ice shelves and margins of the ice sheet have been mapped for 1978 to 1987 from satellite passive microwave data (Zwally and Fiegles 1994). The occurrence of surface melting is indicated by a marked increase in microwave brightness temperature (T_b), which is caused by moisture in the near-surface firn. A surface-melt index (duration · area) is calculated as the sum of the areas in which melting is indicated by T_b incrases more than 30 K above the annual mean T_b. The regional surface-melt indices correlate well with the monthly average temperatures measured at Antarctic stations. Most Antarctic surface melting occurs during December and January (Fig. 6).

The melt index calculated for Antarctica is 24×10^6 days · km^2, averaged over nine summers. The observed interannual and regional variability is large, varying by more than a factor of 2 from year to year. Surface melting was most intensive during the 1982/1983 summer (36×10^6 days · km^2) and least intensive during the 1985/1986 summer (16×10^6 days · km^2). Linear fits (Fig. 6) indicate a decline in surface melting over the 8 years. Although definitive changes in surface melting have been measured, inferences regarding trends in surface melting are not meaningful because of the large interannual variability.

3.4 Ice Sheet Mass Balance and Sea Level

The Antarctic ice sheet, as large in area as 4% of the world's oceans, would be sufficient to raise global sea level by more than 60 m if it melted. Each year about one part in 10^4 of the ice mass, or about 6 mm of global sea level equivalent, is exchanged with the ocean through the processes of evaporation, atmospheric moisture transport, snowfall on the ice sheet, and iceberg discharge. Uncertainty in the current mass balance of the Antarctic ice sheet

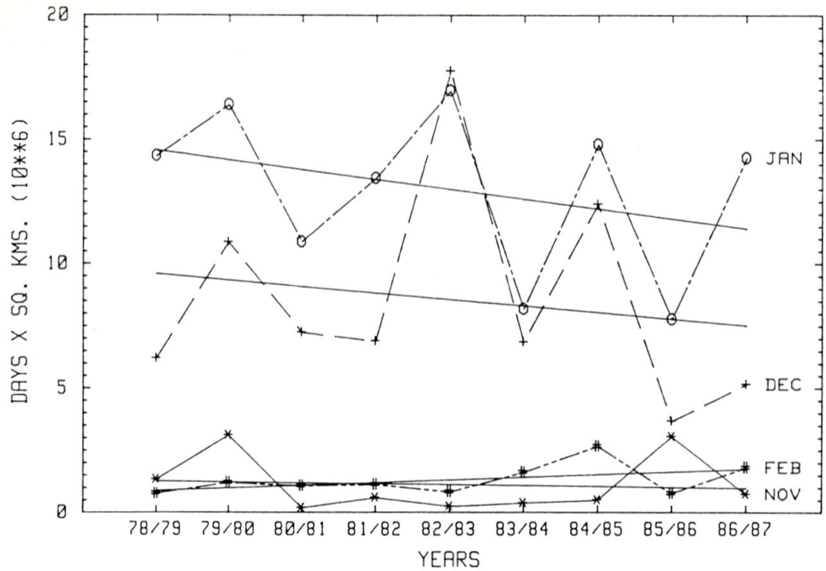

Fig. 6. Surface melting (duration · area) on Antarctic ice-shelves and ice sheet margins, derived from Nimbus-7 passive microwave data from 1978 to 1987. Although the linear fits indicate a decrease in surface melting, the interannual variability is large, so inferences regarding a trend are not meaningful

is probably at least ±30%, or ±2 mm/year in sea level equivalent. Recent estimates of accumulation and ice outflow fluxes have suggested a positive mass balance in the range of 2–25%, which is equivalent to 0.1 to 1.1 mm/year of sea level fall (Bentley and Giovinetto 1991). On the other hand, estimates of iceberg discharge (Orheim 1985) and basal melting from ice shelves (Jacobs et al. 1985), along with net accumulation rate estimates, imply a significant negative mass balance, and a contribution to sea level rise. At present, global sea level appears to be rising by 2.4 ± 0.9 mm/year (Peltier and Tushingham 1989) or 1.8 ± 0.1 mm/year (Douglas 1991). Although both thermal expansion of the ocean and melting of small glaciers contribute to sea level rise, the major source of water for the current sea level rise is undetermined.

Over century time scales, the primary effect of ice sheet interactions with climate changes is vertical growth or shrinkage of the ice in response to changes in precipitation and surface heat flux. Changes in precipitation and melting have immediate effects on the mass balance, whereas changes in ice flow and discharge occur on century time-scales. While the possibility of enhanced ice-sheet melting in a warmer climate has been of concern (e.g. Meier et al. 1985), more attention has been given recently to a well-known positive correlation between precipitation and temperature in polar regions, as discussed in Zwally (1989) for example. Modeling of the Antarctic precipitation and evaporation in a greenhouse-warmed climate, including the effects of sea ice reduction and the consequent enhanced oceanic moisture source, suggest an accumulation increase as large as 68% or 4 mm/year of sea level lowering (Budd and Simmonds 1991). From analysis of ice cores within about 300 km of the coast of East Antarctica, Morgan et al. (1991) found an increase in accumulation of about 30% since 1960. Therefore, increasing precipitation in a warmer climate might cause the ice sheet to grow more, or shrink less.

Satellite altimeter measurements of surface elevation change offer a unique means of determining the ice sheet mass balance. Radar altimeter measurements of the southern 40% of the Greenland ice sheet indicated a growth rate of about 23 cm/year from 1978 to 1985, implying a positive mass balance of 25–45% for that period and a sea level depletion of a few tenths of a millimeter per year (Zwally et al. 1989). However, the accuracy of current altimeter systems has been limited, mainly by errors of the radar method over the steeper slopes near the margins and by errors in the calculation of the satellite's orbital position.

Recent analysis of 3 years of data from the exact repeat mission period of Geosat (November 1986 to September 1989) shows changes in the surface elevation of the East Antarctic ice sheet. An important aspect of the analysis is correction of the orbital heights of the satellite by referencing of the measurements over the ocean south 20° S to a common surface derived from altimeter data. The analysis of elevation change consists of constructing a time series using the height differences at orbital crossovers (cf. Zwally et al. 1989). All the crossovers between orbital paths in the first 90-day period and the second 90-day period are averaged to obtain the average elevation change between the two periods, and so forth for each successive 90-day interval. Similar series starting with the second period and so forth are combined to form the series shown in Fig. 7.

The crossover data in the elevation band from 2700 to 3300 m in the longitudinal sector 140–160° E (Amery ice shelf drainage basin) suggest an elevation decrease of 16 cm/year during the 3-year period. In contrast, the data in the band from 2000 to 2700 m in the sector 140–160° F suggest an increase of 0.07 m/year. In each case, similar values are obtained after the corresponding time series on the coastal sea ice is subtracted. The variations of these elevation series about the linear fit may be a combination of residual time-dependent errors in the measurement and real seasonal elevation changes, perhaps related to seasonal changes in surface accumulation. Although additional analysis is required to confirm these changes, these examples demonstrate the potential of monitoring the surface elevation changes by satellite altimetry. In particular, seasonal and interannual variations in surface accumulation due to changes in precipitation will be observable. Also, improvements in accuracy from planned satellite laser altimeters will increase the spatial coverage toward the margins and provide useful measurements in the low accumulation regions.

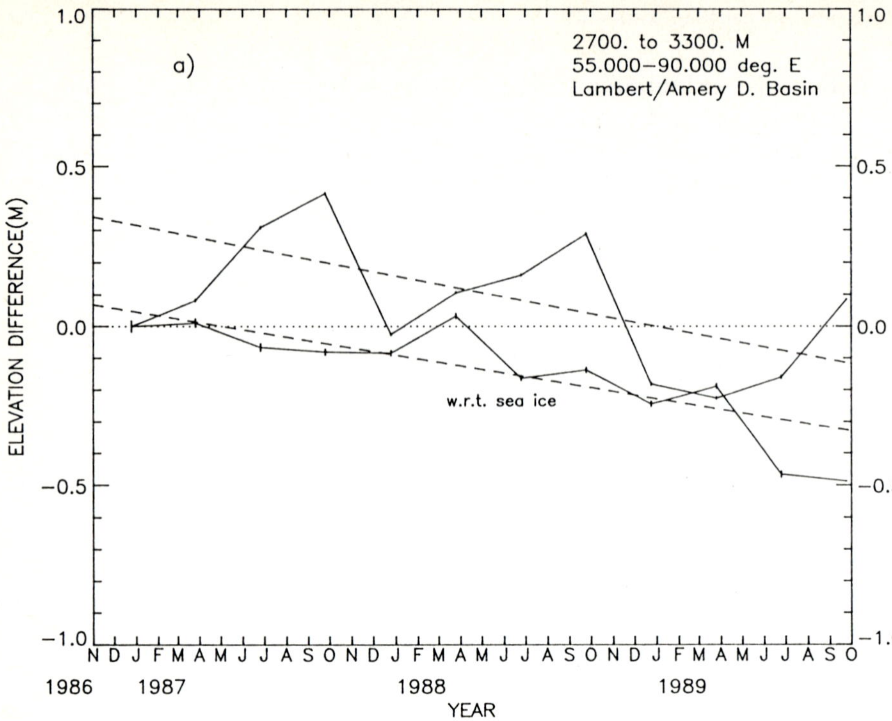

2700. to 3300. M
55.000–90.000 deg. E
Lambert/Amery D. Basin
a)
w.r.t. sea ice
ELEVATION DIFFERENCE(M)
1.0
0.5
0.0
−0.5
−1.0
N D J F M A M J J A S O N D J F M A M J J A S O N D J F M A M J J A S O
1986 1987 1988 1989
YEAR

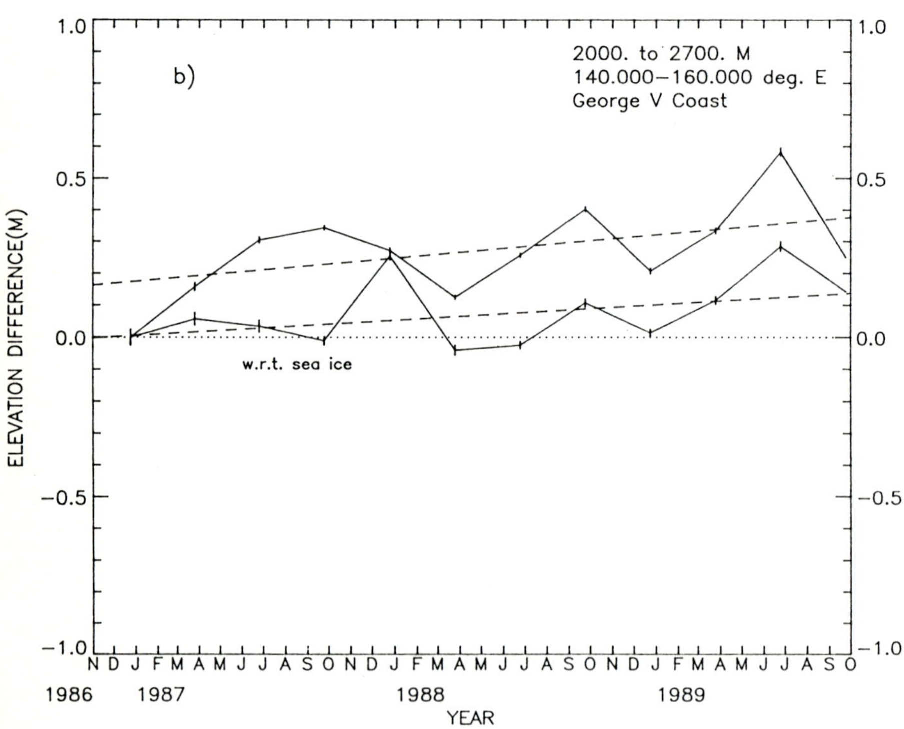

2000. to 2700. M
140.000–160.000 deg. E
George V Coast
b)
w.r.t. sea ice
ELEVATION DIFFERENCE(M)
1.0
0.5
0.0
−0.5
−1.0
N D J F M A M J J A S O N D J F M A M J J A S O N D J F M A M J J A S O
1986 1987 1988 1989
YEAR

3.5 Ice-Shelf and Ice Sheet Stability

Much attention has also been given to the question of stability of the West Antarctic ice sheet, because of the uncertainties regarding possible ongoing changes in response to past changes in climate and sea level, and because of the possibility that greenhouse warming might cause major changes in the future. The West Antarctic ice sheet is grounded mainly on a submarine bed and most of the ice discharges through ice streams into large floating ice shelves. The ice shelves are important in restraining and, perhaps, stabilizing discharge from the continent. Therefore, recent measurements relevant to these questions include observations of ice-shelf disintegration in the Antarctic peninsula and measurements of changes in ice stream velocities.

Using satellite imagery, Doake and Vaughan (1991) have shown that the Wordie ice shelf on the Antarctic peninsula has been slowly breaking up over the past few decades, during a period of warming in the region. However, the much larger and more southerly Ross and Filchner-Ronne ice shelves, which help stabilize the West Antarctic ice sheet, lie in a climate regime that is at least 10 K colder. Nevertheless, observation of ice-shelf breakup provides information on the processes involved in thinning and fracturing ice shelves, which may be taking place on other ice-shelves in the Peninsula or around the East Antarctica coast. In recent years, large tabular icebergs (about 30 000 km^2 total area) have calved from the Ross and Filchner-Ronne ice shelves (which cover about 1 000 000 km^2), but such events are likely to be episodic and unrelated to the Wordie ice-shelf breakup.

The West Antarctic ice streams flowing into the Ross ice shelf appear to be undergoing significant changes (Shabtaie and Bentley 1987; Stephenson and Bindschadler 1988). Also, Ferrigno et al. (1993) have measured an 8% increase between 1984 and 1990 in the velocity of the Thwaites Glacier, which discharges from West Antarctica into the Amundsen Sea where an ice-shelf probably existed in the past. However, there is apparently no consistent overall pattern of change, because the ice flow appears to be slowing down in some places and speeding up in others. Improved techniques for deriving velocity fields on ice streams from sequential high-resolution satellite imagery (e.g. Bindschadler and Scambos 1991) will play an important role in monitoring the rate of ice discharge from the continent.

Fig. 7. Time series of surface elevation change from crossover analysis of Geosat radar altimeter data, showing a decreasing elevation of 16 cm/year between 2700 and 3300 m surface elevations in the Lambert/Amery drainage basin (**a**) and an increasing elevation of 7 cm/year between 2000 and 2700 m elevations in the George V coast area (**b**) of East Antarctica. *Lower curves* in each case have the corresponding time-series over the coastal sea ice region subtracted

3.6 Phytoplankton Distribution

A parameter that is related to marine biology and is capable of large-scale monitoring by satellite is the concentration of phytoplankton pigment in the oceans. Large areas of elevated phytoplankton pigment concentrations were observed near the sea ice edge in the Weddell Sea using Coastal Zone Color Scanner (CZCS) data from Nimbus-7 (Comiso et al. 1990). Phytoplankton blooms near the sea ice edge during spring and summer may contribute significantly to the bioproductivity of the Southern Ocean. An overview of Southern Ocean data sets and discussion of the nutrient, physical, and illumination factors influencing the pigment concentrations is given in McClain et al. (1991). Figure 8 shows a composite of pigment concentration in the Southern Ocean derived from CZCS data. In the future, satellite monitoring of the pigment concentration in the Southern Ocean will provide the basis for a better understanding of the bioproductivity of the Southern Ocean and the global carbon cycle, and possibly for the detection of changes in these parameters.

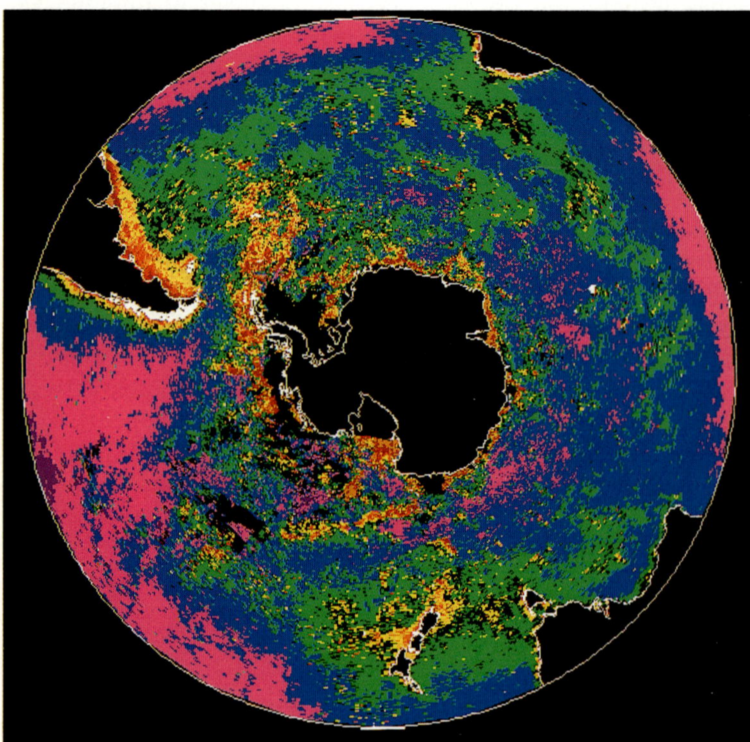

Fig. 8. Distribution of pigment concentration (chlorophyll-a plus pheaophytin) in the Southern Ocean, derived from data from the Coastal Zone Color Scanner on Nimbus-7 (McClain et al. 1991)

4 Conclusion

Research on detection of change in Antarctica is likely to be an important part of the international program on global change. A number of important measurements and scientific studies have begun, largely as a result of the continuing research program conducted by many nations in the Antarctic as well as the interests of scientists working on Antarctic science.

Nevertheless, to accomplish the desired goals of a global change detection program, conventional measurements at selected locations need to be expanded, improved temporal and spatial sampling strategies should be developed and implemented, and additional key satellite measurements should be initiated or expanded. A new emphasis on the systematic collection of well-calibrated data and the attention to these factors in the data analysis is especially important. Some of the work may require considerable efforts and may not yield much new information for some time. However, failure to commence better measurements will insure that our inability to determine current trends of key environmental parameters, due to a paucity of well-calibrated data records with adequate spatial and temporal sampling, will continue. Today's routine measurements will provide the baseline data sets for the future.

References

Alley RB, Bentley CR (1988) Ice-core analysis on the Siple Coast of West Antarctica. Ann Glaciol 11:1–7

Anderson JG, Brune WH, Proffitt WH (1989) Ozone destruction by chlorine radicals within the Antarctic vortex: the spatial and temporal evolution of C10-O_3 anticorrelation based on in situ ER-2 data. J Geophys Res 94(D9):11465–11479

Bentley CR, Giovinetto MB (1991) Mass balance of Antarctica and sea level change. Proc Role of Polar Regions in Global Change. University of Alaska, Fairbanks, AK, pp 481–488

Bindschadler RA, Scambos TA (1991) Satellite-image-derived velocity field of an Antarctic ice stream. Science 252:242–246

Budd WF, Simmonds I (1991) The impact of global warming on the Antarctic mass balance and global sea level. Proc Role of Polar Regions in Global Change. University of Alaska, Fairbanks, AK, pp 489–494

Cavalieri D, Parkinson CL (1981) Large-scale variation in observed Antarctic sea ice extent and associated atmospheric circulation. Mon Weather Rev 109:2323–2336

Comiso JC, Maynard NG, Smith WO, Sullivan CW (1990) Satellite ocean color studies of Antarctic ice edges in summer and autumn. J Geophys Res 95(C6):9481–9496

Dickinson RE, Meehl GA, Washington WM (1987) Ice-albedo feedback in a CO_2 doubling simulation. Clim Change 10:241–248

Doake CSM, Vaughan DG (1991) Rapid disintegration of the Wordie Ice Shelf in response to atmospheric warming. Nature 350:328–330

Douglas B (1991) Global sea level rise. J Geophys Res 96(C4):6981–6992

Farman JC, Gardner BG, Shankin JD (1985) Large losses of total ozone in Antarctica reveal seasonal ClO_x/NO_x interaction. Nature 315:207–210

Ferrigno JG, Lucchitta BK, Mullins KF, Allison AL, Allen RJ, Gould WG (1993) Velocity measurements and changes in position of Thwaites Glacier/iceberg tongue from aerial photography, Landsat images and NOAA AVHRR data. Ann Glaciol 17:239–244

Giovinetto MB (1964) Distribution of diagentic snow facies in Antarctica and Greenland. Arctic 17:32–40

Gloersen P, Campbell WJ (1991) Recent variations in Arctic and Antarctic sea ice covers. Nature 352:33–36

Jacka TH, Budd WF (1991) Detection of temperature and sea ice extent changes in the Antarctic and southern ocean. Proc Role of Polar Regions in Global Change. University of Alaska, Fairbanks, AK, pp 63–70

Jacobs SS, Fairbanks RG, Horibe Y (1985) Origin and evolution of water masses near the Antarctic continental margin from $H_2{}^{18}O/H_2{}^{16}O$ ratios in seawater. In: Jacobs S (ed) Oceanology of the Antarctic Continental Shelf. Antarct Res Ser 43:59–85

Kukla G, Gavin J (1981) Summer ice and carbon dioxide. Science 214:497–503

McClain CR, Koblinsky CJ, Firestone J, Darzi M, Yeh E, Beckley BD (1991) Examining several southern ocean data sets. EOS Trans 72:345, 351

Meier et al. (1985) Glaciers, ice sheets, and sea level: effect of a CO_2-induced climatic change. NAS Press, Washington, DC, pp 210–215

Mercer JH (1978) West Antarctic ice sheet and CO_2 greenhouse effect: a threat of disaster. Nature 271:321–325

Morgan VI, Goodwin ID, Etheridge DM, Wookey CW (1991) Evidence from Antarctic ice cores for recent increases in snow accumulation. Nature 354:58–60

Newman P, Stolarski R, Schoeberl M, McPeters R, Krueger A (1991) The 1990 Antarctic ozone hole as observed by TOMS. Geophys Res Lett 18:661–664

Orheim O (1985) Iceberg discharge and the mass balance of Antarctic. In: Meier et al. Glaciers, ice sheets, and sea level: effect of a CO_2-induced climatic change. NAS Press, Washington, DC, pp 210–215

Parkinson CL, Bindschadler RA (1984) Response of Antarctic sea ice to uniform atmospheric temperature increases. Climate processes and climate sensitivity. Geophys Monogr 29:254–264

Paterson WSB (1984) Present and future melting on Antarctic ice shelves. In: environment of West Antarctica: potential CO_2-induced changes. NAS Press, Washington, DC, pp 140–154

Peltier WR, Tushingham AM (1989) Global sea level rise and the greenhouse effect: might they be connected? Science 244:806–810

Raper SCB, Wigley TML, Mayes PR, Jones PD, Salinger MJ (1984) Variations in surface air temperatures, 3 the Antarctic, 1957–1982. Mon Weather Rev 112:1341–1353

Shabtaie S, Bentley CR (1987) West Antarctic ice streams draining into the Ross ice shelf: configuration and mass balance. J Geophys Res 92(B2):1311–1336

Stephenson SN, Bindschadler RA (1988) Observed velocity fluctuations on a major Antarctic ice stream. Nature 334:695–697

Stolarski RS, Krueger AJ, Schoeberl MR, McPeters RD, Newman PA, Alpert JA (1986) Nimbus 7 SBUV/TOMS measurements of the springtime Antarctic ozone decrease. Nature 322:808–811

Stuiver M, Denton GH, Hughes TJ, Fastook JL (1981) History of the marine ice sheet in West Antarctica during the last glaciation: a working hypothesis. In: Denton GH, Hughes TJ (eds) The last great ice sheets. Wiley, New York, pp 319–436

Westherly JM, Walsh JE, Zwally HJ (1991) Antarctic sea ice variations and seasonal air temperatures. J Geophys Res 96 C9:15119–15130

Weller G, Budd W, Gordon A, Hempel G, Lorius C (1989) The role of Antarctica in global change. ICSU Press/SCAR, Cambridge

Zwally HJ (1989) Growth of the Greenland ice sheet: interpretation. Science 246:1589–1591

Zwally HJ, Fiegles SG (1994) Extent and duration of Antarctic surface melting. J Glaciol (in press)

Zwally HJ, Gloersen P (1977) Passive microwave images of the polar regions and research applications. Polar Rec 18:43–50

Zwally HJ, Parkinson CL, Comiso JC (1983) Variability of Antarctic sea ice and changes in carbon dioxide. Science 220:1005–1012

Zwally HJ, Brenner AC, Major JA, Bindschadler RA, Marsh JG (1989) Growth of the Greenland ice sheet: measurement. Science 246:1587–1589

Antarctic Krill – Changing Perceptions of Its Role in the Antarctic Ecosystem

Stephen Nicol[1]

1 Introduction

The stature of Antarctic krill within the Antarctic ecosystem is reflected in their scientific name – *Euphausia superba*, however, they have been called a variety of other terms most of which are somewhat inaccurate. The word krill, itself, is a misnomer since it is derived from the word *kril* which refers to small fish in Norwegian. This term was then used by North Atlantic whalers to describe the crustaceans found in the stomachs of baleen whales and so has been adopted as a general term for the 85 or so species of euphausiid. Various species of krill have been referred to as: squillae, small red insects, animalcules, pelagic prawns, opposum shrimps and skeleton shrimps, but perhaps the most often used phrase to describe Antarctic krill is "tiny shrimp-like crustaceans". This phrase is misleading since Antarctic krill are by no means tiny, either considered as crustaceans or as animals in general (Fig. 1). An animal which weighs 1 g as an adult is a medium-sized animal and its size actually makes krill quite large for a crustacean. This is not necessarily a trivial point since there appears to be a direct relationship between the size of an animal and the conservation efforts that are expended on its behalf.

All species of krill (euphausiids) are pelagic or free-swimming, generally occurring in the open ocean rather than being associated with the sea floor. Some species are large – up to 14 cm long – and deep living; most of the others are surface-dwelling organisms, inhabiting the top 300 m of the water column. Antarctic krill are the largest of those krill which are found in the upper layers of the ocean and are certainly the most abundant.

So what are krill and why is Antarctic krill the superb krill? There are several notable features which set the euphausiids apart from other crustaceans: the gills are exposed unlike those of most other advanced crustaceans which are sheltered under the carapace; they possess a series of luminous organs which produce a blue light; and there is no statocyst or balance organ. The general body plan is, however, similar to many familiar

[1] Australian Antarctic Division, Channel Highway, Kingston, Tasmania 7050, Australia

Hempel (Ed.) Antarctic Science

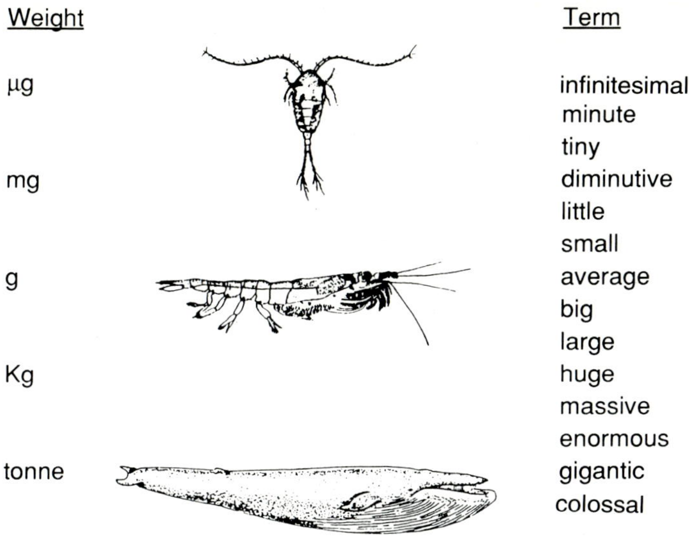

Fig. 1. The size spectrum of metazoan (multi-celled animal) life ordered by weight with a semi-arbitrary listing of descriptive terms for each size category and an illustrative selection of animals

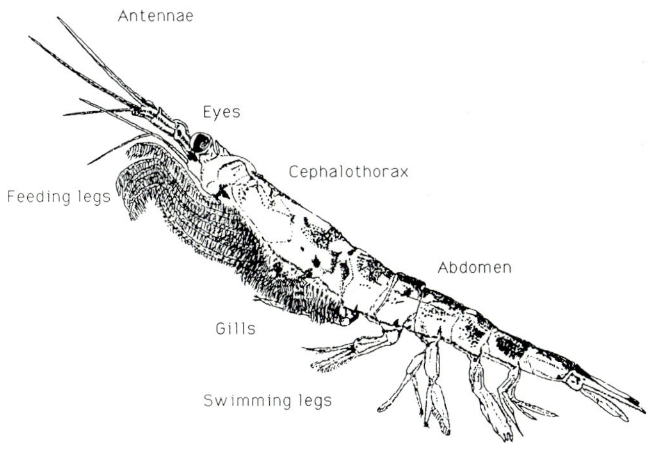

Fig. 2. Antarctic krill, *Euphausia superba*; an outline of its body plan and general anatomic features. Adult krill reach a maximum length of ~65 mm measured from the tip of the tail to the centre of the eye

crustaceans (Fig. 2). The cephalothorax contains most of the internal organs; the digestive gland, stomach, heart and gonads, and houses most of the animal's sensory appendages; the large eyes, the antennae. The limbs of the cephalothorax are modified into highly specialized feeding appendages. The muscular abdomen carries the swimming legs which paddle the animal through the water in a smooth metachronal rhythm.

The role of krill in the Antarctic food web became apparent as soon as the first studies into the natural history of the more obvious elements of the ecosystem began. Almost all of the species of penguins, seals and baleen whales which fed in the Antarctic seas in summertime were revealed to be feeding to some degree on krill. The Southern Ocean Ecosystem became known as the "krill based ecosystem" and for a long period was held up as an illustration of a direct, efficient food web connecting the diatoms of the phytoplankton to the great whales and other vertebrate predators through one vital link – Antarctic krill (see, for example, Murphy 1962). Since the adoption of this model of the Antarctic ecosystem 30 plus years ago there have been a number of fundamental changes in our knowledge of the biology of Antarctic krill and of how the ecosystem itself functions. The way in which we view krill is now fundamentally different from our perceptions of it even 20 years ago and our knowledge of its biology is greatly improved. This review is a synthesis of some of the important developments in the field of krill biology that have occurred recently and is an attempt to see how well the concept of a krill-based ecosystem has stood the test of time.

2 A Brief History of Krill Research

Krill are hugely abundant in the Southern Ocean and are frequently to be found at the surface colouring the water a deep reddish-brown. It is therefore surprising that early records of voyages to Antarctic waters rarely remark on the presence of krill and it was not until 1840 that *Euphausia superba* was scientifically described by Dana. The most comprehensive account of early records of the existence of krill occurs in the classic monograph on Antarctic krill by Marr (1962). Marr also provides a useful summary of all studies of Antarctic krill up to the early 1960s and it is worth noting that most of the studies he lists report on either the distribution of krill or some aspect of the anatomy of *E. superba*. Exceptions to this pattern include studies of krill as the food of vertebrates, or which examined the dietary habits of krill themselves. Marr's study extended this tradition of basically descriptive studies of krill distribution and natural history and by the accuracy and breadth of his observations he laid the foundations for the explosion in krill research that occurred in the 1970s and 1980s.

One of the earliest reported laboratory investigations on krill was the determination of their chemical composition (Heyerdahl 1932, quoted in Mauchline and Fisher 1969) and possibly the first known experimental results from work on live Antarctic krill published were those of Fraser (1936) who observed spawning by captive animals on board the RRS *Discovery II*. Mackintosh (1967) who reported some results on moulting intervals of krill kept at South Georgia was the first to provide advice on how to keep this species alive for experimental purposes and his experiments may actually have preceded those of Fraser as they were carried out some 37 years prior to publication!

The true dawn of experimental research into the biology of *E. superba* occurred with the work of McWhinnie in the early 1960s which began by reporting the first measurements of oxygen consumption by krill made on board RV *Eltanin* (McWhinnie and Markiniak 1964) and followed by a series of experiments in the laboratories at Palmer Station at the Antarctic Peninsula. The success of experimental studies on krill has depended on the ability to maintain them alive for appropriate periods in the laboratories of research ships, or in laboratories on the Antarctic continent or on sub-Antarctic islands. The increasing sophistication of the laboratory facilities available on research vessels and on Antarctic stations has been a major factor in our burgeoning knowledge of the biology of living krill. A more recent trend has been to transport live Antarctic krill to laboratories far from the Antarctic (Ikeda et al. 1980) and this has resulted in the publication of a variety of studies of living *E. superba* which have been carried out Australia, Japan, South Africa and Germany.

By the middle of the 1970s a general picture had emerged of the life history of Antarctic krill which was based largely on the findings of the Discovery Expeditions. The works of Bargman (1945), Marr (1962) and Mackintosh (1972, 1973) depicted a species which was confined to waters south of the polar front but which was most abundant in the area of the Southern Ocean subject to the seasonal advance and retreat of the pack ice. Within this vast area (12.5 million km^2), krill were unevenly distributed both regionally and on a much smaller scale occurring in swarms or schools of uncertain size. Krill were seen as planktonic animals which arose from eggs which sank as they developed until they reached the first swimming stage at which point they began the long slow ascent into the euphotic zone where they lived for the rest of their lives. Krill grew rapidly through a bewildering variety of larval stages before becoming adults in their second year and senescent sometime early in their third summer.

The Discovery Investigations were prompted by the need to put the management of the whale stocks of the Southern Ocean on a more scientific footing and focussed on krill as the primary food of most of the important commercial species of baleen whales. Between the 1930s when the Discovery Investigations were initiated and the 1960s and 1970s, when the main Discovery Reports on krill were being published, most of the great whale stocks had been dramatically reduced, some of them almost to the point of extinction and the information collected was of little use in the management context for which it had been intended. As the exploitation potential of the great whales declined, attention was being turned to other resources in the Southern Ocean, particularly krill. One side effect of the Discovery Investigations was to highlight the widespread distribution and great abundance of krill and to pinpoint their focal role in the Antarctic ecosystem. Soviet fishery investigations on krill began in the 1960s (summarized in Everson 1977) and other nations began to take notice of the harvesting potential of the stocks of krill. The size of the circumpolar krill population was estimated to be up to 7000 million tonnes (Moiseev 1970) but due to the central ecological role of krill, the fraction available for harvesting was, like the biomass estimates, subject to considerable uncertainty. The potential for krill to become a resource of global proportions was highlighted by Laws (1977) who calculated the potential "surplus" of krill that might be available to a fishery due to the drastic reduction in numbers in their major predators, the great whales. The "krill surplus" hypothesis, which has been surrounded by controversy since its publication, firmly put krill on the map as one of the world's last great unexploited marine resources with a potential harvest of up to 150 million tonnes per year. At the time the global harvest of all species from all the oceans was around 70 million tonnes a year so it is easy to see why krill were hailed as one of the solutions to the world's perceived protein shortfall.

Commercial harvesting of krill began in the mid-1970s on a small scale and at about this time SCAR began planning for a second generation of investigations into the biology and distribution of Antarctic krill – the BIOMASS (Biological Investigations of Marine Antarctic Systems and Stocks) programme. It became apparent that, despite the wealth of knowledge accumulated during the Discovery Investigations, the detailed information which would be necessary to assess the impact of the developing krill fishery and to manage its growth was almost entirely absent. The main aim of the BIOMASS programme was to assess krill stocks around the Antarctic continent and to study the aggregating nature of the krill populations using new techniques – particularly hydroacoustics. A side effect of the programme was to focus attention on the biology of Antarctic krill resulting in a burst of experimental research activity examining various features of the behaviour and physiology of krill.

As the fishery for Antarctic krill grew, the need for some form of resource management regime for the Southern Ocean became more apparent. Out of these concerns and on the foundations of BIOMASS, the Convention for the Conservation of Antarctic Marine Living Resources (CCAMLR) was born. One result of CCAMLR has been a further focussing of research activities on the aspects of krill biology which must be quantified before scientific management principles can be employed in the assessment of the fishery and of its impacts. The progress towards a comprehensive feedback management plan for krill is going to require intensive research into the basic biology of krill and into its ecological relationships and this is likely to be a major driving force in Antarctic marine biology into the next century.

Although the major research programmes into the biology of Antarctic krill have been driven largely by their resource implications – either krill as food for whales or as food in their own right – there has always been a solid basis of research which has been examining the biology of this uniquely successful crustacean and its role in the Southern Ocean ecosystem. All aspects of the biology of krill are now being subjected to scrutiny as the attention turns to the role of the Southern Ocean in the global carbon cycle. Krill, as the dominant single animal species of the Antarctic ecosystem, obviously plays a role in the biological cycles which may be so important in the ability of the Southern Ocean to take up excess carbon dioxide. How important this role is likely to be is going to be a major focus of research throughout the 1990s.

3 Changes in Perceptions

The beginning of the 1970s can be seen as a turning point in the study of Antarctic krill, as it is from this point on that the focus shifted from the descriptive work of the Discovery Investigations towards a more experimental approach both in the field and in the laboratory. The Discovery Reports presented a wealth of detailed descriptive results which provided the basis for experimental studies which could test their findings. No sooner had the major results on krill been published than the conclusions were being examined and the hypotheses they contained were being tested.

4 Sampling

The information on which the Discovery Reports was based was derived from a sampling programme, the scope of which has not been attempted since. The samples obtained were all collected by nets which were fished at predetermined stations all around the Antarctic continent at a number of depths. Although a general picture of the distribution pattern of Antarctic krill emerged from these studies (Fig. 3) which is still accepted to this day, Marr (1962) was unhappy about his ability to provide quantitative estimates of the distribution or abundance of krill. What he saw as being the fundamental problem in the analysis of krill distributions was the problem of patchiness; krill do not occur randomly on any scale, they aggregate into schools or swarms; these swarms occur in larger groupings or patches and these patches make up larger areas where krill are more abundant. The sampling problems faced by the Discovery Investigations were that they were investigating a discontinuous phenomenon using integrating samplers – nets – which provide maximal information from continuous phenomena. Marr recognized this limitation when he wrote: "(When we can) watch as well as capture the animals we have for so long been blindly sampling . . . we shall obtain a reliable picture of their natural abundance, as well as of much

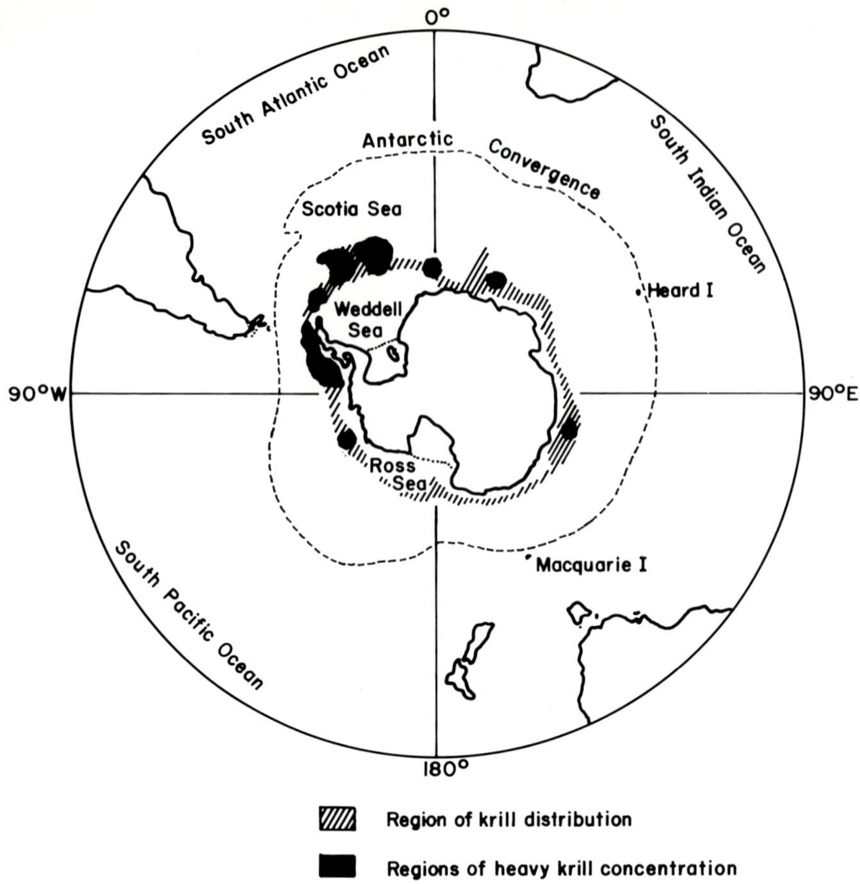

Fig. 3. The generalized distribution of Antarctic krill. Information compiled from a number of sources

else concerning them that at present, remains to us obscure". It was not until scientific echo-sounders began to be used in the Southern Ocean that it was possible to *see* the scales of krill patchiness.

The advantage of echo-sounders is that they provide a continuous two-dimensional record of the distribution of sound scatterers in the water column below the ship's track thus giving insight into the sizes of swarms, their density and their vertical extent. The major problem with this technique is that there has, up to recently, been an imprecise knowledge of the relationship between the amount of sound reflected and the biomass of krill below the ship. Recent experiments have clarified this relationship and this has resulted in the revision of all previous acoustic estimates of krill abundance upwards by a factor of ~5 (Everson et al. 1990; Greene et al. 1991; Hewett and Demer 1991). The hydroacoustic technique still has some significant problems; krill in the surface layer – above the downwards looking transducer – are not sampled and at times much of the population of krill may be present in the top 20 m of the water column. Echo-sounders also work badly in pack ice and the signal may be distorted by the noise of ice hitting the hull or the towed body even in areas where there is little ice cover. There is also, as yet, no way of determining exactly the species of the organisms reflecting the sound and this necessitates the use of nets to identify the animals in the scattering layer and to determine their size distribution.

5 Swarming

Despite the problems of the hydroacoustic technique it has radically altered the ability of researchers to examine the distribution and abundance of krill in particular areas. It allows for rapid (relative to net sampling) surveys of areas and provides a continuous output, so is less affected than nets by the problem of patchiness. Echo-sounders have also provided us with a clearer picture of the size and complexity of krill swarms, some of which can contain millions of tonnes of krill (Shulenberger et al. 1984) and can have extremely detailed internal structure (Fig. 4). Krill swarms can be large complex entities whose formation, maintenance and dispersal have been largely unstudied but which are now becoming the target for intense scrutiny using hydroacoustics and other sophisticated instrumentation. A new generation of scientific echo-sounders is now in use in the Southern Ocean and will allow for the most accurate assessment of abundance of the krill resource yet.

Given the dynamic nature of krill swarms and the vast area over which krill are known to occur, it is unlikely that any attempts will be made to attempt again large-scale synoptic biomass estimates, such as those carried out during BIOMASS. It is more likely that echo-sounding techniques will be used to gain regional information on the scales of occurrence of krill and to build up a picture, through surveys of discrete areas, of the overall distribution and abundance over a number of years. Such surveys are likely to focus first on the areas subject to the fishery (the South Atlantic) and then to move to other areas determined by their potential for exploitation.

Hydroacoustic surveys produce a static picture of krill abundance and distribution but finding out how these patterns relate to the dynamics of the water movements of the Southern Ocean is going to be a major task for the 1990s. CCAMLR (1991) has identified the study of the flux between the krill populations of the three main fishing areas of the South Atlantic – the Antarctic Peninsula, the South Orkneys and South Georgia – as an urgent priority. The current patterns link these three areas and the amount of movement of krill stocks between them is likely to be a complex interplay

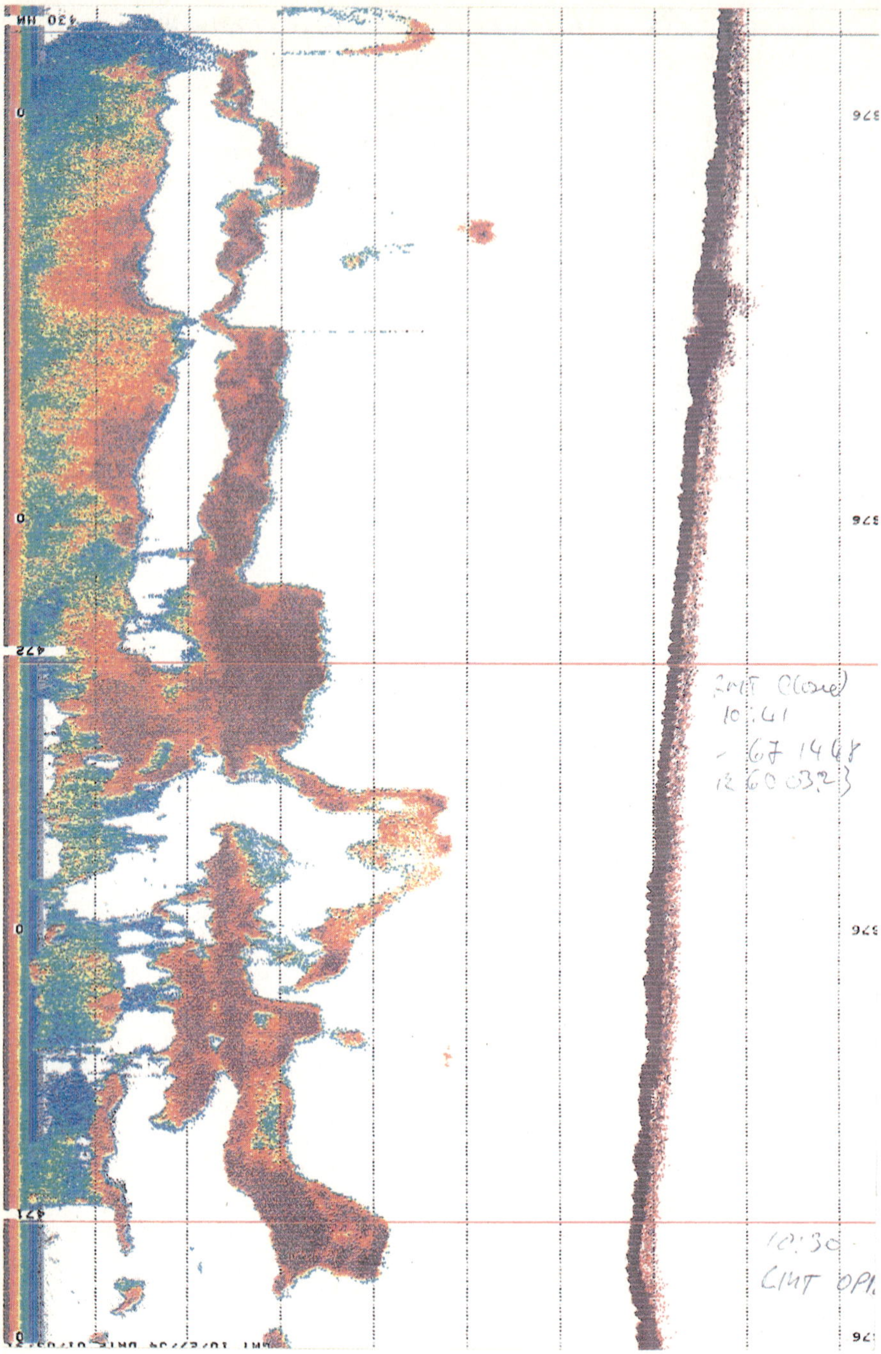

between the hydrography and the behaviour of the krill swarms. Obviously, the fisheries management strategy for each area is dependent on how much of the population in each area is resident and how much has been derived from the other areas. Marr, Mackintosh and early Russian authors (see Miller and Hampton 1989 for review) pointed out the relationship between the krill distribution patterns and the currents around Antarctica and the sub-Antarctic islands but the complex task of moving from these static relationships to understanding the dynamics of the processes underlying these patterns is only just beginning.

The swarming behaviour of krill was eloquently described by Marr (1962) but, although its importance has been recognized, it has been little studied since, except by remote mechanisms such as echo-sounding. Early descriptions of schooling krill arose from examination of surface swarms (Marr 1962) or from scuba divers (Ragulin 1969) but the experimental analysis of the swarming behaviour of krill has been difficult because of their oceanic habitat and because they have proved difficult to maintain in laboratories. Recent studies have indicated the complexity of this type of behaviour (Hamner et al. 1983; Hamner 1984) and the complications that this behaviour add to the study of krill population dynamics (Watkins et al. 1986) but a major gap in our knowledge is still the causes and consequences of the aggregative habit of krill. Krill were initially envisaged as plankton (drifting organisms) and for much of their early life krill are indeed very much at the mercy of water movements at all scales. As adults, however, they have considerable powers of locomotion, being capable of sustained swimming speeds of $25\,\mathrm{cm\,s^{-1}}$ (Hamner 1984) and are thus more accurately described as nektonic or free swimming pelagic animals. Marr refers to krill as "a creature of great agility, powers of locomotion, purposeful intent and not a little awareness"! There are records of krill swimming in a directed fashion against a current for many hours (Gunther in Marr 1962) and swarms moving for days at speeds up to $0.5\,\mathrm{km\,h^{-1}}$ (Kanda et al. 1982) but such observations are rare. The social nature of krill, and their ability to move en masse horizontally for considerable distances, further emphasizes the shortcomings inherent in attempting to describe the distribution of krill on the basis of the water movements alone.

Krill have long been known to be capable of extensive vertical migration, beginning with the "developmental ascent" of the larvae (Marr 1962) and continuing into adulthood as the diurnal movements from the deeper waters

←

Fig. 4. Echogram of a large swarm of Antarctic krill obtained by a Simrad EK500 scientific echo-sounder operating at 120 kHz from the Research vessel *Aurora Australis*. *Dark red area* are where the density of krill is highest, *white areas* where krill are absent and considerable vertical and horizontal structure is apparent. The swarm extended for approximately 10 km and was encountered at 10.30 h (local time) on 1 March 1991 at 67°14.14 S, 60°03.23 E. The sea-floor is visible at the 175 m contour

(~100–200 m) during the day to the surface at night. Often viewed as a simple response to ambient light levels, this behaviour has proved to be much more complex. Marr and all early researchers reported the presence of krill swarms at the surface during daytime, a period when they should, according to theory, be at their maximum depth. Similarly, swarms of krill could also be found at depth during the night also confounding the predictions based on simple light responses. Modern echo-sounders have revealed further complexities to the vertical structure of krill swarms and because of their ability to resolve distributions in the vertical scale so well they will be invaluable in the task of understanding the relationships between krill patches, the thermohaline structure of the water column and the distribution of other organisms in the vertical plane.

6 Feeding

One form of behaviour which has received a great deal of attention is that of feeding by krill. Antarctic krill are mainly herbivorous and this was pointed out early on (Barclay 1940), but the mechanism of their feeding, the range of their food items and their ability to utilize food items other than planktonic diatoms have all been the subject of considerable experimentation and speculation over the last 20 years.

The feeding appendages of krill carry an array of fine hairs or setae, which interlock to form what at first glance appears to be a well-constructed plankton net (McClatchie and Boyd 1983). Early examinations of the feeding behaviour of krill assumed that they swam through the water, either with the "net" open or flapping the sides of the "net", to strain the dilute phytoplankton from the water (Kils 1983). Studies of feeding by filter feeding zooplankton, using high-speed cinematography and mathematical analysis of the hydrodynamics of water flow on the appropriate scales in the 1970s and 1980s, indicated that filter feeding may not be as simple as was first imagined (Koehl and Strickler 1981; Strickler 1982). Antarctic krill feeding, when examined in detail, exhibited all the complexities exhibited by smaller copepods (Hamner 1988). Krill have a battery of sensory appendages including long antennae and large eyes so it was not surprising to learn that they might be capable of discriminating between food sources and of detecting patches of food from a distance. Examination of live krill revealed that motions by the feeding appendages were complex and that krill do not feed continuously or indiscriminately (Hamner 1988). Quantification of food intake rates has almost all been based on experiments which involve incubating krill in small (<1 l) jars for varying periods with various food sources and estimating an ingestion rate from the amount of food removed from the water (see Morris 1984 for a review). These experiments have arrived at "filtration rates" – the amount of water that would have to be filtered to obtain the amount of food removed – which are generally too low – ~1 l per hour – to account adequately for the required energetic input to allow for growth, reproduction and general metabolism given average phytoplankton concentrations (Ross and Quetin 1986). Our understanding of the mechanisms of krill feeding is still incomplete but krill are no longer viewed as mechanistic vacuum cleaners, indiscriminately sweeping the water clear of all that it contains. Whether they exploit layers of phytoplankton which are many times the average density or augment their mainly herbivorous diet with energy-rich zooplankton is still being investigated. What is certain is that their diet must be radically different in the winter time when phytoplankton are absent.

7 Overwintering

The question of where krill go and what they do in the winter has been asked since the earliest studies and remains unanswered (Quetin and Ross 1991). Since the main distributional area of krill is that region between the minimum extent of pack-ice in summer and the maximum extent in winter, it is logical to assume that they would be found in the ice covered waters in winter. Consistent anecdotal evidence suggested that krill might be found associated with the underside surface of the ice especially in the spring when they were observed on floes which had been turned over by passing ships (Marr 1962). As the importance of the sea-ice microbial community became established, it became apparent that this might be a source of food for krill, especially in the spring (Bunt 1963; Garrison et al. 1986). Studies by divers and by remotely operated vehicles did, indeed, reveal krill living in the complex three-dimensional habitat formed by the rafting of ice floes (Garrison et al. 1986; O'Brien 1988). In situ observations and experimental results established the reality of krill feeding on algae which was growing on the ice underside (Marschall 1988; Stretch et al. 1988) but the extent to which this occurs and its importance in the annual energy budget of krill is still a matter of debate.

An alternative hypothesis for the winter behaviour of krill is that they migrate to the hyper-benthic layer and there either feed off detritus and zooplankton or enter a state of semi-hibernation. Krill have been observed in the lowest levels of the water column during winter (Kawaguchi et al. 1986) and have been found in the guts of benthic fish (Duhamel and Williams 1990) so this may be a behaviour practised in continental shelf areas.

Evidence for the over-winter behaviour of krill has been augmented by investigations into their biochemistry. Clarke (1980) showed low levels of lipid buildup in krill at the end of summer and absence of storage lipids, which suggested that krill did not greatly store energy reserves to survive the winter. If they have no overt energy reserves then what might they do to reduce the energy necessary to keep swimming and to provide the necessary fuel?

8 Age and Longevity

Laboratory experiments indicated that krill could survive for over 200 days without food (Ikeda and Dixon 1982). They appeared to survive prolonged starvation by utilizing their structural body protein and by shrinking at each moult. Reducing size also reduces the energy necessary to stay buoyant and the body protein serves as an energy source, thus is doubly efficient. There has been great debate over whether shrinkage is a laboratory artifact but recent results do tend to indicate that shrinkage does occur in the field in winter (Quetin and Ross 1991) so it may be one of the ways in which this adaptable crustacean survives the annual long dark period of food scarcity.

Body shrinkage by adult krill was shown to be accompanied by regression of the sexual characteristics so that after a number of moults, the adults came to resemble juveniles (Thomas and Ikeda 1987). The implications of these allied findings were profound. The age of crustaceans is generally determined from some measure of size and the findings for krill indicated that there was, at best, an imprecise relationship between size and age (Nicol 1991a). Furthermore, animals which appeared young at the beginning of summer may well have been older adults. What this implied is that the studies of the population dynamics and production of krill which assumed a two-year life span – derived from length frequency analysis – were probably in error. How much in error was not apparent until laboratory rearing experiments indicated that krill could live for up to 9 years (Ikeda and Thomas 1987) and extrapolations from laboratory measurements indicated that a longevity of 11 years was a possibility (Ikeda 1985). What this meant was that the productivity of krill, based on a two-year life span, was maybe several times too high. This finding had obvious implications for the fishery.

The debate on the longevity of krill continues and although novel ageing techniques have been tried with varying levels of success (Ettershank 1984; Nicol et al. 1991) and the length frequency analysis technique has been refined (Hosie et al. 1988; Siegel 1988) there is still uncertainty surrounding the number of age classes in the adult krill population. Despite this uncertainty, there is general consensus that the two-year life span as proposed by Mackintosh (1972) is incorrect and that some proportion of the krill population probably survives to some ripe old age. How big this proportion of the population is and how old they get are still a matter for conjecture.

9 Productivity of Krill

The questions of age and growth are but part of the inputs required to determine the productivity of the Antarctic krill population. The development of techniques to keep krill alive in the laboratory gave rise to a number of studies investigating growth rates under a variety of conditions (Ikeda 1985). The demonstration that krill have the potential to shrink led to a re-examination of the length-based growth curves of Mackintosh (1972) by simulation studies of individual growth (Astheimer et al. 1985) and by re-analysis of length composition data from the Discovery expeditions by modern techniques (Rosenberg et al. 1986). These studies all pointed to Antarctic krill being longer lived, slower growing and less productive than was initially envisaged (Ross and Quetin 1988).

At the same time that growth rates were being re-examined and productivity estimates were being lowered, laboratory investigations into the fecundity of krill were producing results which tended to show that krill were, from the reproductive aspect, rather more productive than early studies had shown. The initial estimates of fecundity were based on analyses of the number of eggs in the ovary (Bargmann 1937), the volume of the ovary (Mauchline and Fisher 1969) and on the quantity of ovarian lipid (Clarke 1980). These studies indicated that krill were capable of laying up to 11 000 eggs and the general assumption was that females spawned only once a year. Laboratory studies on live krill indicated that females could release up to 8000 eggs at a time and that spawning episodes might occur as frequently as every 6.7 days during the summer (Ross and Quetin 1983). The suggested potential of krill to spawn many times in one season effectively doubled their estimated fecundity and led to further investigations into their production and to an examination of the energetics required to fuel such productivity (Ross and Quetin 1986). The spawning rate of Antarctic krill is still under investigation but it is a further example of the contribution made to our understanding of the biology of krill by the recently acquired ability to examine live krill. If krill do produce many more eggs than was originally estimated then their larvae may be more abundant and will suffer greater mortality than was previously thought and this markedly affects our view of the population dynamics of this species and of its role in the Antarctic ecosystem.

10 Krill in the Southern Ocean Ecosystem

The ecosystem of the Southern Ocean has been described in the past as being simple, involving short direct links between the phytoplankton (mainly distoms) their grazers – krill – and their predators the vertebrates, such as seabirds, whales and seals. Recent advances have indicated that there are extra dimensions of complexity to the ecosystems of these waters. There is a specialized community associated with the pack-ice which involves krill at certain times of the year (Garrison et al. 1986). There is a "microbial network" which recycles the products of primary production which is driven in many areas by phytoplankton other than diatoms (Smetacek et al. 1990). There are communities of zooplankton which do not include krill or their larvae (Smith and Schnack-Schiel 1990; Hosie 1994) and there are some species such as salps which may actually exclude krill from certain areas by outcompeting them for the available food (Huntley et al. 1989). Finally there may be large sections of the vertebrate populations in certain areas, which have been traditionally regarded as krill eaters, which may depend on other food sources such as fish and squid for much of the year (Whitehead et al. 1990). Despite these revelations about the complexity of the Southern Ocean Ecosystem it is still true to say that krill, as a single species, play a dominant role in the food webs in much of the Antarctic zone. Few pathways exist between the primary producers and the tertiary consumers which do not pass through krill and this funnelling of so much of the oceanic productivity through one species makes the Southern Ocean Ecosystem unique.

The role of krill as consumers and as prey has been long known but their potential as producers is only now being appreciated. The Southern Ocean may, in its brief summer, be responsible for up to 15% of the global primary production and krill are certainly responsible for grazing down a significant part of the biomass thus produced (Huntley et al. 1991). The emphasis has changed recently from considering krill purely as consumers of the plants of the plankton to their role in transferring carbon fixed by the plants from the surface waters to the ocean depths. Krill convert small particles (phytoplankton) into larger particles (body tissue, eggs, faeces and moults) and dissolved material such as carbon dioxide and ammonia. The large particles sink rapidly (Clarke et al. 1988; Nicol and Stolp 1989) and decompose at depth where their carbon is lost to the atmosphere for long periods and the carbon dioxide produced by the krill's metabolic activity enters the vast oceanic pool of dissolved inorganic carbon. Because the Southern Ocean is generally rich in macronutrients, it has been suggested recently that it might be able to absorb a substantial portion of the carbon dioxide of the atmosphere generated by the burning of fossil fuels by increases in primary productivity.

Krill consume an unknown but large – up to 50% – fraction of the primary productivity in the Southern Ocean (Huntley et al. 1991). It has been pointed out recently, however, that because so much of the krill production is consumed by air breathing vertebrate predators such as seals, birds and whales, up to 25% of the carbon fixed by the plants of the Southern Ocean may end up by being respired back into the atmosphere rather than entering the ocean's carbon store. This may make the Southern

Ocean rather less of a carbon sink than was previously imagined. This role of krill as an intermediary in global scale processes is a new twist to the examination of the life cycle of krill in relation to its complex environment.

11 The Krill Fishery and Its Management

The environment in which krill live is a complex and dynamic one subject to massive short- and long-term changes. In recent times, the biotic environment of the Southern Ocean has been dramatically altered by the activities of humanity through the intensive harvesting of selected species. The fisheries of the Southern Ocean initially concentrated on traditional species and, one by one, the stocks of fur and elephant seals, whales and then fish were devastated (Sahrhage 1989).

Right from their first scientific discovery, there has been speculation about using Antarctic krill for food. Some of the first records acknowledging the presence of this species mention attempts to eat it, generally with unfavourable results. Von Drygalski (1989) in the reports of the German South Polar Expedition of 1901–1903, noted catching "huge quantities of shrimps (krill)" close to the ice. "In such quantities that we were able to eat them too; they tasted quite good, but they were rather small and tiresome to peel . . ." thus they anticipated some of the problems later faced by the commercial fishery.

Much of the interest in using krill as a resource arose from the over-exploitation of other fisheries and particularly from the reduction of the whale stocks in the Southern Ocean. The decline of many of the world's fisheries and the establishment of exclusive 200-mile zones in the coastal waters of traditional fishing areas led to a surplus of large factory trawlers and a search for new resources to exploit. The Southern Ocean was ideal in that regard, because the Antarctic Treaty had frozen all territorial claims, and thus there were no exclusive economic zones in the region. Additionally, because of the destruction of the populations of krill's main predators, there was judged to be a vast untapped "surplus krill" population. Couple this with the swarming habit, and hence the ease with which they can be caught, and krill were being touted as the protein source which was to feed the "starving millions". Such simplistic assumptions did not take into account a number of pertinent factors. Ecosystems are dynamic entities and any perturbation such as the removal of a large proportion of the top predators results in shifts in the balance of species present so that there is rarely likely to be a "surplus" of prey for any more than a very short period of time. Several species of krill predators are thought to have increased in numbers during this century, probably as a direct consequence of the extra krill available because of the reduced number of great whales (Croxall et al. 1988). Any commercial harvesting of krill would have to compete with these established predators and would also compete for krill with the remnants of the great whale populations which are struggling for survival, let alone recovery.

Going krill fishing in the Southern Ocean turned out to be a costly exercise; it is a very hostile environment to work in, it is a long way from almost everywhere, and krill, although proving easy to catch, ended up being costly to process and difficult to market. Krill possess some of the most powerful protein digesting enzymes ever encountered, a fact that has made them the source of chemicals for medical use (Anheller et al. 1989). Once krill die these enzymes begin to break down the animal's body rapidly, reducing its proteins to water-soluble forms and turning the krill black and mushy. Crushing krill in a fishing net or on the deck of a trawler only exacerbates this problem and as a result catches have to be kept small (less than 10 tonnes per set of the net when 50 tonnes are probably feasible) and the catch has to be processed immediately after it is landed. Krill trawled in commercial operations are deemed useless for human consumption after they have been on deck for 3 h, and cannot even be used as animal fodder after 10 h (Budzinski et al. 1985). As if this did not put enough constraints on the fishery, a further problem began to become apparent once the fishery was already established. The shells of krill are extremely rich in fluoride (Soevik and Braekkan 1979), so much so that animal feed made from whole krill has four times the European Community allowable fluoride level (Budzinski et al. 1985). In the early to mid-1980s there was a huge drop in the krill catch and many have attributed this decline to the industry adapting to processes which produced low-fluoride krill products. Nowadays much of the krill catch is peeled before being frozen or canned and then is marketed as "Antarctic shrimp", a luxury food which competes with other low-cost shrimp products at the cheaper end of the crustacean market (Nicol 1989). What was originally touted as protein for the Third World has become a luxury product for the rich nations.

The krill fishery, despite its technical problems, has been the largest fishery in the Southern Ocean since 1979. At its peak in the early 1980s the fishery was one of the 30 largest fisheries in the world and constituted the ocean's biggest crustacean catch. The fishery has been concentrated mainly in the South Atlantic sector of the Southern Ocean and, currently, ships from five nations are involved in the harvest (Fig. 5). There has been a recent decline in the annual catches of krill associated with the breakup of the Soviet Union but a number of nations not currently fishing have expressed an interest in entering the fishery and rapid expansion is a constant threat.

When it became apparent that many of the fish stocks of the Antarctic were in a parlous state and it appeared that the krill fishery was about to expand rapidly, the Antarctic Treaty nations convened a conference to try and develop some way of protecting the living resources of the Southern Ocean. The end-product of these negotiations was the Convention for the

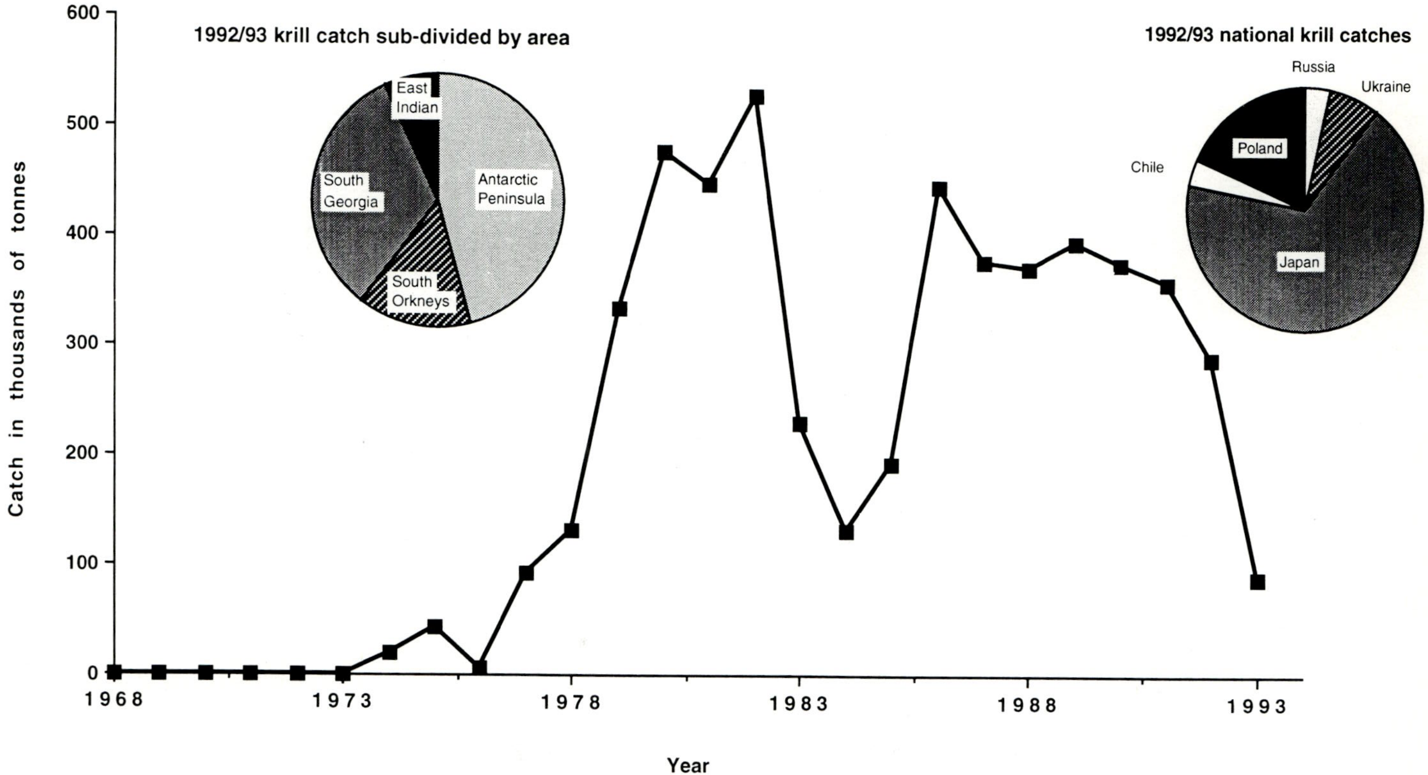

Fig. 5. The fishery for Antarctic krill. The annual catch of Antarctic krill in Antarctic waters, the proportion of the catch taken from each of the fishing regions on 1992/93 and the proportion of the catch taken by each of the fishing nations in 1992/93. Data from CCAMLR database

Conservation of Antarctic Marine Living Resources which came into force in 1981 (Edwards and Heap 1981). This is a far-reaching treaty which lays out the framework for managing the fisheries of the region. It is unique for a variety of reasons. It does not, as most fisheries conventions do, set out only to manage the species of commercial interest; CCAMLR has as its central theme the management of the ecosystem of the Southern Ocean. Although the Convention is concerned with conservation, it also acknowledges the right of signatories to the "rational use" of the resources of the area and it sets out, in some detail, the biological conditions beyond which a fishery should not proceed. The Convention also broke ground in that it came into force before the species which was its principal focus, krill, was in any immediate need of protection. Most fisheries management schemes are only applied when they are necessary because of over-fishing.

The Convention established a Commission which meets annually and regulates the fisheries that take place in the CCAMLR area. At its tenth meeting, in 1991, the Commission passed its first conservation measure for krill which set a precautionary catch limit at 1.5 million tonnes for the South Atlantic thus setting in motion the regulatory process which should ensure the well-being of the Southern Ocean Ecosystem (CCAMLR 1991). Because the concept of managing an ecosystem is such a complex task and because it had never been attempted before, the Commission has spent a great deal of time just trying to work out how it should go about its business. There were also some pressing problems facing other Antarctic fisheries at the time the Convention was signed, most notably the serious depletion of a number of fish stocks. It took most of the 1980s to get some of these fish stocks protected and to establish efficient working groups to deal with the problems of ecosystem management, fish stock assessment and the krill fishery. This explains why it has apparently taken so long for CCAMLR to come to grips with the management of the largest resource in the Convention area – krill (Miller 1991). Another factor which may explain some of the time lag between the formulation of the Convention and the enactment of regulation on the krill fishery is the fact that the krill fishery has consistently been misrepresented as a small operation (Nicol 1991b). Whilst the current catch may be small compared to the astronomical figures originally suggested the average annual catch of 266 500 tonnes over the fishery's 20-year history makes it large by global standards.

We are still in a phase of learning about the Antarctic ecosystem and of the effects of the krill fishery on the other inhabitants of the region; thus we will be unable to reach a stage where we will be able to manage the fishery in a predictive fashion (Nicol and de la Mare 1993). Consequently CCAMLR's decision to impose limits on the rate of expansion of the fishery until we are able to assess more fully the impact of increased harvests on the rest of the ecosystem is in keeping with the spirit of the Convention.

12 Conclusions

Over the last decade we have gone from perceiving krill as short-lived planktonic animals to long-lived, socially organized ones. We are moving from the concept of krill as passive and indiscriminate filterers of the waters of the Southern Ocean to seeing them as active feeders and significant contributors to the flux of organic matter and minerals from the surface waters at high latitudes. Krill are now the focus of a large fishery rather than merely being the staple food of a range of exploited species. The products of the fishery are being developed for the luxury end of the market and are no longer seen as cheap protein for the Third World. Despite all these changes in our knowledge of the biology of krill, our perception of the central role that this organism plays in the Antarctic ecosystem remains largely unchanged. The ecosystem is, if not krill-based, at least krill-centric with this uniquely successful crustacean constituting the lynchpin of the world's only circumglobal ecosystem. The scientific study of krill will continue for a number of reasons; partly because they are the prey of so many diverse predators, partly because they are a resource in their own right, but also because their huge abundance highlights an evolutionary success story which is yet to be told. It is to be hoped that the initiatives begun in 1991 will bear fruit and that the management of krill will in its turn be as successful as the species has been.

References

Anheller J-E, Hellgren L, Karlstam B, Vincent J (1989) Biochemical and biological profile of a new enzyme preparation from Antarctic krill *Euphausia superba* Dana suitable for debridement of ulcerative lesions. Arch Dermatol Res 281:105–110

Astheimer H, Krause H, Rakusa-Suszczewski S (1985) Modelling individual growth of the Antarctic krill *Euphausia superba* Dana. Polar Biol 4:65–73

Barclay E (1940) Nahrung und Filterapparat des Walkrebschens *Euphausia superba* Dana. Z Fisch Hilfswiss 1:65–156

Bargman HE (1945) The development and life history of adolescent and adult krill, *Euphausia superba*. Discovery Rep 23:103–176

Bargmann HE (1937) The reproductive system of *Euphausia superba*. Discovery Repts 14:325–350

Budzinski E, Bykowski P, Dutkiewicz D (1985) Possibilities of processing and marketing of products made from Antarctic krill. FAO Fish Tech Pap 268:1–46

Bunt JS (1963) Diatoms of Antarctic sea ice as agents of primary productions. Nature 199:1254–1257

CCAMLR (1991) Report of the tenth meeting of the Commission. CCAMLR Hobart, 101 pp

Clarke A (1980) The biochemical composition of krill *Euphausia superba* Dana, from South Georgia. J Exp Mar Biol Ecol 43:221–236

Clarke A, Quetin LB, Ross RM (1988) Laboratory and field estimates of faecal pellet production by Antarctic krill, *Euphausia superba*. Mar Biol 98:557–563

Croxall JP, McCann TS, Prince PA, Rothery P (1988) Reproductive performance of seabirds and seals at South Georgia and Signy Island, South Orkney Islands, 1976–1987: implications for Southern Ocean monitoring studies. In: Sahrhage D (ed) Antarctic Ocean and resource variability. Springer, Berlin Heidelberg New York, pp 261–285

Duhamel G, Williams R (1990) Ichthyofaune demersale. Analyse des captures de chalut à perche. Les Rapports des Campagnes à la Mer à bord du "Marion Dufrense", MD 42/SIBEX, No. 85-01, pp 149–172

Edwards DM, Heap JA (1981) The Convention on the Conservation of Antarctic Marine Living Resources: a commentary. Polar Rec 20(127):353–362

Ettershank G (1984) A new approach to the assessment of longevity in the Antarctic krill *Euphausia superba*. J Crustacean Biol 4 Spec Issue 1:295–305

Everson I (1977) The living resources of the Southern Ocean. Southern Ocean Fisheries Research Programme. GLO/SO/77/1. FAO, Rome, 156 pp

Everson I, Watkins JL, Bone DG, Foote KG (1990) Implications of a new acoustic target strength for abundance estimates of Antarctic krill. Nature 345:338–340

FAO (1991) Fisheries statistics. Catches and landings (1989) 68, Food and Agriculture Organisation of the United Nations, Rome, 518 pp

Fraser FC (1936) On the development and distribution of the young stages of krill (*Euphausia superba*). Discovery Rep 14:1–192

Garrison DL, Sullivan CW, Ackley SF (1986) Sea ice microbial communities in Antarctica. Bioscience 36:243–250

Greene CH, Stanton TK, McClatchie S (1991) Acoustic estimation of Antarctic krill. Nature 349:110

Hamner WM (1984) Aspects of schooling of *Euphausia superba*. J Crustacean Biol 4 Spec Issue 1:67–74

Hamner WM (1988) Biomechanics of filter feeding in the Antarctic krill, *Euphausia superba*: review of past work and new observations. J Crustacean Biol 8(2):149–163

Hamner WM, Hamner PP, Strand SW, Gilmer RW (1983) Behaviour of Antarctic krill, *Euphausia superba*: chemoreception, feeding, schooling and molting. Science 220: 433–435

Hewitt RP, Demer DA (1991) Krill abundance. Nature 253:310

Hosie GW (1994) Structure and distribution patterns of macrozooplankton communities in the Prydz Bay region, Antarctica. In: Proc of the 1991 BIOMASS Colloquium (in press)

Hosie GW, Ikeda T, Stolp M (1988) Distribution, abundance and population structure of the Antarctic krill (*Euphausia superba* Dana) in the Prydz Bay region, Antarctica. Polar Biol 8:213–224

Huntley ME, Sykes PF, Marin V (1989) Biometry and trophodynamics of *Salpa thompsoni* Foxton (Tunicata: Thaliacea) near the Antarctic Peninsula in Austral summer, 1983–1984. Polar Biol 10:59–70

Huntley ME, Lopez MDG, Karl DM (1991) Top predators in the Southern Ocean: a major leak in the biological carbon pump. Science 253:64–66

Ikeda T (1985) Life history of Antarctic krill *Euphausia superba*; a new look from an experimental approach. Bull Mar Sci 37:599–608

Ikeda T, Dixon P (1982) Body shrinkage: a possible overwintering strategy of the Antarctic krill (*Euphausia superba* Dana). J Exp Mar Biol Ecol 62:143–151

Ikeda T, Thomas PG (1987) Longevity of the Antarctic krill (*Euphausia superba* Dana) based on a laboratory experiment. Proc Natl Inst Polar Res Symp Polar Biol 1:56–62

Ikeda T, Mitchell AW, Carleton JH, Dixon P (1980) Transport of living Antarctic zooplankton to a tropical laboratory: a feasibility study. Aust J Mar Freshwater Res 31:271–274

Kanda K, Takagi K, Seki Y (1982) Movement of the larger swarms of Antarctic krill *Euphausia superba* population off Enderby Land during 1976-1077 season. J Tokyo Univ Fish 68:25–42

Kawaguchi K, Ishikawa S, Matsuda O (1986) The overwintering of Antarctic krill *Euphausia superba* Dana under the coastal fast ice off the Ongul Islands in Lutzow-Holm Bay, Antarctica. Mem Natl Inst Polar Res Spec Issue 44:67–85

Kils U (1983) Swimming and feeding of Antarctic krill *Euphausia superba* – some outstanding energetics and dynamics, some unique morphological details. Ber Polarforsch 4:130–155

Koehl MAR, Strickler JR (1981) Copepod feeding currents: food capture at low Reynolds number. Limnol Oceanogr 26(6):1062–1073

Laws RW (1977) Seals and whales in the Southern Ocean. Philos Trans R Soc Lond B 279:81–96

Mackintosh NA (1967) Maintenance of living *Euphausia superba* and frequency of moults. Norsk Hvalfangstitt 5:97–102

Mackintosh NA (1972) Life cycle of Antarctic krill in relation to ice and water conditions. Discovery Rep XXXVI:1–94

Mackintosh NA (1973) Distribution of post-larval krill in the Antarctic. Discovery Rep 36:1–94

Marr JWS (1962) The natural history and geography of the Antarctic krill (*Euphausia superba* Dana). Discovery Res 32:33–464

Marschall HP (1988) The overwintering strategy of Antarctic krill under the pack-ice of the Weddell sea. Polar Biol 9:129–135

Mauchline J, Fisher LR (1969) The biology of euphausiids. Adv Mar Biol 7:1–454

McClatchie S, Boyd CM (1983) Morphological study of sieve efficiencies and mandibular surfaces in the Antarctic krill *Euphausia superba* Dana. Can J Fish Aquat Sci 40:955–967

McWhinnie MA, Markiniak (1964) Temperature responses and tissue respiration in Antarctic crustacea with particular reference to the krill *Euphausia superba*. In: Lee MO (ed) Biology of the Antarctic seas. Antarct Res Ser 1:63–72

Miller DGM (1991) Exploitation of Antarctic marine living resources: a brief history and a possible approach to managing the krill fishery. S Afr J Mar Sci 10:321–339

Miller DGM, Hampton I (1989) Biology and ecology of the Antarctic krill (*Euphausia superba* Dana): a review. BIOMASS 9:1–166

Moiseev PA (1970) Some aspects of the commercial use of the krill resources in the Antarctic. In: Holgate MW (ed) Antarctic Ecology, vol 1. Academic Press, London, pp 213–216

Morris DJ (1984) Filtration rates of *Euphausia superba*: under or overestimates. J Crustacean Biol 4 Spec Issue 1:185–197

Murphy RC (1962) The oceanic life of the Antarctic. In: Oceanography, readings from Scientific America. W.H. Freeman, San Fransisco, pp 287–299

Nicol S (1989) Who's counting on krill? New Sci 1690:38–41

Nicol S (1991a) The age-old problem of krill longevity. Bioscience 40(11):833–836

Nicol S (1991b) CCAMLR and its approaches to management of the krill fishery. Polar Rec 27(162):229–236

Nicol S, de la Mare W (1993) Ecosystem management and the Antarctic krill. Am Sci 81(1):36–47

Nicol S, Stolp M (1989) Sinking rates of cast exoskeletons of Antarctic krill (*Euphausia superba* Dana) and their role in the vertical flux of particulate matter and fluoride in the Southern Ocean. Deep Sea Res 36(11):1753–1762

Nicol S, Stolp M, Hosie GW (1991) Accumulation of fluorescent age pigments in a laboratory population of Antarctic krill (*Euphausia superba* Dana). J Exp Mar Biol Ecol 146:153–161

O'Brien DP (1987) Direct observations of the behaviour of *Euphausia superba* and *Euphausia crystallarophias* (Crustacea, euphausiacea) under the pack ice during the Antarctic winter of 1985. J Crustacean Biol 7:437–438

Quetin L, Ross RM (1991) Behavioural and physiological characteristics of Antarctic krill *Euphausia superba* Dana. Am Zool 31(1):49–63

Ragulin AG (1969) Underwater observations of krill. Trudy, Vses Nauchno-issled Ryb. Khoz Okeanogr 66:231–234 MAFF (UK) Translation NS92

Rosenberg AA, Beddington J, Basson M (1986) Growth and longevity of krill during the first decade of pelagic whaling. Nature 324(6093):152–154

Ross RM, Quetin LB (1983) Spawning frequency and fecundity of the Antarctic krill *Euphausia superba*. Mar Biol 77:201–205

Ross RM, Quetin LB (1986) How productive are Antarctic krill? Bioscience 36(4):264–269

Ross RM, Quetin LB (1988) *Euphausia superba*: a critical review of annual production. Comp Biochem Physiol 90B(3):499–505

Sahrhage D (1989) Antarctic krill fisheries: potential resources and ecological concern. In: Caddy JF (ed) Marine invertebrate fisheries, their assessment and management, pp 13–33

Shulenberger E, Wormuth JH, Loeb VJ (1984) A large swarm of *Euphausia superba*: overview of patch structure and composition. J Crustacean Biol 4 Spec Issue 1:75–95

Siegel V (1988) A concept of seasonal variation of krill (*Euphausia superba*) distribution and abundance west of the Antarctic peninsula. In: Sahrhage D (ed) Antarctic Ocean and resources availability. Springer, Berlin Heidelberg New York, 304 pp, pp 218–230

Smetacek V, Scharek R, Nöthig, E-M (1990) Seasonal and regional variation in the pelagial and its relationship to the life history of krill. In: Kerry KR, Hempel G (eds) Antarctic Ecosystems. Ecological changes and conservation. Springer, Berlin Heidelberg New York, pp 103–115

Smith SL, Schnack-Schiel SB (1990) Polar zooplankton. In: Smith WO (ed) Polar oceanography Part B. chemistry, biology and geology. Academic Press, London, pp 527–598

Soevik T, Braekkan OR (1979) Fluoride in Antarctic krill (*Euphausia superba*) and Atlantic krill (*Meganyctiphanes norvegica*). J Fish Res Board Can 36(11):1414–1416

Stretch JJ, Hamner PP, Hamner WM, Michel WC, Cook J, Sullivan CW (1988) Foraging behaviour of Antarctic krill *Euphausia superba* on sea ice microalgae. Mar Ecol Prog Ser 44:131–139

Strickler JR (1982) Calanoid copepods, feeding currents, and the role of gravity. Science 218:158–160

Thomas PG, Ikeda T (1987) Sexual regression, shrinkage, re-maturation and growth in female *Euphausia superba* in the laboratory. Mar Biol 95:357–363

Von Drygalski E (1990) The Southern ice continent. The German South Polar Expedition aboard the Gauss, 1901–1903. Bluntisham Books, Erskine Press, UK

Watkins JL, Morris DJ, Ricketts C, Priddle J (1986) Differences between swarms of Antarctic krill and some implications for sampling krill populations. Mar Biol 93:137–146

Whitehead MD, Johnstone GW, Burton HR (1990) Annual fluctuations in productivity and breeding success of Adelie penguins and fulmarine petrels in Prydz Bay, East Antarctica. In: Kerry KR, Hempel G (eds) Antarctic ecosystems. Ecological change and conservation. Springer, Berlin Heidelberg New York, pp 214–223

Impact of Shelf and Sea Ice on Water Mass Modifications and Large-Scale Oceanic Circulation in the Weddell Sea

Eberhard Fahrbach[1], Ernst Augstein[1] and Dirk Olbers[1]

1 Introduction

The thermohaline circulation of the global ocean is driven to a large extent by water mass modifications that take place in the northern and southern Atlantic Ocean and its adjacent seas. In the Arctic and northern Atlantic Oceans, lateral and vertical mixing leads to the formation of North Atlantic Deep Water, a layer whose characteristics can be traced throughout the deep layers of most of the world ocean. South of the Antarctic Circumpolar Current, however, density layers influenced by the North Atlantic are shallow and they consequently lose large quantities of heat to the atmosphere. This occurs both through direct contact of the upper ocean/sea ice layer with the atmosphere or indirectly by conduction through the quasi-permanent ice shelves. Such heat losses initiate processes that produce cold and dense water masses, with the most important being formed in the Weddell Sea which in turn recirculate back toward the north at abyssal depths as various forms of Antarctic Bottom Water.

Oceanic water masses are typically identified on the basis of temperature and salinity, and by their concentrations of dissolved substances such as nutrients and gases. These characteristics can be strongly modified at the sea surface by evaporation or precipitation and by heat or gas exchanges with the atmosphere. Such modifications may lead to increased densities of the surface water, causing it to sink and be replaced by less dense water. Once a water parcel has descended into the oceanic interior, its conservative characteristics such as potential temperature and salinity can be further modified only through mechanical mixing with other water types; molecular diffusion is of minor importance.

In terms of annual averages, the polar and sub-polar oceans lose large quantities of heat to the atmosphere. Because the density of typical sea water increases continuously with decreasing temperature, unlike fresh-water which has a density maximum above the freezing point, this cooling leads to sinking motions. Formation of sea ice further induces sinking motions be-

[1] Alfred-Wegener-Institut für Polar- und Meeresforschung, Columbusstr., D-27568 Bremerhaven, Germany

Hempel (Ed.) Antarctic Science

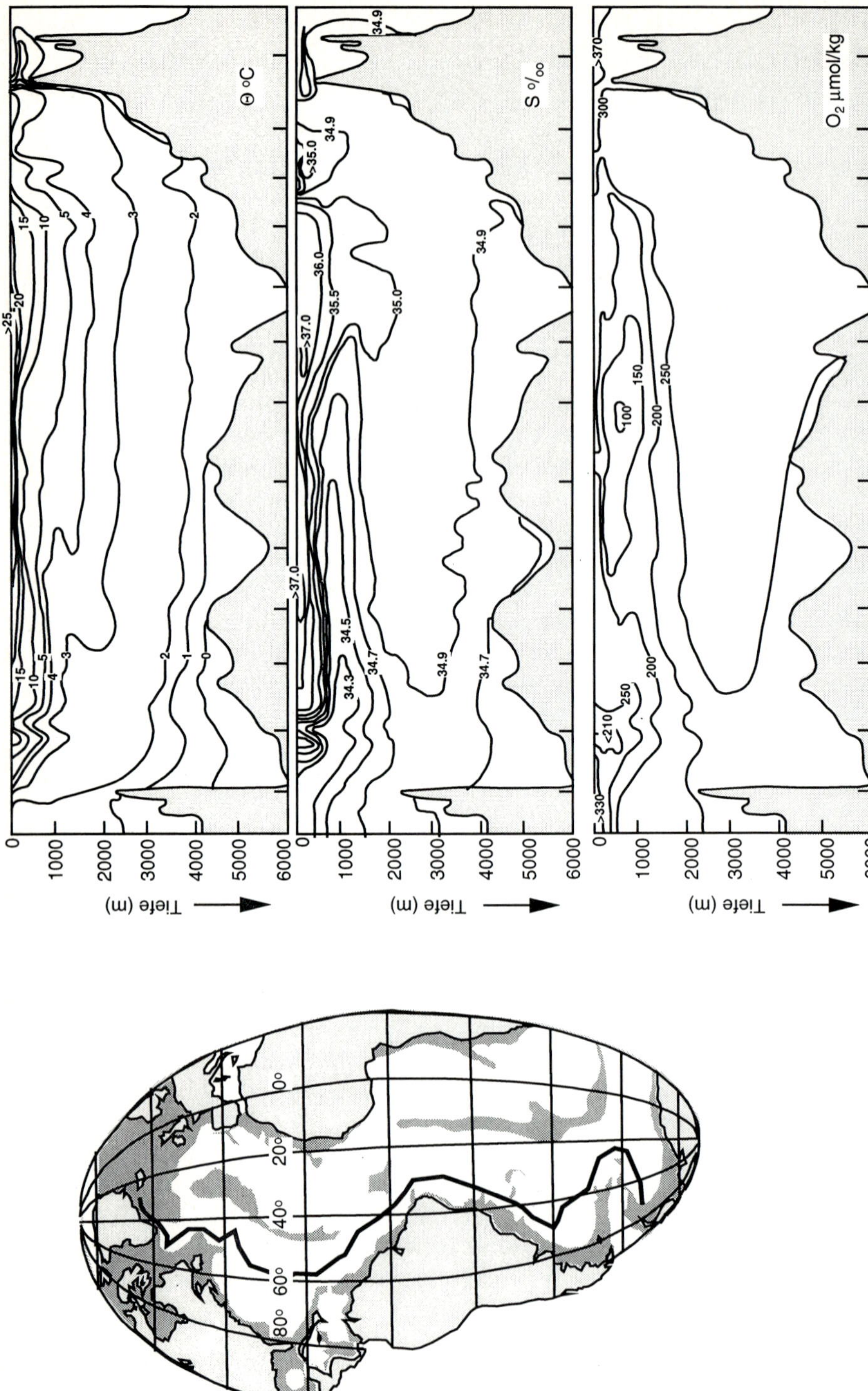
Θ °C
S ‰
O_2 µmol/kg
Tiefe (m)
0
1000
2000
3000
4000
5000
6000
50° S 40° 30° 20° 10° 0° 10° 20° 30° 40° 50° 60° N 70°

cause only a small part of the salt remains in the ice, thus causing an increase in density of the underlying water into which the excess salt is released. The vertical overturning induced by ice formation, called haline convection, enhances the thermal convection. But the ocean–atmosphere heat exchange is strongly controlled in turn by the thermal characteristics and concentration of ice, which combines with the internal oceanic circulation to create an equilibrium between ice formation and heat loss to the atmosphere. As the insulating effect of the ice cover increases, ice growth and haline convection become inhibited and eventually stopped (see e.g. Untersteiner 1986).

The characteristics of ice cover are determined not only by thermodynamic processes but also by mechanical forces of the wind and current fields. These often cause ridges and leads in the ice, which then affect the ocean–atmosphere heat flux. Around Antarctica, the coastal region is of particular interest because the prevailing easterly surface winds have an offshore component which generates a belt of open water along the coast, even in winter, the so-called coastal polynyas. Since the extremely cold outflow from the continent can lead to temperature differences between air and water of more than 20 K, the upward heat fluxes may rise to $500\,W\,m^{-2}$. On average, the heat fluxes over the coastal polynyas are an order of magnitude higher than those in the offshore areas (Comiso and Gordon 1987; Gordon and Comiso 1988; Smith et al. 1990; Kottmeier and Engelbart 1992).

In most parts of the Southern Ocean south of the Antartic Circumpolar Current, warm water which partly originates in the North Atlantic is advected in from the north underneath a cold surface layer. If thermohaline convection penetrates down to the warm layer, known in the Weddell Sea as Warm Deep Water, it results in a significant vertical heat flux which may prevent further sea ice formation or even cause the ice to melt (Bagriantsev et al. 1989; Gordon and Huber 1990; Martinson 1990). Consequently, leads and polynyas can be formed and maintained by this mechanism near the coast as well as in offshore areas. A particularly large and long-lived open-ocean polynya in the area of Maud Rise in the eastern Weddell Sea was detected from satellites during the period of 1974 to 1976, and it is thought that vertical convection down to the Warm Deep Water maintained it (Carsey 1980; Comiso and Gordon 1987).

The vertical distribution of potential temperature along a meridional section in the Atlantic Ocean (Fig. 1) show a top layer warmer than 8 °C, the Warm Water Sphere, which covers about 75% of the sea surface but occupies only 25% of the oceanic volume. The Cold Water Sphere underneath comes into contact with the atmosphere only in high latitudes. There-

←

Fig. 1. Meridional sections of potential temperature, salinity and dissolved oxygen along the western side of the Atlantic Ocean (after Bainbridge 1976). The *map on the left* shows the location of the transect

fore, the polar oceans can be viewed as gateways through which the water mass characteristics of the deep ocean are controlled, and these high-latitude controls of the Cold Water Sphere are thought to define the primary role of the ocean in modulating global climate change.

Processes leading to the formation of deep water occur in both hemispheres, accounting for the vertical layering in Fig. 1. At high northern latitudes, North Atlantic Deep Water is formed from a mixture of waters originating in the Nordic and Labrador seas and then propagates southward as a tongue of water having relatively high salt and oxygen contents to the Southern Ocean, from where it spreads into the Pacific and Indian Oceans (Warren 1981; Swift 1984). In the south, Antarctic Bottom Water is formed in regions near the Antarctic continent. Through most of its range, this water can be identified by its low values of temperature, salinity and oxygen (Brennecke 1921; Wüst 1933; Mosby 1934; Deacon 1937). It spreads over most of the world's sea floor (Fig. 2) while filling about 30% of the oceanic volume (Worthington 1981). In the North Atlantic it can be found up to 53 °N (Mantyla and Reid 1983), where it is identified by its high levels of silica.

Deep and bottom water formation in the Southern Ocean is restricted to a few rather small areas around Antarctica (Fig. 3; Killworth 1983; Carmack 1990). According to present estimates, about 70% of all the Antarctic Bottom Water is produced by processes taking place in the Weddell Sea (Carmack 1977). About 7% comes from the Ross Sea and the rest from areas of lesser importance. Deep water, which occupies the depth range of 1000 to 2000 m and thus does not reach the bottom, is formed in Prydz Bay

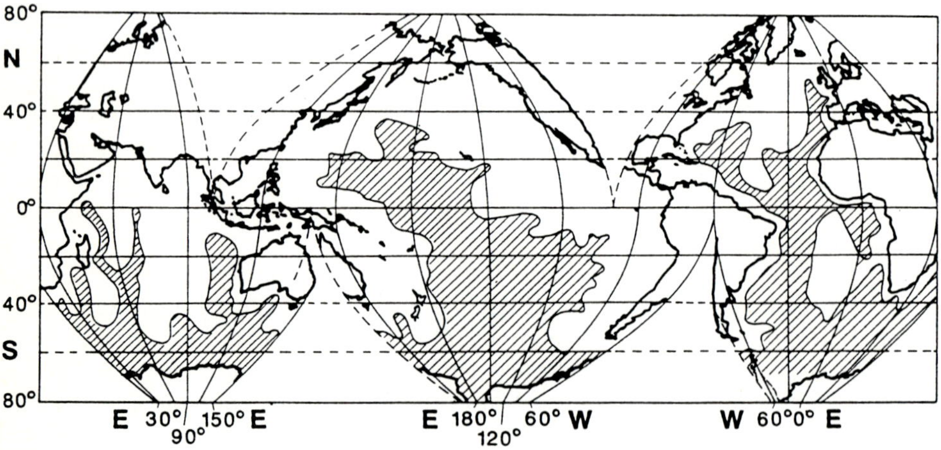

Fig. 2. Distribution of Antarctic Bottom Water (*hatched*) in the abyssal World Ocean. (After Emery and Meincke 1986). (Reprinted from Deep-Sea Res., Vol. 23, Foster TD and Carmack EC, Frontal zone mixing and Antarctic Bottom Water formation in the southern Weddell Sea, Page 314, Copyright 1976, with kind permission from Pergamon Press Ltd, Headington Hill Hall, Oxford 0X3 0BW, UK.)

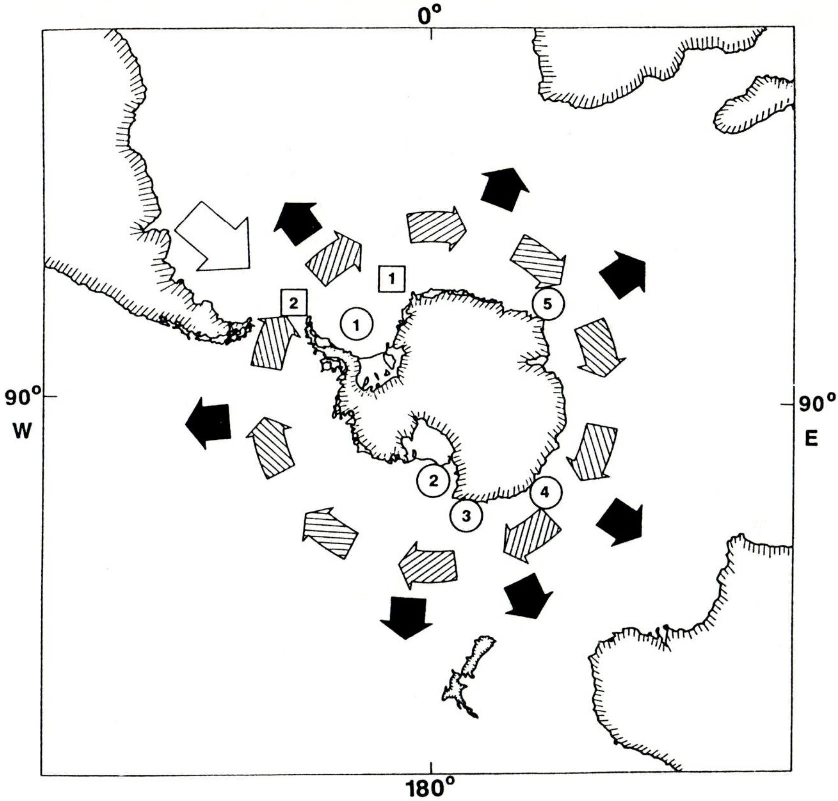

Fig. 3. Schematic representation of water mass formation and circulation of bottom and deep waters around Antarctica. Inflowing North Atlantic Deep Water is indicated as the *open arrow*, outflow of Antarctic Bottom Water as *black arrows* (after Gordon 1991), and the Antarctic Circumpolar Current as *hatched arrows*. Areas where formation occurs through near boundary-processes are shown by *circles* (*1* Weddell Sea; *2* Ross Sea; *3* Wilkes Land; *4* Adelie Coast; *5* Enderby Land) and through open-ocean convection by *squares* (*1* Weddell Sea; *2* Bransfield Strait)

(75° E) and in the open Weddell Sea. Both water masses are transported by sub-polar gyres into the Antarctic Circumpolar Current where they mix with Circumpolar Deep Water and propagate to the north through gaps in the mid-ocean ridges (Fig. 3; see e.g. Gordon 1991). There is a high degree of uncertainty in how much Antarctic Bottom Water is exported to the north, but at present it is estimated as being between 15 and 35 $\times$ $10^6\,m^3\,s^{-1}$ in the annual mean (Stommel and Arons 1960; Gordon 1971; Gill 1973; Broecker and Peng 1982). In order to achieve greater certainty about this climatically important quantity, an increased level of new research is now being concentrated on the Weddell Sea, the major source of the world's abyssal waters.

2 Water Mass Modifications in the Weddell Sea

Observations made during the *Challenger* Expedition of the 1870s revealed the existence of an extremely cold bottom layer in the western South Atlantic that was surmised to originate near Antarctica (Buchan 1895). With succeeding expeditions in the early twentieth century, the role of the Weddell Sea began to emerge (Brennecke 1921; Mosby 1934; Deacon 1937). More recent data from the *Glacier* and *Eltanin*, together with those from the "International Weddell Sea Oceanographic Expedition 1973", have been evaluated by Foster and Carmack (see e.g. 1976a,b). They found a tongue of cold water in the western Weddell Sea whose lowest temperatures were near the northern tip of the Antarctic Peninsula (Fig. 4). At present, the question to which extent this cold water is being advected along the continental slope from the Filchner-Ronne shelf or formed along the east coast of the Antarctic Peninsula is one of the major goals of the ice station drift which occurred in the southern winter of 1992 in the framework of the ANZONE Programme[2] (Gordon et al. 1993). The ice station, manned by American and Russian scientists, drifted from south to north along the coast of the Antarctic Peninsula and permitted to carry out measurements under the perrenial ice for the first time. In the southern summer of 1992/1993 "Polarstern" could complement the winter data by a transect from the Larsen Shelf Ice into the deep Weddell Gyre.

The mechanisms for deep mixing and the resulting formation of Antarctic Bottom Water are basically understood (for a review see Carmack 1986). In a bivariate graphical representation, the potential temperature and salinity of a particular water type is defined by a single point. Incomplete mixing between two different water types produces a series of points lying on a straight line. However, due to variations in the initial conditions, one does not find in the real ocean clearly defined water types, but water classes instead which cover an area in such a temperature-salinity diagram. Foster and Carmack (1976a) identified three classes which mix over the upper part of the continental slope to form Weddell Sea Bottom Water (Fig. 5, left). Warm Deep Water is advected in from the north and mixes at the shelf break with overlying locally formed Winter Water to produce Modified Warm Deep Water. This class then mixes with Western Shelf Water, which has a high density due to accumulated salt from ice formation over the broad western shelves, to produce Weddell Sea Bottom Water. This in turn is modified through mixing with warm Deep Water to finally yield Antarctic Bottom Water.

A second pathway of mixing, which does not require the involvement of Modifed Warm Deep Water, is possible in the southernmost Weddell Sea

[2] ANZONE = Multinational programme of oceanographic investigations of the Antarctic Zone.

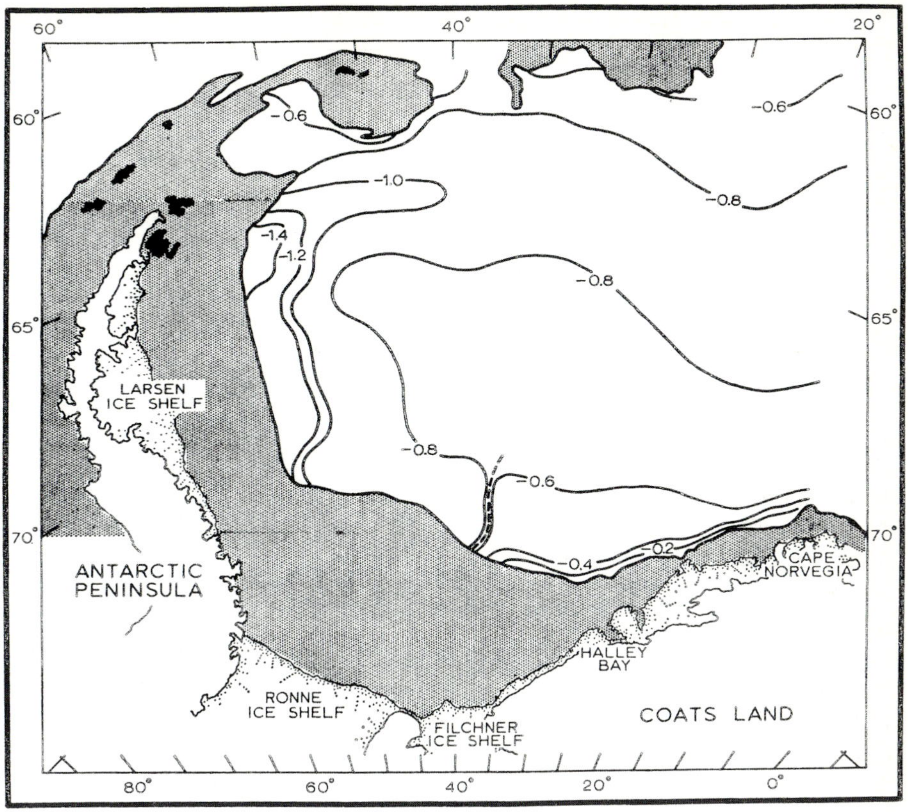

Fig. 4. Distribution of potential temperature at the bottom of the Weddell Sea. (Foster and Carmack 1976a)

because of interactions taking place underneath the vast and floating Filchner-Ronne Ice Shelf (Carmack and Foster 1975a; Weiss et al. 1979; Foldvik et al. 1985). The ice shelf, mainly consisting of freshwater ice, is 400 to 1500 m thick, and because of the pressure at its base the local freezing point is significantly lower than the surface values of −1.8 to −1.9 °C. A flow of Western Shelf Water is subducted under the ice shelf and loses heat to the ice, causing the base of the ice to melt. The water emanating from under the ice shelf, cooled to −2.2 °C is termed Ice Shelf Water, and when mixed with Warm Deep Water over the continental slope the result is Weddell Sea Bottom Water (Fig. 5, right). According to measurements of the stable oxygen isotope O^{18} concentrations, which admit of distinguishing between the freshwater input from melting shelf ice and sea ice, the bottom water receives a significant amount of meteoric meltwater from the base of the ice shelf. This supports opinions that significant part of the Weddell Sea Bottom Water is formed by this process (Weiss et al. 1979; Schlosser et al. 1990).

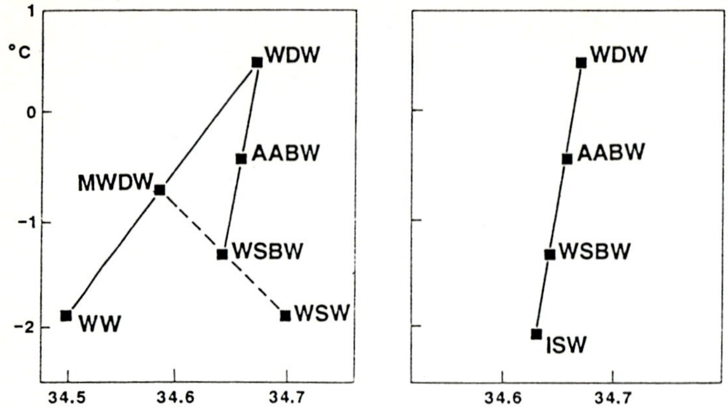

Fig. 5. Potential temperature-salinity characteristics of water masses (*filled squares*) and mixing pathways (*lines*) leading to the formation of Weddell Sea Bottom Water: *On the left*, through formation of Modified Warm Deep Water (after Foster and Carmack 1976a), and *on the right* through mixing with Ice Shelf Water (Foldvik et al. 1985); *WW* Winter Water; *ISW* Ice Shelf Water; *WSW* Western Shelf Water; *WDW* Warm Deep Water; *MWDW* Modified Warm Deep Water; *WSBW* Weddell Sea Bottom Water; *AABW* Antarctic Bottom Water. The *mixing line* between MWDW and WSW is *broken* to indicate that mixing occurs in two steps

3 The Weddell Gyre Study

Because of the global importance of the Antarctic Bottom Water and our poor understanding of its role in climate change, the "Weddell Gyre Study 1989–1993" was carried out as a part of the World Ocean Circulation Experiment (WOCE). The primary goals of this study were to directly measure the modifications to and formation rates of water masses in the southern and western Weddell Sea, as well as the surface forcing conditions. The study complements the simultaneous ice station drift in the framework of the ANZONE project, which is focussed upon similar processes in the interior of the Weddell Sea. Also, observations from the Weddell Gyre Study are used in support of numerical model experiments.

For this project, data concerning the surface forcing conditions are being obtained from a variety of sources. The European Centre for Medium Range Weather Forecasting (ECMWF) is providing results from a diagnostic model that assimilates all available synoptic observations. Satellite passive microwave sensors are providing data on sea-ice cover and its concentration. Examples of the seasonal and interannual variations in the Weddell Sea ice cover for February and September 1988 and 1989, are shown in Fig. 6 (Lemke and Viehoff 1991). The power of satellite remote sensing for the polar regions is demonstrated in more detail by Comiso (1991) or Zwally

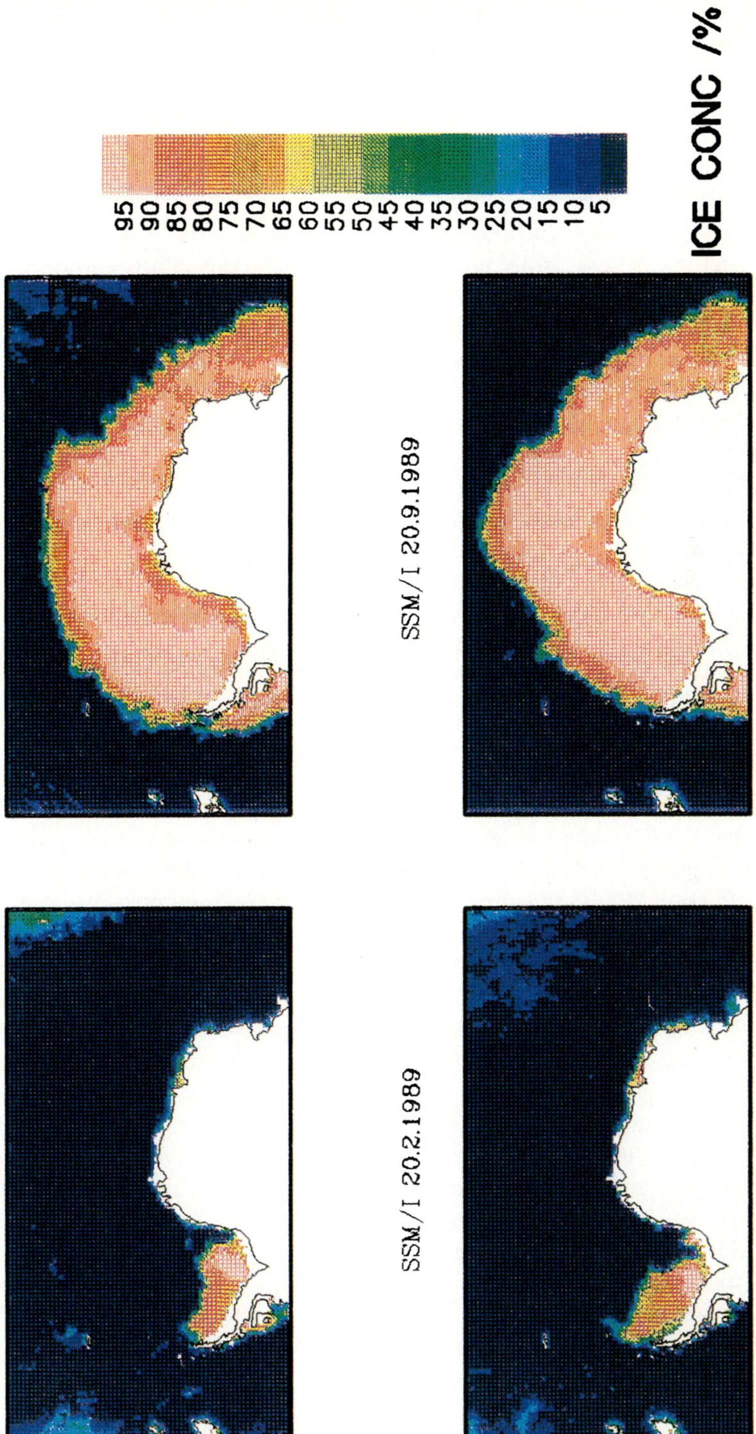

Fig. 6. Sea-ice cover in summer (*left*) and winter (*right*) in 1988 (*top*) and 1989 (*bottom*) obtained from satellite microwave images (SSM/I). (Lemke and Viehoff 1991)

(this Vol.). Sea-ice drift, atmospheric parameters at the surface and upper ocean temperatures can also be measured with satellite tracked buoys deployed on ice floes. Monthly mean ice drift vectors in the Weddell Sea from such buoys (Fig. 7) have been compiled by Kottmeier et al. (1992), and these clearly show the gyral circulation pattern. Ice thicknesses, which were determined primarily by direct observations using drill holes (Wadhams, this Vol.), are the poorest known part of the sea-ice mass budget. First time series of ice thickness from moored upward-looking echo sounders were obtained from the Weddell Gyre Study.

Oceanographic in situ measurements can be made either by lowering instruments into the water column from a ship or by instruments moored to the ocean bottom that are deployed and recovered by a ship. The moored instruments measure various parameters, typically at hourly intervals, and automatically record them for operating periods of up to 1 or 2 years. During winter expeditions to the Weddell Sea in 1986, 1989 and 1992, the ice-breaking research vessel *Polarstern* enabled us to carry out systematic measurements along cruise tracks through the fully developed sea ice (Schnack-Schiel 1987; Augstein et al. 1991). Growing experience makes it possible for us to conduct multi-year programs using moored instruments in areas covered by ice while maintaining a satisfactory rate of instrument recovery.

The 1989 phase of the Weddell Gyre Study was carried out jointly by the two ships *Akademik Fedorov* and *Polarstern* (Fig. 8, from Augstein et al. 1991). The latter operated along a transect from the northern tip of the Antarctic Peninsula to Kapp Norvegia. This transect represents a contribution to the WOCE Repeat Hydrography Programme and was in between repeated four times: in 1989 in late winter, 1990 in early spring, 1992 and mid-winter and 1992/93 in summer. Full depth measurements of temperature and salinity were carried out along this transect at about 50 sites spaced 20 to 60 km apart. The temperature field observed across the Weddell Basin from *Polarstern* during the first winter survey in 1989 is shown in Fig. 9 (Fahrbach et al. 1991). The near surface layer consists of Winter Water colder than −1.8 °C, which occupies the upper 50 to 100 m in the centre of the Weddell Gyre but extends to larger depths near the coasts due to an onshore Ekman transport (Sverdrup 1954) and deep convective mixing (Fahrbach et al. 1992). In summer Winter Water is covered by a shallow layer of warmer and less saline surface water. Beneath this, to roughly 1500-m depth, is an intermediate layer of Warm Deep Water. This water mass is most evident near the continental edges of the gyre, particularly in the east where it first enters the gyre as a relatively warm inflow from the north. This layer is less warm by the time it reaches the western side of the cyclonic Weddell Gyre, thus indicating that there is a pronounced heat loss from the Warm Deep Water while it moves through the southern Weddell Sea. Weddell Sea Bottom Water, having potential temperatures of less than −0.8 °C, is observed throughout the basin except along the eastern con-

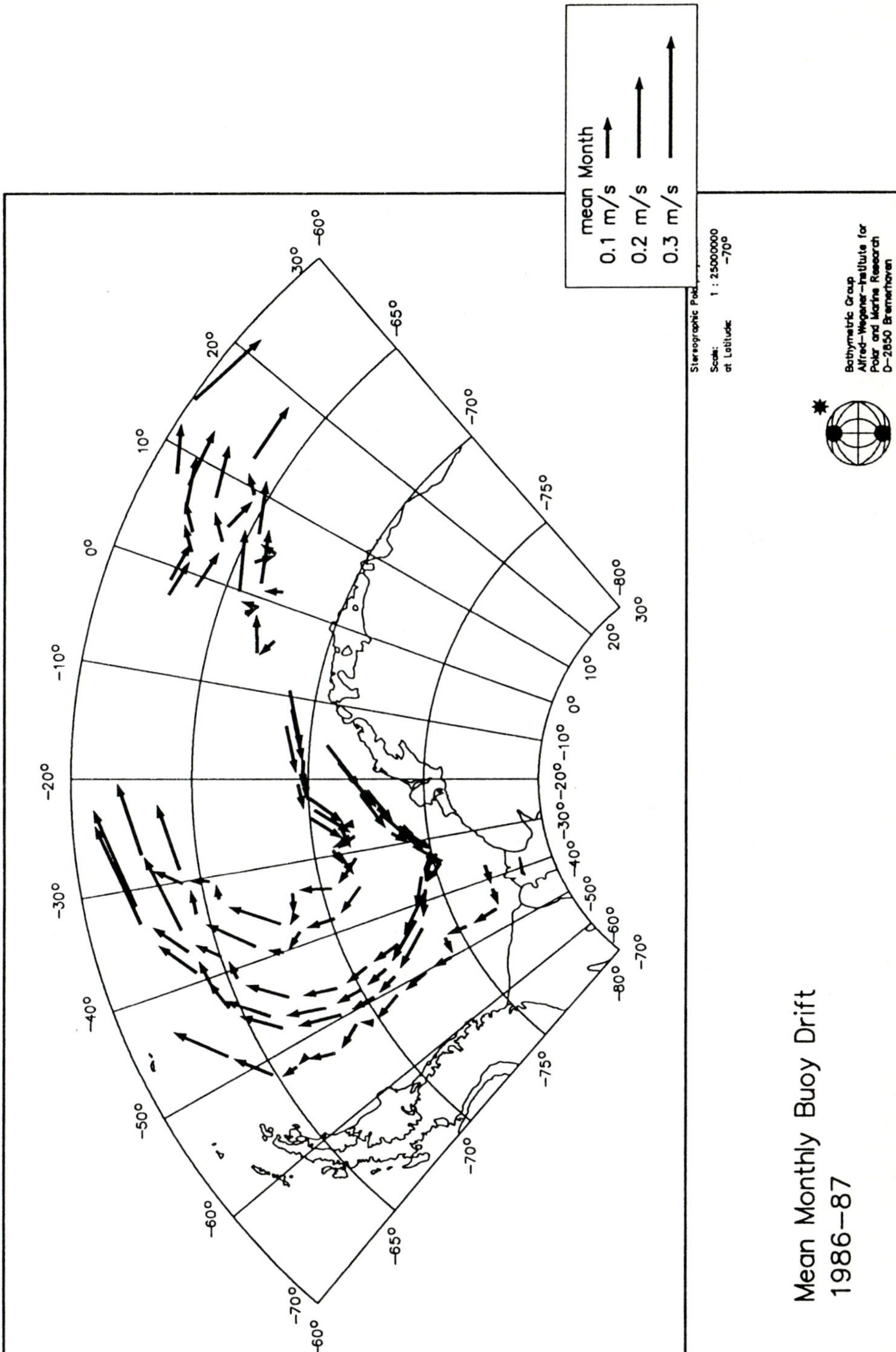

Fig. 7. Monthly means of sea-ice drift measured by satellite-tracked drifting buoys. (Kottmeier et al. 1992)

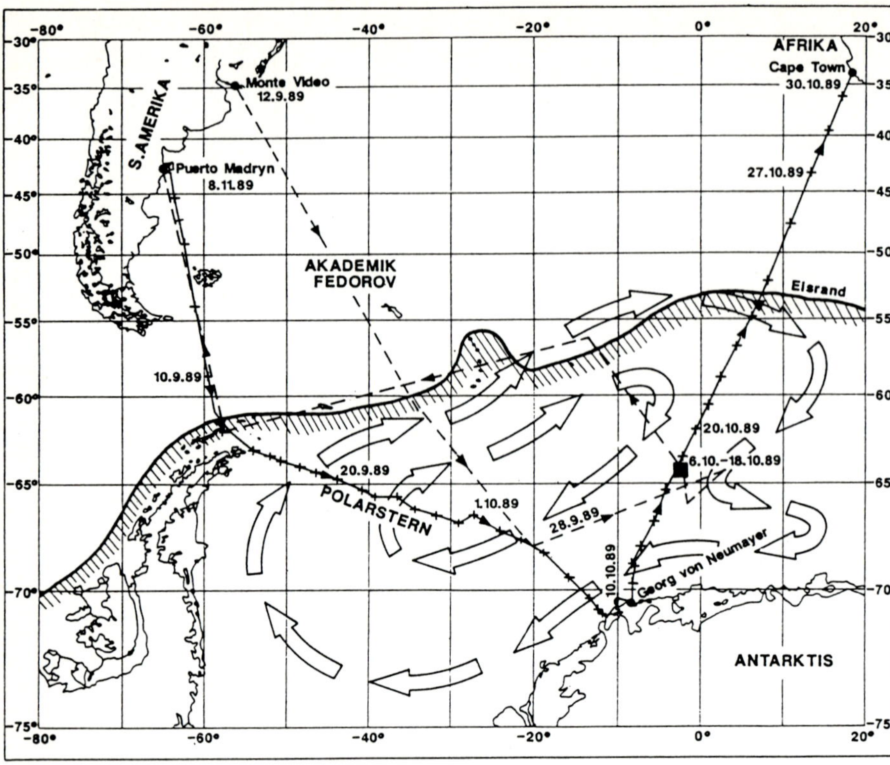

Fig. 8. Schematic representation of the Weddell Gyre and cruise tracks of R.V. *Akademik Fedorov* and *Polarstern* during the Winter Weddell Gyre Study 1989 (Augstein et al. 1991). The edge of the sea ice is indicated by *hatching*

tinental slope. This confirms the hypothesis that Bottom Water is formed in either the southern or western portions of the Weddell Sea. Recent measurements during the ANZONE drift of the Ice Station Weddell and the Weddell Gyre Study indicate that different types of bottom water are formed along the southern and western rims. But the volume of this water mass is relatively small; the largest volume is contained in the layer between the Warm Deep Water and the Weddell Sea Bottom Water. This is the Antarctic Bottom Water or otherwise called Weddell Sea Deep Water (e.g. Orsi et al. 1993), which originates partly through mixing between the adjacent water masses and partly through deep convection (Gordon 1991).

Information about the large-scale velocity field has been obtained from 27 current meters on 7 moorings deployed in September and October 1989 and recovered in November and December 1990 (Bathmann et al. 1992). During that cruise a second set of 21 moorings was deployed. The observed mean currents perpendicular to the transect are shown in Fig. 10. While the structures of the boundary currents are represented fairly well, the interior

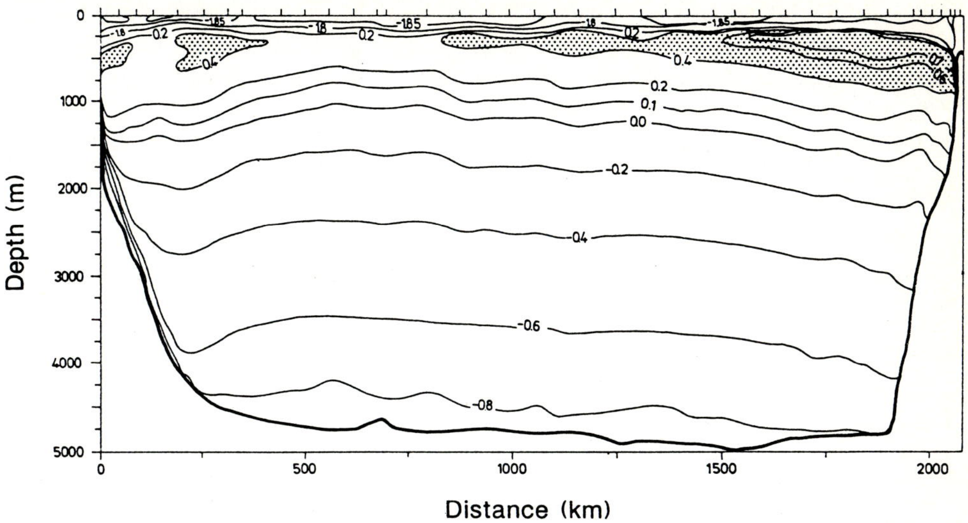

Fig. 9. Vertical section of potential temperature in °C across the Weddell Basin from the northern tip of the Antarctic Peninsula to Kapp Norvegia as measured during the Winter Weddell Gyre Study 1989. The core of the Warm Deep Water is *shaded*

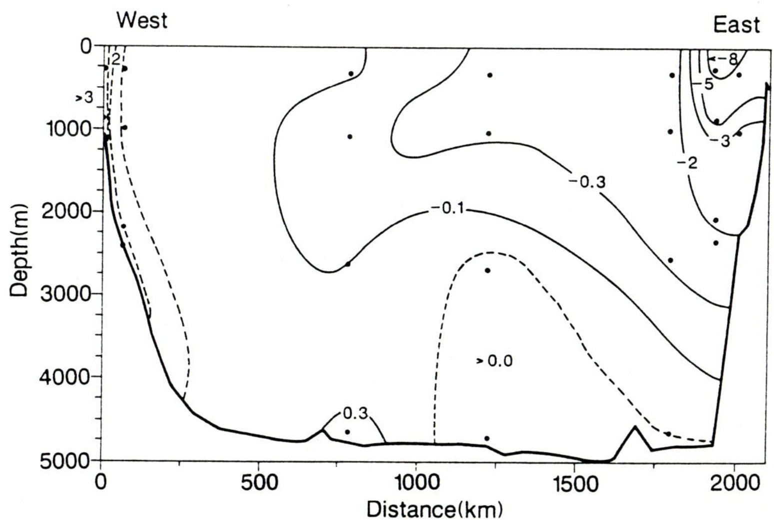

Fig. 10. Annual mean currents in $cm\,s^{-1}$ normal to the transect from the northern tip of the Antarctic Peninsula to Kapp Norvegia as measured with moored instruments from October 1989 to November 1990. *Positive values* and *dashed contours* represent northeastward flow

flow is much less certain. It was measured by only a few instruments and the velocities were extremely low, which sometimes caused the rotors to stall for lengthy periods. From the second deployment 18 moorings were recovered in the southern summer of 1992/93 and the records are at present processed. The much higher data density will permit to describe the structure of the gyre in more detail.

Our current meter records confirm earlier suspicions of there being a significant barotropic component of velocity in the Weddell Gyre, i.e. the depth-independent velocity that would occur in a homogeneous ocean from variations in height of the sea surface. We find the total volume transport of the gyre to be $30 \pm 10 \times 10^6\,m^3\,s^{-1}$, significantly less than previous estimates such as that by Carmack and Foster (1975b) of $97 \times 10^6\,m^3\,s^{-1}$; this particular discrepancy is most likely due to the short duration and small number of current meter measurements available to them.

If one computes the baroclinic component of velocity, i.e. the deviation from the vertical average due to the density field as observed from shipborne temperature and salinity measurements, the flow of the individual water masses across the transect can be estimated. Water mass modifications cause differences between the in- and outflow of the individual water masses, and from our observations we find that Warm Deep Water is transformed to surface layer water, Antarctic Bottom Water and Weddell Sea Bottom Water at a mean rate of $6 \times 10^6\,m^3\,s^{-1}$. The horizontal heat transport across the transect is equivalent to a long-term mean heat loss of $19\,W\,m^{-2}$ from the ocean to the atmosphere in the area south of the transect. The observed transport of salt by the Ocean currents suggests that the salt gain of the water columne due to the formation of sea ice is almost compensated by precipitation over the entire southern area and glacial ice melt.

4 Comparisons Between Observations and Numerical Simulations

Even with the most powerful computers it is not now possible to run models over large horizontal domains with spatial resolutions comparable to oceanographic observations. The most advanced model of the Antarctic region, the Fine Resolution Antarctic Model (FRAM), is run by a team of researchers at universities in the UK and at the Institute of Oceanographic Sciences in Wormley (Webb et al. 1991). Because of the enormous amount of computation time needed for even a specific regional study, one is however usually not able to run experiments with that model. Special models are therefore required. From the large number of regional models presently being used in various institutions, we have selected three which are particularly well suited for addressing specific questions about the Weddell Sea.

The large-scale circulation of the Southern Ocean has been numerically investigated with a coarse grid primitive equation model confined to the Atlantic and Indian ocean sectors (Olbers and Wübber 1991). The horizontal resolution is 1.5° in longitude and 1.0° in latitude, and in the vertical there are 18 layers. The model is formulated so that the two zonal boundaries are continous with one another. Results from two experiments run with different surface forcing conditions are depicted in Fig. 11. If the model is driven only by the mean wind field and thermal fluxes across the sea surface the circulation pattern in the top of Fig. 11 is obtained. But according to the current meter records, the strength of the Weddell Gyre is underestimated by a factor of 2 or more. However, when freezing and melting rates from a sea-ice model (Fig. 12, from Lemke et al. 1990) are included in the forcing terms, the transport of the Weddell Gyre increases from 10 to $40 \times 10^6\,m^3\,s^{-1}$ (Fig. 11, bottom). This is due to additional density gradients that now appear and are related to the differential input of salt- and freshwater. In addition, the transport of the Antarctic Circumpolar Current is intensified throughout the model domain. Our estimates of the volume transport of the Weddell Gyre from the current meter measurements agree more closely with the model experiment that includes the effects of sea-ice formation and melting than with the simpler model.

Numerical experiments with a special ice-ocean model of Lemke et al. (1990) result in an annual mean heat flux of $2\,W\,m^{-2}$ and an ice formation rate of 0.6 m/year for the area south of the transect. Although the heat fluxes are smaller and the ice formation rate is larger in the model than what we have observed, these discrepancies are not too surprising because both model and observational studies are still rather crude. Effects, such as melting ice shelves, are not included in the model, but affect the observed fluxes.

The influence of the Filchner-Ronne Ice Shelf on the formation of water masses in the southern Weddell Sea has been studied with a third model, the ice-ocean model of Hellmer and Olbers (1991). In it, the two-dimensional flow beneath the Filchner-Ronne Ice-Shelf is simulated from the Filchner Depession to the Ronne Ice Shelf (Fig. 13). The circulation under the ice shelf is controlled by boundary conditions at the ice front and by freezing and melting under the ice shelf. This model appears to confirm observational evidence that a significant amount of the shelf ice is melted near the grounding line and that the ice-shelf is maintained by marine ice formation and precipitation (Jenkins and Doake 1991; Determann 1991). In the ice-ocean model of Hellmer and Olbers (1991) two circulation cells interact. One of them is centered under the Filchner and the other under the Ronne side of the ice-shelf. The intensities of these cells vary significantly with small variations in boundary conditions at the ice front. The two cases depicted in Fig. 14 are quite different with respect to intensities of flow, but result from the use of only slightly different salinities at the Ronne ice front: 34.65 PSU (top) and 34.68 PSU (bottom). Also, the sense of the circulation

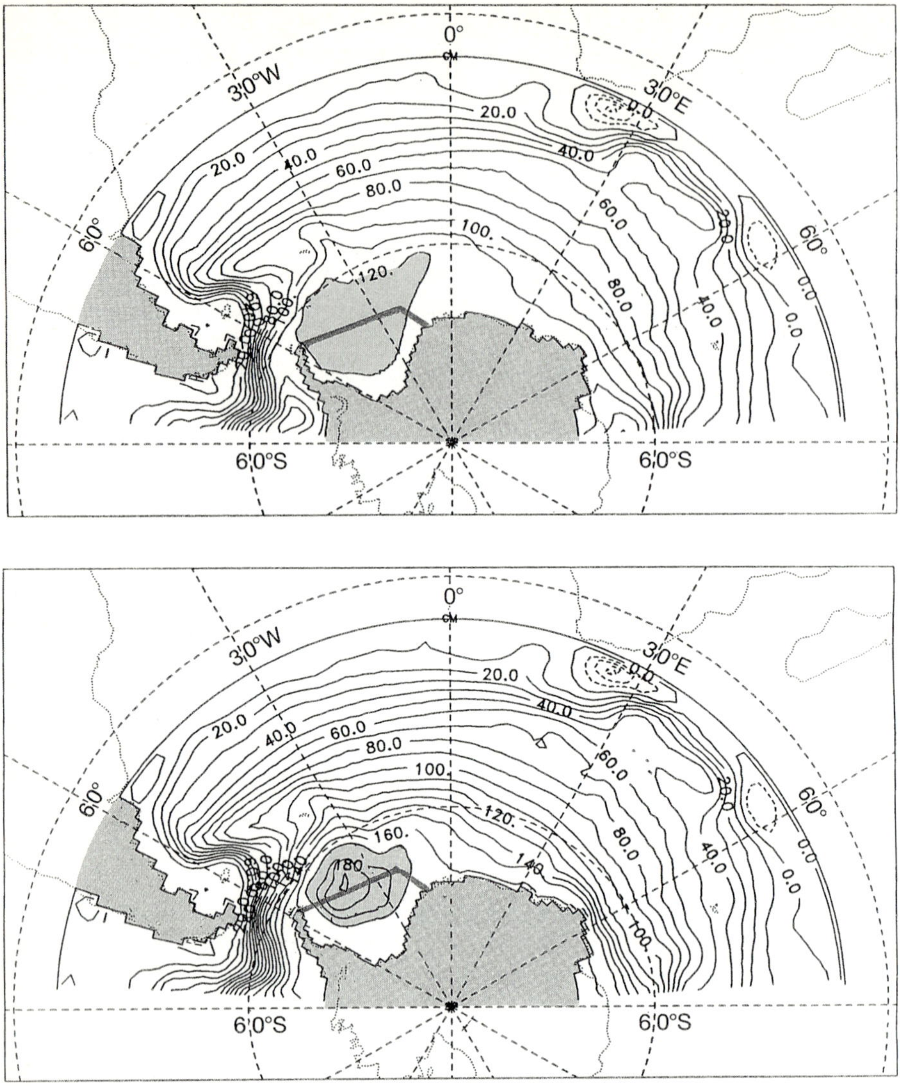

Fig. 11. Stream function of the total mass transport in $10^6\,m^3\,s^{-1}$, in the Atlantic-Indian sector of the Southern Ocean from model simulations (Olbers and Wübber 1991). The *upper field* results from neglecting effects of salt flux beneath the seasonally varying ice cover, whereas the *lower field* accounts for this by using the results of the model shown in Fig. 12

is changed, from one initially having a net flow from Filchner to Ronne to one in the opposite direction. The temperature field and the freezing and melting rates are affected by these small changes in salinity as well, variations that might be present in the seasonal cycle.

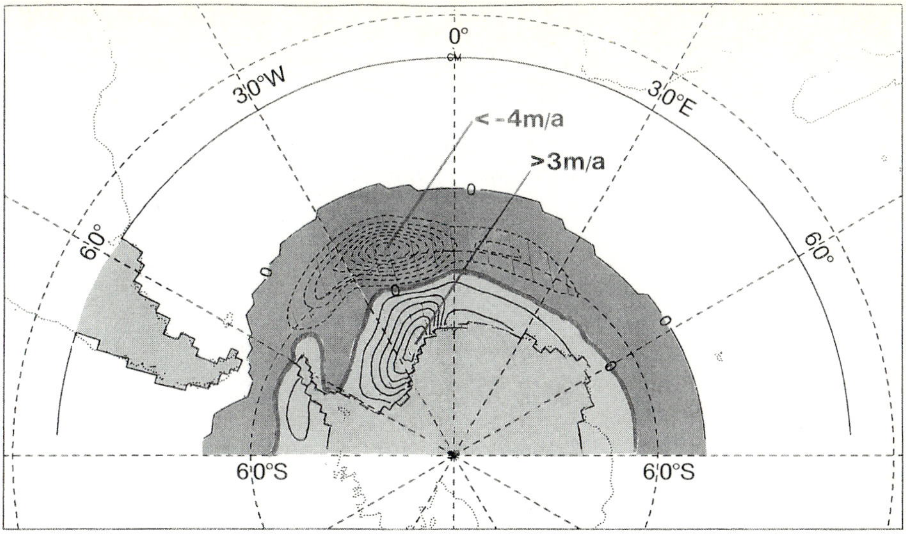

Fig. 12. Mean annual freezing and melting rates in cm/year, in the Weddell Sea obtained with a coupled ice-ocean-atmosphere model. (Lemke et al. 1990)

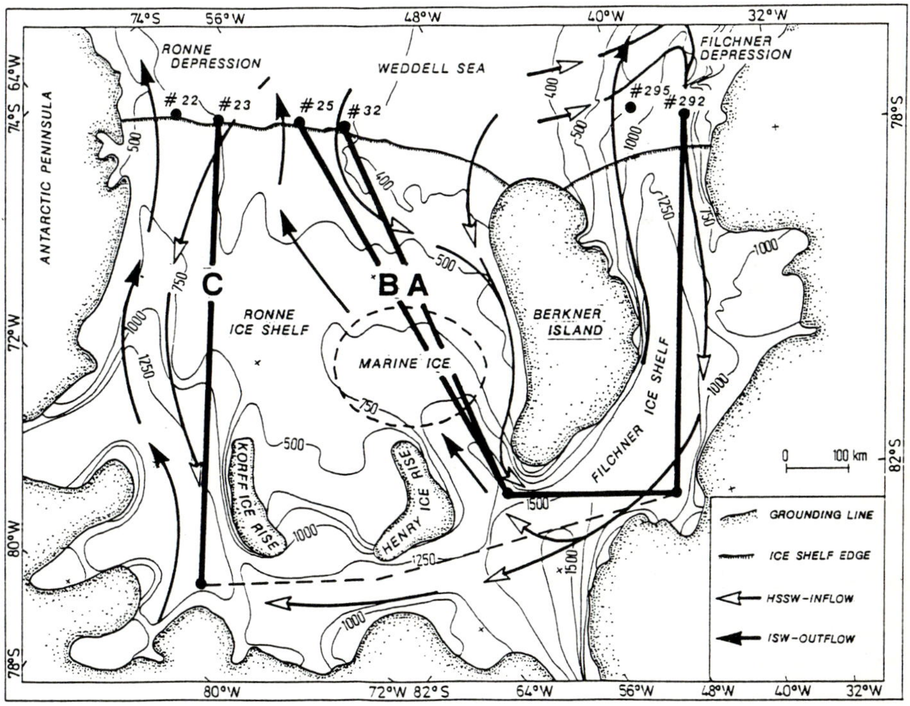

Fig. 13. Locations of transects under the Filchner-Ronne Ice Shelf along which the oceanic circulation was simulated by a two-dimensional numerical model. (Hellmer and Olbers 1991)

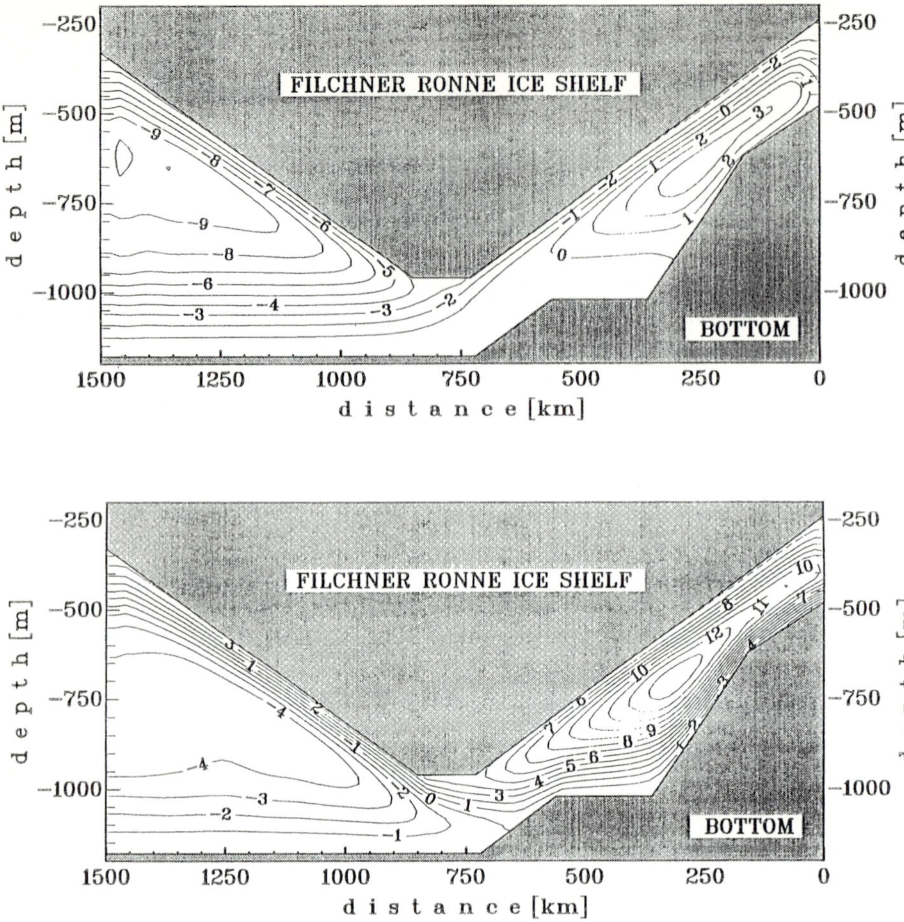

Fig. 14. Stream function of the circulation along path *C* (Fig. 13) under the Filchner-Ronne Ice Shelf obtained for different boundary conditions with a numerical model by Hellmer and Olbers (1991). In the *upper panel* the salinity of the inflow at the Ronne side (*on the right*) is given as 34.65 PSU, whereas in *the lower* it is 34.68 PSU. Contour interval $1\,m^2\,s^{-1}$

5 Conclusions and Outlook

At present, we have attained a rather satisfactory level of qualitative knowledge about the processes relevant to water mass formation and circulation in the Weddell Sea. More quantitative information on these topics will come from the evaluation of the Weddell Gyre Study, the field phase of which ended in early 1993 with the exception of a reduced set of moorings to monitor longer-term variability. The data are sufficient for determining annual mean values as well as seasonal variations in transports and water mass formation rates. Upward looking sonars, active and passive satellite microwave instruments, and automatic surface drifters on ice floes will altogether provide a comprehensive data set from which the sea-ice budget can be evaluated in detail. Additionally, attempts are being made to measure the temperature, salinity and current fields under the Filchner-Ronne Ice Shelf and to observationally investigate interactions between the ocean and ice shelf (Nicholls et al. 1991).

A more realistic simulation of the large scale circulation of the Southern Ocean will be achieved with a coupled ocean–sea-ice model presently being tested in Bremerhaven; interactions between the ocean and ice shelf will be included in the model in a later stage. For this purpose the above mentioned two-dimensional concept has to be fully extended to three dimensions. Similarly, further studies must be designed for fully investigating the oceanic vertical mixing in the coastal zones near Antartica as well as in the open ocean regime of the full sea-ice belt.

As we have described, considerable efforts are being made on both the observational and modelling fronts, and these efforts will lead to a significant increase in our ability to quantify processes taking place in the Weddell Sea, and hence its roles in the general oceanic circulation and long-term global change.

Acknowledgements. We are grateful to Dr. R. Peterson who significantly improved this manuscript with helpful comments. This is AWI-contribution No. 507.

References

Augstein E, Bagriantsev N, Schenke HW (1991) The Expedition ANTARKTIS VIII/1–2, 1989, with the Winter Weddell Gyre Study of the research vessels "Polarstern" and "Akademik Fedorov". Ber Polarforsch 84:134

Bagriantsev NV, Gordon AL, Huber BA (1989) Weddell Gyre: temperature maximum stratum. J Geophys Res 94:8331–8334

Bainbridge AE (1976) GEOECS Atlantic Expedition, vol 2. Sections and profiles. National Science Foundation, Washington, 198 pp

Bathmann U, Schulz-Baldes M, Fahrbach E, Smetacek V, Hubberten HW (eds) (1992) The expedition ANTARKTIS IX/1–4, 1990/91. Ber Polarforsch 100:403

Brennecke W (1921) Die ozeanographischen Arbeiten der Deutschen Antarktischen Expedition 1911–1912. Arch Dtsch Seewarte 39:1–214

Broecker WS, TH Peng (1982) Tracers in the ocean. Eldigio, Palisades, New York, 690 pp

Buchan A (1895) Report on oceanic circulation, based on the observations made on board H.M.S. Challenger, and other observations. In: Report on the scientific results of the voyage of H.M.S. Challenger during the years 1872–76, a summary of the scientific results, second part, Appendix (Physics and Chemistry Part VIII), Her Majesty's Government, London, 38 pp

Carmack EC (1977) Water characteristics of the Southern Ocean south of the Polar Front. In: Angel M (ed) A voyage of discovery, George Deacon 70th Anniversery Vol. Pergamon Press, Oxford, pp 15–41

Carmack EC (1986) Circulation and mixing in ice-covered waters. In: Untersteiner N (ed) The geophysics of sea ice. NATO ASI Ser B, Physics, vol 146. Plenum, New York, pp 641–712

Carmack EC (1990) Large-scale physical oceanography of polar oceans. In: Smith WO Jr (ed) Polar oceanography, part A, physical science. Academic Press, San Diego, pp 171–222

Carmack EC, Foster TD (1975a) Circulation and distribution of oceanographic properties near the Filchner Ice Shelf. Deep-Sea Res 22:77–90

Carmack EC, Foster TD (1975b) On the flow of water out of the Weddell Sea. Deep-Sea Res 22:711–724

Carsey FD (1980) Microwave observation of the Weddell polynya. Mon Weather Rev 108:2032–2044

Comiso JC (1991) Satellite remote sensing of the Polar Oceans. J Mar Syst 2:395–434

Comiso JC, Gordon AL (1987) Recurring polynyas over the Cosmonaut Sea and the Maud Rise. J Geophys Res 92:2819–2834

Deacon GER (1937) The hydrology of the Southern Ocean. Discovery Rep 15:3–122

Determann J (1991) Numerical modelling of ice shelf dynamics. Antarct Sci 3:187–195

Emery WJ, Meincke J (1986) Global water masses: summary and review. Oceanol Acta 9:383–391

Fahrbach E, Knoche M, Rohardt G (1991) An estimate of water mass transformation in the southern Weddell Sea. Mar Chem 35:25–44

Fahrbach E, Rohardt G, Krause G (1992) The Antarctic Coastal Current in the southeastern Weddell Sea. Polar Biol 12:171–182

Foldvik A, Gammelsröd T, Törresen T (1985) Circulation and water masses on the southern Weddell Sea shelf. In: Jacobs SS (ed) Oceanology of the Antarctic Continental Shelf. Antarct Res Ser 43:5–20

Foster TD, Carmack EC (1976a) Frontal zone mixing and Antarctic Bottom Water formation in the southern Weddell Sea. Deep-Sea Res 23:301–317

Foster TD, Carmack EC (1976b) Temperature and salinity structure in the Weddell Sea. J Phys Oceanogr 6:36–44

Gill AE (1973) Circulation and bottom water production in the Weddell Sea. Deep-Sea Res 20:111–140

Gordon AL (1971) Antarctic Polar Front Zone. In: Reid JL (ed) Antarctic oceanology I. Antarct Res Ser 15:205–221

Gordon AL (1991) Two stable modes of Southern Ocean winter stratification. In: Chu PC, Gascard JC (eds) Deep convection and deep water formation in the oceans, Elsevier, Amsterdam, pp 17–35

Gordon AL, Comiso JC (1988) Polynyas in the Southern Ocean. Sci Am 256:90–97

Gordon AL, Huber BA (1990) Southern Ocean winter mixed layer. J Geophys Res 95:655–11672

Gordon AL, Huber BA, Hellmer HH, Ffield A (1993) Deep and Bottom Water of the Weddell Sea's Western rim. Science 262:95–97

Hellmer HH, Olbers DJ (1991) On the thermohaline circulation beneath the Filchner-Ronne Ice Shelves. Antarct Sci 3:433–442

Jenkins A, Doake CSM (1991) Ice-Ocean interaction on Ronne Ice Shelf, Antarctica. J Geophys Res 9:791–813

Killworth PD (1983) Deep convection in the world ocean. Rev Geophys Space Phys 21:1–26

Kottmeier C, Engelbart D (1992) Generation and atmospheric heat exchange of coastal polynyas in the Weddell Sea. Boundary-layer Meteorol 60:207–234

Kottmeier C, Olf J, Frieden W, Roth R (1992) Wind forcing and ice motion in the Weddell Sea region. J Geophys Res 97:20373–20383

Lemke P, Viehoff T (1991) Anwendungen der Fernerkundung in der Meereisforschung. Geowissenschaften 9:266–271

Lemke P, Owens WB, Hibler WD III (1990) A coupled sea ice-mixed layer-pycnocline model for the Weddell Sea. J Geophys Res 95:9513–9526

Mantyla AW, Reid JL (1983) Abyssal characteristics of the World Ocean waters. Deep-Sea Res 30:805–833

Martinson DG (1990) Evolution of the Southern Ocean winter mixed layer and sea ice: open ocean deepwater formation and ventilation. J Geophys Res 95:11641–11654

Mosby H (1934) The waters of the Atlantic Ocean. Scientific Results of the Norwegian Antarctic Expedition 1927–1928. Norske Videnskaps-Akademi, Oslo 1:1–131

Nicholls KW, Makinson K, Robinson AV (1991) Ocean circulation beneath the Ronne Ice Shelf. Nature 354(6530):221–223

Olbers D, Wübber C (1991) The role of wind and buoyancy forcing of the Antarctic Circumpolar Current. In: Latif M (ed) Strategies for future climate research, Max-Planck-Institut für Meteorologie, Hamburg, pp 161–192

Orsi AH, Nowlin WD Jr, Whitworth T III (1993) On the circulation and stratification of the Weddell Gyre. Deep-Sea Res 40:164–203

Schlosser P, Bayer R, Foldvik A, Gammelsröd T, Rohardt G, Münnich KO (1990) Oxgen 18 and helium as tracers of ice shelf water and water/ice interaction in the Weddell Sea. J Geophys Res 95:3253–3264

Schnack-Schiel S (ed) (1987) Die Winter-Expedition mit FS "Polarstern" in die Antarktis (ANT V/1–3). Ber Polarforsch 39:259

Smith SD, Muench RD, Pease CH (1990) Polynyas and leads: an overview of the physical process and environment. J Geophys Res 95:9461–9480

Stommel H, Arons AB (1960) On the abyssal circulation of the World Oceans. Deep-Sea Res 6:140–154

Sverdrup HU (1954) The Currents off the Coast of Queen Maud Land. Nor Geograf Tidsskr 14:239–249

Swift JH (1984) The circulation of the Denmark Strait and Iceland-Scotland overflow waters in the North Atlantic. Deep-Sea Res 31:1339–1355

Untersteiner N (ed) (1986) The geophysics of sea ice. NATO ASI Ser B, Physics; vol 146. Plenum, New York, 1196 pp

Warren BA (1981) Deep circulation of the World Ocean. In: Warren BA, Wunsch C (eds) Evolution to physical oceanography. MIT Press, Cambridge, pp 6–41

Webb DJ, Killworth PD, Coward AC, Thompson SR (1991) The FRAM atlas of the Southern Ocean. Natural Environment Research Council, Swindon, 67 pp

Weiss RF, Ostlund HG, Craig H (1979) Geochemical studies of the Weddell Sea. Deep-Sea Res 26A:1093–1120

Worthington LV (1981) The water masses of the world ocean: some results of a fine-scale census. In: Warren BA, Wunsch C (eds) Evolution of physical oceanography. MIT Press, Cambridge, pp 42–69

Wüst G (1933) Das Bodenwasser und die Gliederung der Atlantischen Tiefsee. Wiss Ergeb Dtsch Atlant Exped "Meteor" 1925–1927, 6 Walter de Gruyter, Berlin, pp 1–106

Evolutionary History of Antarctica

J. A. Crame[1]

1 Introduction

One of the most important breakthroughs in Antarctic geological research over the last three decades has been the elucidation of the continent's fossil record. Although fossils have been known since the very earliest days of scientific exploration in the south polar regions, it is only comparatively recently that their study has been placed within a firm scientific framework. Detailed taxonomic studies of many key groups have been completed and it is now possible, for the first time, to take a broad perspective of the history of life on our southernmost continent. A surprising diversity of fossil plants and animals has now been found in Antarctica; the story they tell is every bit as important to the development of our understanding of the broad patterns of the history of life on Earth as that from other continents.

Latest Proterozoic (Vendian; approximately 610–570 Ma) microfossils are known from the Watts Needle Formation of the Shackleton Range (Buggisch and Kleinschmidt 1991), and probably occur also within the basement complex of Northern Victoria Land (Playford 1989). However, our main concern in this study is with the Phanerozoic history of Antarctica, that is, the last 570 Ma (Harland et al. 1989). Fossiliferous Early Palaeozoic rocks (principally Cambrian in age; 570–510 Ma) are known from a variety of localities along the length of the Transantarctic Mountains (TAM) and within the Ellsworth Mountains (Fig. 1). Late Palaeozoic fossils (principally Devonian; 408–367 Ma) have been found at four main sites within the TAM and also in western Marie Byrd Land. Classic Gondwana fossils, such as *Glossopteris* and *Lystrosaurus* are linked to the Late Palaeozoic-Early Mesozoic (i.e. Permian-Triassic; 290–208 Ma) sedimentary rocks within the Beacon Supergroup of the central TAM; an important exposure also occurs within the Prince Charles Mountains of East Antarctica (Fig. 1). Late Mesozoic taxa (i.e. Jurassic and Cretaceous; 208–65 Ma) are particularly common in the Antarctic Peninsula region and Cenozoic (65–0 Ma) ones at peripheral localities such as the Vestfold Hills and Larsemann Hills, East Antarctica, and King George Island, South Shetland Islands (Fig. 1).

[1]British Antarctic Survey, Natural Environment Research Council, High Cross, Madingley Road, Cambridge CB3 0ET, UK

Hempel (Ed.) Antarctic Science

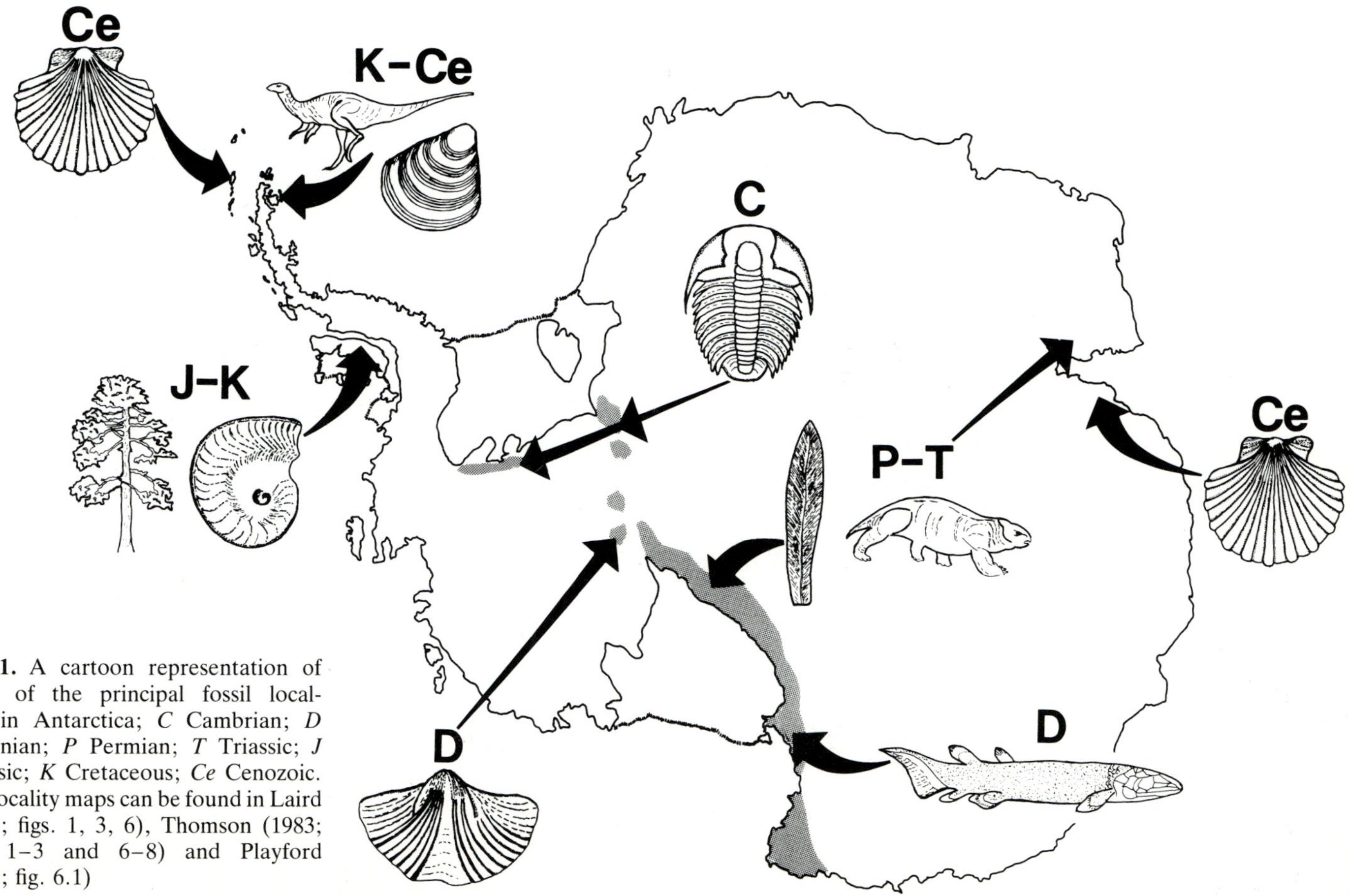

Fig. 1. A cartoon representation of some of the principal fossil localities in Antarctica; *C* Cambrian; *D* Devonian; *P* Permian; *T* Triassic; *J* Jurassic; *K* Cretaceous; *Ce* Cenozoic. Full locality maps can be found in Laird (1981; figs. 1, 3, 6), Thomson (1983; figs. 1–3 and 6–8) and Playford (1989; fig. 6.1)

2 Fossil Discoveries

In the following review, a selection of the principal fossil discoveries from each of these localities will be highlighted. It is the intention then to go on and demonstrate what the broader implications of some of these finds might be. In particular, what can the development of polar floras and faunas through time tell us about global ecological and biogeographical processes?

2.1 The Palaeozoic Record

Our story really begins some 570 million years ago with the worldwide diversification of skeletonized marine fossils that marked the base of the Cambrian period. However, before discussing the Antarctic biotas, it is necessary to place them within a palaeogeographical framework. There is currently a lively academic debate as to the disposition of continents at the Proterozoic-Phanerozoic transition.

The Cambrian palaeocontinental reconstruction shown in Fig. 2 is very much the traditional one, based on sources such as Scotese et al. (1979) and Ziegler et al. (1979) (see also refinements in Scotese and McKerrow 1990). In it, the proto-supercontinent of Gondwana occupies the right-hand (or eastern) flank of the globe and it can be seen that, within it, the familiar component continents (Africa, South America etc.) are actually upside-down. On the left-hand (western) flank of proto-Gondwana are the China, Kazakhstan and Siberia continents, and farther west still is the large continent Laurentia (North America) and the smaller one, Baltica. All these continental blocks are essentially equatorial in their distribution and it is apparent that there was no land above 60°; free interconnections existed between the two polar oceans.

In marked contrast to this standard view of largely dispersed continents, other authors have suggested that, in the Late Precambrian (i.e. latest Proterozoic or Neoproterozoic), there was just a single, massive supercontinent (e.g. Bond et al. 1984). The latest of these hypotheses, as proposed by Moores (1991), Dalziel (1991) and Hoffman (1991), is based upon the premise that the North American craton was once contiguous with the East Antarctica-Australia one; particularly striking is the possible continuation of the Grenville Front tectonic belt from eastern North America into western Dronning Maud Land (e.g. Dalziel 1991, fig. 3). If this correlation is correct, then it is likely that the outermost margin of the East Antarctica-Australia block formed a conjugate rift pair with the Laurentian margin that is now coincident with the west coast of North America. In the Neoproterozoic, localities such as the TAM and the Flinders Ranges of South Australia were the sites of deposition of very thick sequences of turbiditic sedimentary rocks. It is likely that they accumulated as deep-water fans within a major intracontinental rift basin.

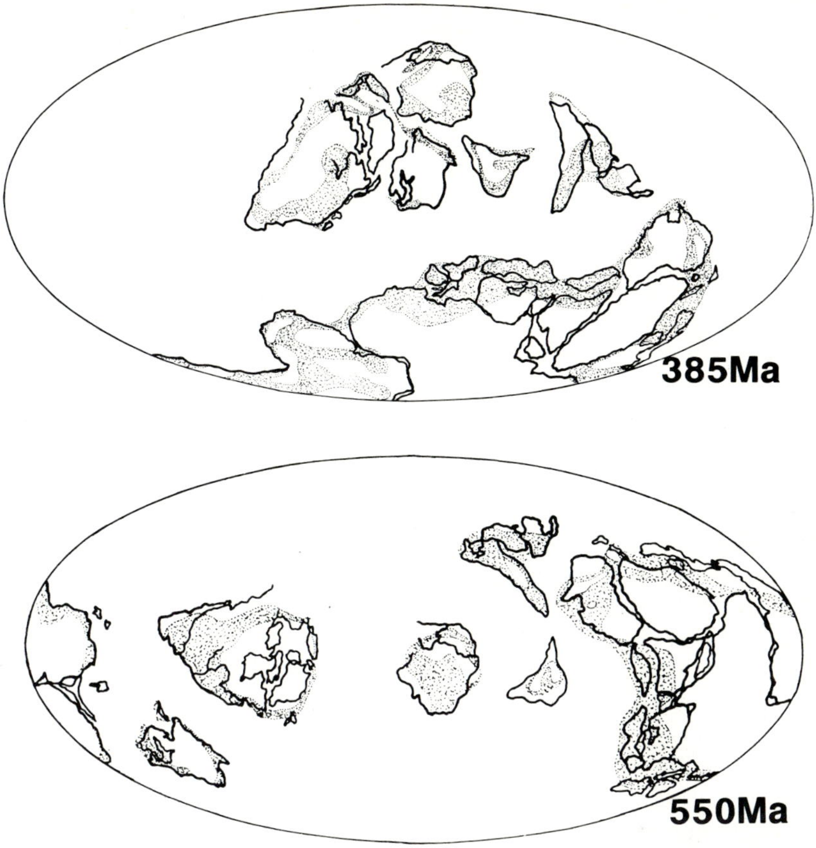

Fig. 2. Continental reconstructions for the Early Cambrian (550 Ma) and Middle Devonian (385 Ma) periods. Sources for the reconstructions are given in the text. *Stippled areas* indicate shallow, epicontinental seas

According to the latest model, Laurentia broke out of the Neoproterozoic continent at approximately 750 Ma and the Pacific Ocean developed along the site of the intracratonic rift between East Antarctica – Australia and Laurentia (Dalziel 1992; Davidson 1992). Between this time and the end of the Cambrian period (510 Ma), a Pacific Ocean basin some 9000 km wide was formed and the Gondwana supercontinent assumed its now-familiar shape. Although detailed palaeomagnetic data are urgently required to substantiate fully this new hypothesis, it would appear likely that a dispersing Neoproterozoic supercontinent would have been contained, essentially, within the Early Cambrian low-latitude regions (Dalziel 1992).

Perhaps the most striking feature to emerge from all the various late Proterozoic – early Phanerozoic reconstructions to date is that Antarctica is placed consistently in an equatorial position. In the dispersed continents model it is actually within the northern hemisphere (10°–30° N) (Bambach et al. 1980; Fig. 2), and in the single supercontinent one it straddles the equator and protrudes 10°–20° S (Dalziel 1992, fig. 3). Such a low-latitude position is reflected in the tropical-subtropical seas that extended across the Antarctic craton during the Early Cambrian period. At this time, thick sequences of limey muds, sand shoals and reefs were deposited, and these are now preserved within lithological units such as the Shackleton Limestone and its equivalents (Laird 1981; Rees and Rowell 1991). The formation of reefs is particularly interesting, for this is the one and only time that they are found within the Antarctic geological record. They were formed by archaeocyathans, small cup-like organisms with affinities to sponges (Fig. 3l), together with encrusting microbial microfossils assigned to the genera *Renalcis*, *Epiphyton* and *Girvanella* (Rees et al. 1989). The taxonomic positions of the latter taxa are uncertain, but they could well be calcified cyanobacteria.

Outboard of the stable carbonate platform with reefs, the accumulation of a thick sequence of shallow water volcanic rocks and carbonates through the Middle Cambrian period is recorded in rock units such as the Liv Group of the Central TAM and the Nelson Limestone, Gambacorta Formation and Wiens Formation of the Pensacola Mountains (Laird 1981; Rowell and Rees 1989). Indeed, the persistence of shallow, open-marine sedimentation into the Late Cambrian is shown by the presence of further thick sedimentary sequences in both the Ellsworth Mountains and Northern Victoria Land (Stump 1991). A rich marine invertebrate fauna from the Early, Middle and Late Cambrian rocks of Antarctica includes types such as trilobites, brachiopods, molluscs and echinoderms (Fig. 3j,k; e.g. Shergold et al. 1976; Clarkson et al. 1979; Laird 1981; Rowell and Rees 1989). Many taxa have close biogeographical ties with Australia and China, and there is growing evidence to suggest that there was a Cambrian Gondwanan faunal province (Cocks 1989).

Concomitant with the deposition of Middle and Upper Cambrian sedimentary rocks, a major tectonic change was taking place along the East

Fig. 3a–l. Antarctic and southern high-latitude marine invertebrate taxa through time. **a** Pliocene pectinid bivalve, *Chlamys*; **b** latest Cretaceous struthiolariid gastropod, *Conchothyra*; **c** living buccinid gastropod, *Chlanidota*; **d** Cretaceous (Albian) dimitobelid belemnite, *Dimitobelus*; **e** Jurassic (Tithonian) buchiid bivalve, *Australobuchia* (NB taxonomically, the austral *Australobuchia* is thought to be very close to the boreal *Buchia*); **f** Late Cretaceous kossmaticeratid ammonite, *Gunnarites*; **g** Devonian trilobite, *Burmeisteria*; **h** Devonian brachiopod, *Pleurothyrella*; **i** Triassic bivalve of unknown affinities, *Hokonuia*; **j** Cambrian trilobite, *Lyriaspis*; **k** Cambrian monoplacophoran mollusc; **l** Cambrian archaeocyathan. Sources for illustrations: **a** Hennig (1911) pl. 2, fig. 1; **b** Trechmann (1917) pl. 20, fig. 3a; **c** Powell (1951) pl. 8, fig. 35; **d–f** BAS photographic archives; **g,h** Doumani et al. (1965), pl. 3, fig. 3 and pl. 17, fig. 1; **i** Cox (1969) fig. C96, 4b; **j** Soloviev and Grikurov (1979) pl. 3, fig. 5; **k** Webers (1982) fig. 78.2, 6; **l** Hill (1965) pl. 2, fig. 1

a
b
c
d
e
f
g
h
i
j
k
l

Antarctic-Australia continental margin. From an original passive (or rifted) state it was being transformed into one of intense compression and magmatic activity as a subduction zone developed along the craton edge. This is the so-called Ross-Delamerian event which may be linked to the rapid opening of the Pacific Ocean (Dalziel 1992). The change explains why further development of early Palaeozoic shelf sequences in Antarctica was so limited. Some Ordovician strata are probably represented within rock units such as the Crashsite Quartzite of the Ellsworth Mountains, Neptune Group of the Pensacola Mountains and uppermost part of the Robertson Bay Group in northern Victoria Land. However, the Ordovician period as a whole is very poorly represented in the Antarctic sedimentary record, and there are virtually no Silurian strata (Rowell et al. 1987). There are almost no proven Ordovician or Silurian fossils (Cocks 1989).

As major new oceans formed during the Early Palaeozoic era, the Gondwana supercontinent moved a vast distance around the globe. By the end of the Ordovician period (approximately 440 Ma) North Africa had crossed the contemporary south pole and an ice cap stretched from Morocco to Arabia (Scotese and McKerrow 1990). In the succeeding Silurian period the pole was located in central Argentina and by the early Middle Devonian (385 Ma) Antarctica had come to occupy a mid- to high-latitude position in the southern hemisphere; it has essentially maintained this ever since (Bambach et al. 1980; Scotese and Barrett 1990). By now, the component Gondwana continents can be seen to be the "right way up" (Fig. 2).

Broad, shallow epicontinental seas were a feature of the Devonian and during the early part of the period a large marine embayment covered much of the Horlick Mountains – Pensacola Mountains – Ellsworth Mountains region. These seas supported a variety of marine invertebrate taxa and exposures such as those in the Lower Devonian Horlick Formation of the Ohio Range in particular have yielded brachiopods, bivalves, gastropods, bryozoans, tentaculitids and a trilobite (Fig. 3g,h; Doumani et al. 1965). However, in comparison with faunas from other continents, the species diversity of this assemblage is low and there is a marked absence of warm-water biotic elements such as corals, stromatoporoids and conodonts (Blodgett et al. 1990). The Antarctic assemblage belongs within the extensive southern hemisphere cold-water Malvinokaffric Province. It is likely that a subpolar gyre brought cold waters from the south polar region (i.e. southern central Argentina) along the Antarctic margin, and perhaps as far as New Zealand (Bradshaw and McCartan 1983).

Shorewards of the Devonian marine sediments, a sequence of terrestrial and freshwater deposits began to accumulate in a series of intermontane basins along the site of the present day TAM. These formed the earliest constituents of the temporally extensive Beacon Supergroup, whose lowest levels have yielded a series of fossil fish faunas. The most important of these is the Givetian (late Middle Devonian) one in the Aztec Siltstone of Southern Victoria Land (Young 1989). In fact, this is one of the most diverse vertebrate faunas known from the entire Antarctic fossil record, with over 30 taxa assigned to four major groups of jawed fishes (the placoderms, acanthodians, chondrichthyans and osteichthyans) and one major group of agnathans (primitive jawless vertebrates). The biogeographic affinities of the Aztec fauna are strongest with those of eastern Australia, and a distinctive East Gondwana faunal province can be recognized through the Early-Middle Devonian (Young 1990). It is possible that representatives from two of the groups, the placoderms and acanthodians, may be regarded as vertebrate equivalents of the Malvinokaffric Province (Young 1989).

Devonian plant fossils (including both macroplants and spores) are known from at least four localities in the TAM, as well as from Marie Byrd Land (Edwards 1989; Playford 1989). The most prolific and best preserved assemblage comes from a locality in the latter area and includes taxa such as *Drepanophycus*, *Haplostigma* and ?*Protolepidodendron*. It can be concluded that Antarctic Middle-early Late Devonian floras were dominated by herbaceous lycopods and were linked closely to those known from Argentina, the Falkland Islands and South Africa (Edwards 1989). Contemporaneous spore assemblages appear at first sight to have been cosmopolitan, although there are some indications of endemism at species level (Playford 1989).

During the succeeding Early Carboniferous period, Gondwana continued to move across the south pole and, as it did so, rotated clockwise so that its eastern regions came to lie in increasingly higher latitudes. By approximately the mid-Visean epoch (340 Ma), the northern fringes of the supercontinent had fused with the major northern landmass of Laurasia to form the embryonic megacontinent of Pangea (Hurley and Van der Voo 1987; Veevers 1988) (see below; Fig. 4). Initially, this collision is thought to have led to a period of high-latitude warming during which cosmopolitan faunas and floras flourished (Raymond et al. 1989). However, by the mid-Carboniferous (Namurian) there is evidence of widespread glaciation in both southern South America and eastern and central Australia (Powell and Veevers 1987). A cool-water, marine Gondwana province, identified in both Argentina and Australia, counterbalanced a contemporaneous northern high-latitude (Siberian) one (Runnegar and Campbell 1976).

The very poor Carboniferous sedimentary record of Antarctica is due to this being a period of intense uplift and glaciation over much of the continent. At the glacial culmination, in the latest Carboniferous-earliest Permian (300–290 Ma), a vast ice sheet (or, more likely, series of interconnected ice sheets) covered southern South America, South Africa, India, Antarctica and Australasia (e.g. Powell and Veevers 1987, fig. 2). It had an areal extent far greater than that of the Pleistocene one at its maximum, and it is clear that, although glaciation was polar in Antarctica and Australia, it ranged from 60°–20° S over the rest of Gondwana.

Late Carboniferous(?) – Permian glacial, periglacial and postglacial sedimentary environments are recorded in Beacon Supergroup sequences such as the Victoria Group of the TAM and Amery Group of the Prince Charles

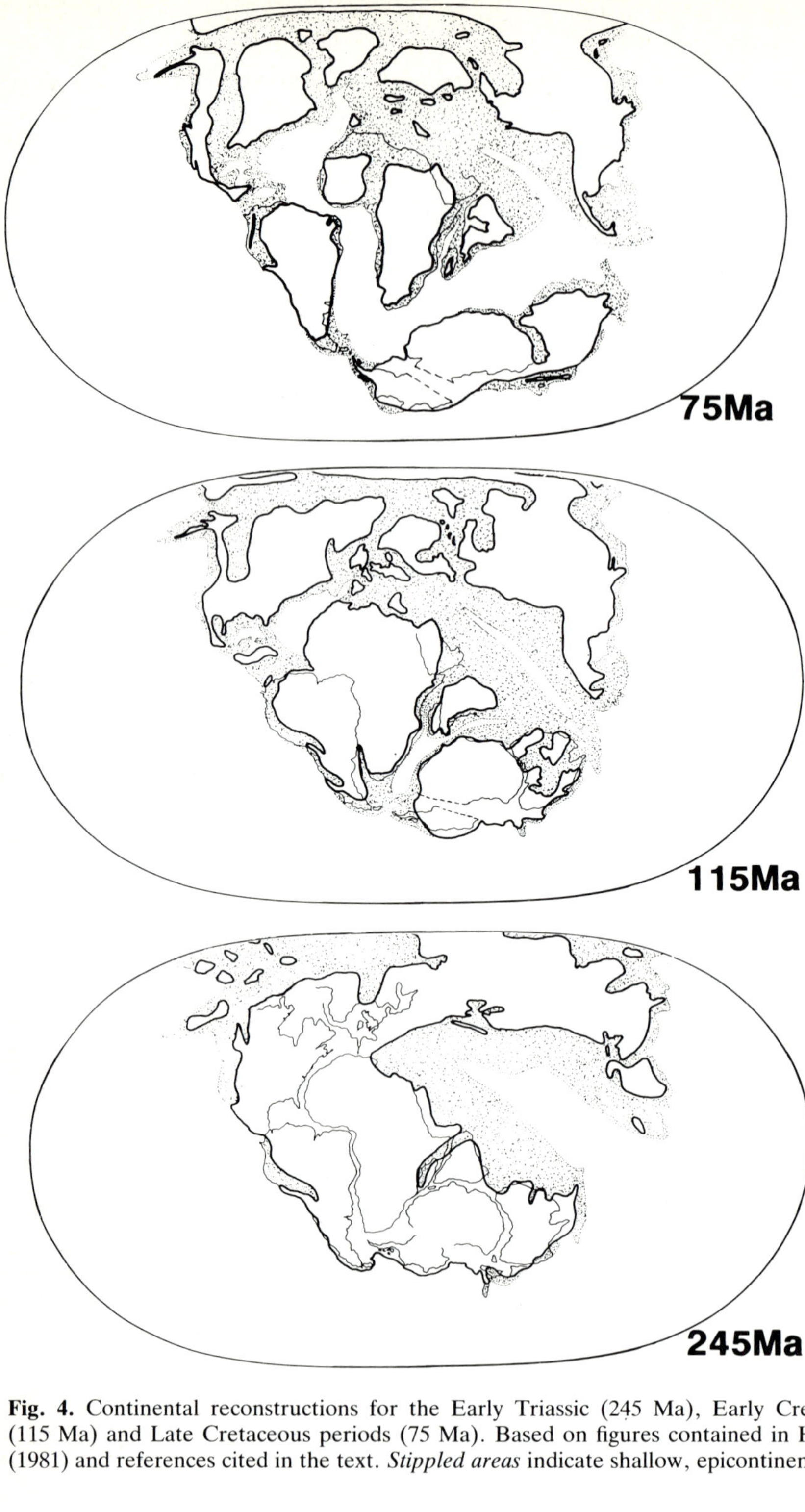

Fig. 4. Continental reconstructions for the Early Triassic (245 Ma), Early Cretaceous (115 Ma) and Late Cretaceous periods (75 Ma). Based on figures contained in Howarth (1981) and references cited in the text. *Stippled areas* indicate shallow, epicontinental seas

Mountains (e.g. Collinson 1991). These rock units are of particular importance for, in conjunction with a number of other key Gondwana localities, they demonstrate a major change in terrestrial vegetation. It was at this time that a number of older (Carboniferous) plant groups became extinct and were replaced by a variety of new forms, the most important of which was the Glossopteridales (Archangelsky 1989). Early Permian floras were distinctly latitudinally zoned, with a tropical flora (0–30° S), a temperate one (30°–60° S), comprising a mixed-*Glossopteris* assemblage, and a glossopterid-dominated polar one (Archangelsky 1989; Ziegler 1990).

The Permian *Glossopteris* flora is one of the most characteristic fossil assemblages of the Gondwana supercontinent. It has been found on all the component continents, and its occurrence at certain African localities and in Turkey, with Cathaysian and Euramerican taxa, suggests strongly that it spread northwards across Pangea (Archangelsky 1989). The most conspicuous elements of the flora were deciduous trees (glossopterids) and herbaceous sphenophytes; it is likely that these types grew together in swamp-forest communities. During the later Permian, *Glossopteris* (and its close allies) remained the most characteristic floral element of cool- and cold-temperate south polar environments. Glossopterids were counterbalanced in northern hemisphere high-latitude localities such as Siberia and Kazakhstan by cordaitids of the Angaran Realm. Although unrelated, these two major groups show a number of similar ecological adaptations (Chaloner and Lacey 1973; Ziegler 1990).

2.2 The Mesozoic Record

Intermittent subduction of Pacific Ocean seafloor beneath the Antarctic craton continued through the Late Palaeozoic era and into the Early Mesozoic (<245 Ma) (Fig. 4). The line of subduction was marked by an immense volcanic arc which seems to have stretched from what is now Marie Byrd Land, through Thurston Island, to the Antarctic Peninsula and Scotia arc. On the fore-arc (Pacific) margin of this major topographic feature a thick pile of deep-sea sediments accumulated as they were scraped off the subducting plate. Collectively, they formed a complex rock unit, known as an accretionary prism, which is exposed today in the Le May Group of Alexander Island, the Trinity Peninsula Group of the northern Antarctic Peninsula, the Miers Bluff Formation of the South Shetland Islands and the Greywacke Shale Formation of the South Orkney Islands (Thomson 1983; Storey and Garrett 1985). Subjected to at least low grade metamorphism and in places intensely deformed, accretionary prism lithologies are dominated by feldspathic sedimentary rocks of turbiditic origin, together with subsidiary basaltic lavas, tuffs, phyllites and cherts. They are very largely unfossiliferous, but radiolarians have been obtained from the Greywacke Shale Formation and these suggest a Late Triassic age (Dalziel et al.

1981). Shallower-water sandstones and shales of the Legoupil Formation (Trinity Peninsula Group) have yielded an early-middle Triassic bivalve fauna which includes *Bakevelloides* and *Neoschizodus* (Thomson 1975). More extensive shallow marine Triassic faunas from New Zealand, New Caledonia, Papua New Guinea and Argentina indicate that a distinct Maorian Province can be traced within the southern high latitudes (Fig. 3i; Stevens 1989).

On the cratonward side of the Triassic orogenic belt, sedimentary rocks of the Beacon Supergroup continued to accumulate in a huge foreland basin running the length of the TAM (Collinson 1991). In the Beardmore Glacier region, a thick sequence of terrestrial sediments was deposited by a major fluvial system draining both the orogen and the Antarctic craton. The Lower-Middle Triassic Fremouw Formation, comprising interbedded braided stream channel sandstones and finer grained flood plain deposits, is famous for its large and varied vertebrate fauna. Near the base, the *Lystrosaurus* fauna has yielded synapsid (mammal-like) reptiles such as *Lystrosaurus*, *Myosaurus*, *Thrinaxodon* and *Ericiolacerta*, together with other reptile groups, amphibians and thecodonts (an order ancestral to the dinosaurs) (Hammer 1989). *Lystrosaurus*, perhaps the most successful of all the mammal-like reptiles, provides firm stratigraphic correlations with the Early Triassic (Scythian) of both South Africa and India. However, it is now apparent that it may also have occurred as far north as 30° N, and it is likely that at least some faunal dispersal was possible across the equatorial barrier. A younger *Cynognathus* fauna, from the upper Fremouw Formation, includes the large carnivorous cynodont reptile, *Cynognathus*, a large kannemeyeriid and three types of temnospondyl amphibian (Hammer 1989).

Fossil plants, ranging in size from tiny rootlet structures to giant silicified logs, occur throughout the Fremouw Formation and its stratigraphical equivalents. In addition, the uppermost levels of the unit have yielded some exquisitely preserved silicified peats and these may prove to be an invaluable source of new information on the early Mesozoic botanical record of the high southern latitudes (Taylor et al. 1986). It is now apparent that, by the end of the Palaeozoic era, the glossopterid flora had been reduced to a few isolated relicts and was replaced in the earliest Triassic by a flora of transitional affinities (Schopf and Askin 1980). However, evidence from fine-grained floodplain deposits and coal seams in the uppermost Fremouw and succeeding Falla formations indicates that a uniform *Dicroidium* flora was present in Antarctica by the latest Early Triassic (Collinson 1991). *Dicroidium* is another of the classic Gondwana fossils (Veevers 1988).

Accretionary prism rocks continued to accumulate along the western margin of at least part of the Antarctic Peninsula during the Jurassic and Cretaceous periods. Associated with them, in the Lully Foothills of central Alexander Island, is a sequence of volcaniclastic sedimentary rocks and lava flows that represents an exotic seamount. The sedimentary rocks have yielded a fauna which includes the ammonite *Epophioceras*?, starfish *Pro-*

tremaster, gastropods, bivalves and solitary corals. The assemblage, of probable Sinemurian age (200 Ma), constitutes the first proven Early Jurassic marine fauna from Antarctica (Thomson and Tranter 1986). Within the foreland basin, a significant influx of volcanic tuffs and volcaniclastic debris in the latest Triassic rocks presaged a major phase of volcanic activity in the Early and Middle Jurassic. At this time voluminous flows of tholeiitic basalts were emplaced along the length of the TAM (Kirkpatrick Basalt Group), as well as in both the Theron Mountains and western Queen Maud Land. These flows were intimately associated with mafic intrusive sills and dykes (Ferrar Dolerite Group), as well as the huge layered gabbro of the Dufek intrusion (Pensacola Mountains) (e.g. Elliot 1991).

The major outpourings of continental flood basalts and related mafic intrusive rocks can be related directly to the disintegration of the Pangean supercontinent. Extensive continental rifts, initiated at approximately the Middle-Late Triassic boundary (230 Ma), propagated extensively through the Early and Middle Jurassic (208-157 Ma) (Veevers 1989). Rifting was particularly prominent in the Gondwana regions with two of the most important rifts being between East and West Gondwana (i.e. between Africa and India-Antarctica), and East and West Antarctica (the Transantarctic Rift of Schmidt and Rowley 1986). A global sea level rise accompanied these major tectonic changes and, for the first time, the Antarctic Peninsula region took on the form of a volcanic archipelago.

On both flanks of the Antarctic Peninsula volcanic arc, extensive sedimentary basins developed through the Jurassic period. The earliest of these was probably that on the south-eastern margins that was to become the site of accumulation of the Middle-Late Jurassic Latady Formation. Sedimentary rocks comprising this unit were deposited in a comparatively broad, shallow shelf environment; Middle Jurassic (Bajocian) fossils collected from them include the ammonites *Megasphaeroceras* and *Teloceras*, and bivalves such as *Oxytoma*, *Entolium*, *Pholadomya* and *Pleuromya* (Quilty 1983). Sedimentation continued in this area into the Late Jurassic, when a somewhat narrower and deeper basin became more prominent on the opposite (western) flank of the peninsula. This was to become the site of accumulation of the Late Jurassic–Early Cretaceous Fossil Bluff Group of eastern Alexander Island (e.g. Butterworth and Macdonald 1991). At this time much of the northern end of the peninsula was blanketed by deep-sea radiolarian-rich muds that formed part of a huge anoxic basin within the proto-South Atlantic region.

Late Jurassic fossils common to all the foregoing regions include ammonites such as *Virgatosphinctes*, *Aulacosphinctoides* and *Kossmatia*, and various representatives of the belemnite family Belemnopseidae. However, of particular interest are certain representatives of the benthic marine invertebrate fauna from the two latest (Kimmeridgian and Tithonian) Jurassic stages. Bivalves, for example, include genera such as *Jeletzkiella* and *Malayomaorica*, which are apparently unique to the southern high latitudes, and others such as *Retroceramus*, *Anopaea*, *Buchia* (Fig. 3e) and *Arctotis*, which form striking cosmopolitan and bipolar distributions (Crame 1986, 1993a). Contrary to the widely held belief that the Jurassic was a period of low marine faunal provinciality, it is now becoming increasingly apparent that temperate faunal provinces were at least intermittently developed in both polar regions (Damborenea and Mancenido 1992; Crame 1993a). With the extension of rift zones and formation of new equatorial seaways, Gondwana began to be isolated once more towards the bottom end of the earth (Fig. 4).

It is possible to trace the Antarctic marine fossil record into the Early Cretaceous period (145-97 Ma) using strata exposed in sedimentary basins flanking both the Antarctic Peninsula and the Scotia arc. Prolific faunas, from localities such as Alexander Island, the James Ross Basin, the South Shetland Islands and South Georgia, have again yielded a range of invertebrate taxa with austral affinities (Fig. 3d). These are perhaps strongest in the Aptian-Albian stages (124-97 Ma), which also appear to have been another time of pronounced bipolarity. The oxytomid bivalve *Aucellina* shows a striking bipolar pattern at this time, and it is likely that the phenomenon is also demonstrated by certain other bivalves, brachiopods and gastropods (Crame 1986, 1993a, and references therein). The same localities, together with certain offshore drill and dredge sites, have also yielded a diversity of plant macro- and microfossils (Truswell 1989). Collectively, these suggest that, throughout the Late Jurassic–Early Cretaceous, much of the Antarctic continent was covered by open-canopied rainforests comprising podocarp and araucarian conifers, associated with ginkgos, taeniopterids and bennettitaleans, and an understorey of ferns, lycopods and bryophytes (Jefferson 1983). By the Aptian-Albian, it is apparent that there was a strong latitudinal gradient between the austral podocarp-araucarian dominated assemblages of Antarctica and Australasia, and the northern Gondwana cheirolepidiacean ones of southern South America and the Falkland Plateau. Several lines of evidence are now indicating that cool-temperate, moist climates prevailed in Antarctica throughout the Early Cretaceous (Dettmann 1989; Crame 1992a).

Intensive studies in the back-arc basin exposed in the James Ross Island area have demonstrated the presence of a nearly complete sequence of Late Cretaceous sediments (97-65 Ma). Indeed, this is proving to be one of the most important Cretaceous successions in the entire southern hemisphere and is yielding a constant stream of important new information about life in the high latitudes at the end of the Mesozoic era. Austral marine invertebrate taxa are again present, being particularly well represented in Santonian-Maastrichtian strata (86-65 Ma) by types such as kossmaticeratid ammonites (Fig. 3f), dimitobelid belemnites, bivalves (e.g. *Lahillia*, *Nordenskjoldia*), aporrhaid gastropods and planktic foraminifera (e.g. *Heterohelix*, *Hedbergella* and *Rugoglobigerina*) (Crame 1992a and references therein). Such occurrences are augmented by the presence of a characteristic *Isabelidinium* dinocyst florule (Dettmann 1989).

It should be emphasized that there is also evidence to indicate the presence of a distinctive southern high-latitude fauna of Cretaceous terrestrial vertebrates. A prolific assemblage from the Aptian-Albian Otway and Strzelecki groups of south-east Australia is dominated by small hypsilophodontid dinosaurs, which occur together with theropods such as *Allosaurus* and a probable labyrinthodont (Rich and Rich 1989). The fauna is characterized by a high proportion of both endemic and relict taxa, and it is interesting to note that hypsilophodontids have now been recorded from the Campanian-Maastrichtian of both New Zealand and Antarctica (Hooker et al. 1991). Mammals too, may have been differentiated in the southernmost latitudes during the Late Cretaceous period (Benton 1991).

Both macro- and microfossil evidence indicate that a number of "southern" plants were added to the austral podocarp-araucarian forests through the later parts of the Cretaceous (approximately 83-65 Ma). These include gymnosperms such as *Lagarostrobus* (Huon Pine) and *Dacrydium*, together with a range of early angiosperms. By the mid-Campanian (80 Ma), the latter include *Nothofagus* (Southern Beech) and certain Myrtaceae in the Antarctic Peninsula region; these were joined in the later Campanian-Maastrichtian by representatives of the Gunneraceae, Epacridaceae, Winteraceae and Trimeniaceae (Dettmann 1989; Truswell 1989; Crame 1992a and references therein). By the terminal (Maastrichtian) Cretaceous stage (74-65 Ma), a podocarp-*Nothofagus* rainforest covered much of southern Gondwana. Palaeobotanical evidence suggests strongly that it grew under cool–temperate, highly seasonal climates, and this has been supported by stable isotope palaeotemperature studies on both Foraminifera and Mollusca (Pirrie and Marshall 1990). There is growing evidence of a substantial cooling in global climates at the end of the Cretaceous period.

2.3 The Cenozoic Record

Unlike many other parts of the world, the Cretaceous-Tertiary boundary in Antarctica does not appear to have been marked by a mass extinction (see below). Earliest Cenozoic (Paleocene, 65-56 Ma) biotas do seem, nevertheless, to have been somewhat reduced in diversity, and this is generally inferred to be a response to continued global cooling. Palaeobotanical studies within units such as the Sobral, Cross Valley and La Meseta formations of Seymour Island, and Fildes, Peninsula and Ezcurra Inlet groups of King George Island (South Shetland Islands), indicate that podocarp-*Nothofagus* forests still covered the northern Antarctic Peninsula region during the Paleocene-Eocene (65-35 Ma) (Birkenmajer and Zastawniak 1989). Other angiosperms recorded during this interval include representatives of the Proteaceae, Loranthaceae, Myrtaceae, Casuarinaceae, Ericales (?Epacridiaceae) and Liliaceae, and a steady increase in taxonomic diversity through the Eocene has been linked to progressive climatic amelioration (Truswell 1989). Such is the distinctive nature of the Eocene floras, together with the associated terrestrial vertebrate fauna which included primitive marsupials, that they have been referred to the high-latitude Weddellian Province (Case 1989; see below).

Some insight into the post-Eocene botanical record is provided by a series of ODP and other drill cores in the Ross and Weddell seas, and Prydz Bay (Amery Ice Shelf, East Antarctica). Palynological information from these cores indicates that pockets of rainforest persisted, at least around the continental margins, until approximately the late Oligocene (23 Ma). *Nothofagus* still dominated assemblages, along with other shrubby angiosperms, gymnosperms such as *Podocarpus* and *Dacrydium*, and occasional ferns (Truswell 1989). From this point in time onwards the sedimentary rock record is largely one of glacigenic strata in which the recovery of both macro- and microfossils is usually poor. Nevertheless, there does appear to have been a progressive diminution of plant taxa through the Oligocene and Miocene epochs, and this was almost certainly due to the rapid growth in continental ice sheets. That pockets of vegetation persisted until the late Pliocene (2.5 Ma) has been demonstrated by the spectacular discovery of *Nothofagus* (in the form of beds of leaves, shrubby twigs and pollen) in the Sirius Group of the TAM (86° S; Carlquist 1987).

Seymour Island has been referred to as the "Rosetta Stone" of Antarctic palaeontology, for it has also yielded prolific invertebrate and vertebrate faunas. The middle-late Eocene (and ?early Oligocene) (approx. 42-33 Ma) La Meseta Formation is the principal fossiliferous unit, having provided representatives of the following types: many different bivalve and gastropod molluscs, asteroid and crinoid echinoderms, decapod crustaceans and brachiopods amongst the invertebrates; sharks, teleost fish, turtles, plesiosaurs, birds (including "giant" penguins), marine mammals (including whales) and primitive marsupials amongst the vertebrates (Feldmann and Woodburne 1988). The nature of the molluscan fauna in particular prompted Zinsmeister (1982) to establish a Weddellian Province in the latest Cretaceous-earliest Cenozoic of the southern high latitudes; characteristic taxa include gastropods of the families Struthiolariidae (Fig. 3b) and Aporrhaidae, and bivalves of the Lahilliidae. To these "palaeoaustral" taxa can be added a range of "neoaustral" ones that were known previously only from lower-latitude localities and much later in the geological record. Examples here would included bivalves such as *Eurhomalea*, *Aulacomya* and *Gaimardia*, asteroids such as *Ctenophoraster* and *Zoroaster*, and decapods such as *Lyreidus* and *Chasmocarcinus* (Zinsmeister and Feldmann 1984).

The La Meseta Formation shows the very first compositional elements of the modern marine fauna. These are apparent from the presence of vertebrates such as penguins and whales, and general traits in the molluscan fauna, such as a strong representation of nuculid, nuculanid and limopsid bivalves, and naticid, struthiolariid and buccinid gastropods (Clarke and Crame 1989). Rich assemblages of late Oligocene-early Miocene (25-20 Ma) marine fossils are known from both offshore drillcores and the glacio-marine Polonez Cove and Cape Melville formations of King George Island (e.g. Webb 1990). They comprise types such as bryozoans, brachiopods, bivalves, gastropods, crinoids, ophiuroids, polychaetes and solitary corals, many of which show surprisingly close resemblances to living taxa (Clarke and Crame 1989). However, it is noticeable that these assemblages also contain certain groups, such as the pectinid bivalves (Fig. 3a) and crabs, that are absent from the living Antarctic fauna. Pectinid bivalves are still well represented in Pliocene (5.2-1.6 Ma) shallow marine biofacies preserved at peripheral coastal localities such as Cockburn Island, southern McMurdo Sound, Vestfold Hills and Larsemann Hills. Such occurrences can be taken as further evidence of Pliocene climatic warming (Webb 1990).

3 Synopsis: A Polar Perspective on the History of Life

One of the most important features to emerge from the Antarctic fossil record is that distinctive high-latitude (or austral) biotas can be traced back over very considerable periods of time. Ever since Antarctica, within the Gondwana supercontinent, achieved a mid- to high-latitude position in the southern hemisphere it has been the site of significant faunal and floral differentiation. Such biotas have been detected in the Devonian (the marine Malvinokaffric Province and possible terrestrial equivalents), Permian (*Glossopteris*-dominated assemblages), Triassic (*Lystrosaurus* and *Cynognathus* faunas; *Dicroidium* flora), Jurassic and Cretaceous (austral marine invertebrate assemblages; distinctive terrestrial floras and faunas) periods, Elements of the living Antarctic biota can be traced back to the late Cretaceous-early Cenozoic Weddellian Province.

Such evidence can be added to that garnered from other high palaeolatitude localities, in both arctic and antarctic regions, to suggest that temperate high-latitude biotas are geologically persistent features. Of course, it must be emphasized that they were not necessarily continually present and the climatic gradients they reflect were almost certainly substantially less than those seen at the present day. Nevertheless, it is striking how, even during intervals of so-called greenhouse warmth (such as the Ordovician and Cretaceous periods), the Earth bears signs of distinct high-latitude biotas (Crame 1992a,b).

The temporal persistence of high-latitude biotas in turn poses a number of important questions. First and foremost, where did they come from?; are we seeing here groups of organisms that arose exclusively in the polar regions or, perhaps, taxa that originated elsewhere and subsequently accumulated at the poles? If, as is generally recognized, most polar biotas are of demonstrably lower taxonomic diversity than lower-latitude counterparts, can we assume that polar-equatorial diversity gradients have been a constant feature through time? Finally, how have polar biotas fared during times of global mass extinction? It is questions such as these that are providing much of the current framework for palaeobiological research in the polar regions.

3.1 Origin of Polar Biotas

There has been a tendency in the literature to regard polar biotas as restricted entities that for long periods of time have been separated from the rest of the world. This is particularly so in Antarctica with progressive isolation of the continent (together with Australasia) through the Jurassic and Cretaceous periods leading to a situation where, in the early Cenozoic (60-50 Ma), it was attached by only a narrow volcanic archipelago to southern South America. Final separation occurred in the late Oligocene (25 Ma) with the establishment of the deep-water psychrosphere (Kennett 1977).

However, despite these tenuous physical connections with the rest of the world, it is apparent that Antarctica was in a state of dynamic biotic interchange with lower-latitude regions over very considerable periods of time (Crame 1993b); indeed, it is by no means completely isolated at the present day. To demonstrate what these processes of interchange may have been, and how they could have contributed to the buildup of polar faunas and floras through time, three simple evolutionary models have been devised (Crame 1992b). It should be emphasized that these are by no means the only models that could be established to demonstrate the evolution of polar biotas, nor are they necessarily mutually exclusive; they simply represent three contrasting ways of approaching the problem.

In the first model (Fig. 5A) it is envisaged that most major groups of plants and animals originated in tropical/subtropical regions. In time, a number of these were displaced into higher latitudes and, ultimately, some of them reached the poles. Here they became relicts, and the latter regions can be regarded as refugia for adaptively anachronistic taxa. Classic examples in the marine realm would include the brachiopods, which were widespread in many shallow-water Palaeozoic and Mesozoic assemblages but today have Antarctica as one of their last strongholds, together with the hexactinellid sponges and bivalve genera such as *Limopsis*, *Astarte* and *Thracia* (Crame 1992b, and references therein). Gastropod relicts in the Antarctic and sub-Antarctic include volutes such as *Guivillea* and *Provocator*, and certain large marginellids (Powell 1965). Probable Cenozoic relicts in the Arctic include certain starfish and bivalve taxa in the White Sea, together with three monotypic whale genera: the Greenland whale (*Balaena*), narwhale (*Monodon*) and white whale (*Delphinapterus*) (Briggs 1974).

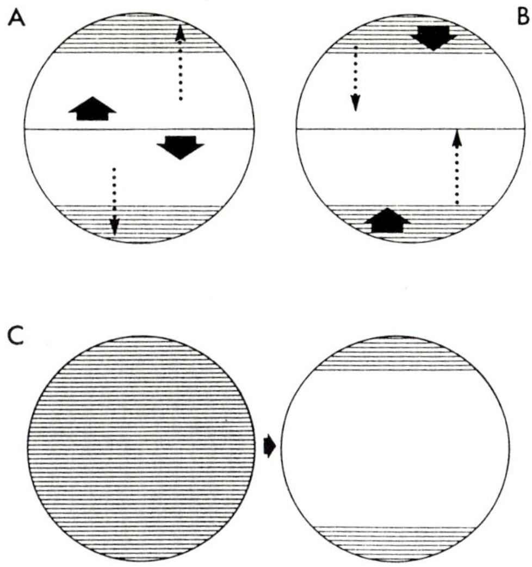

Fig. 5A–C. Evolution of polar biotas: three simple, global models. *Circles* represent the globe, with the *horizontal mid-line* in **A** and **B** being the equator. Polar biotas depicted by *horizontal shading*. (Crame 1992b; figs. 1–3)

There is also a rather striking accumulation of primitive terrestrial plants and animals at the present day in the southern high latitudes (principally southern South America and Australasia). Examples here would include: plants such as lycopods, ferns, araucarians and podocarps; vertebrates such as the Tuatara (*Sphenodon*), frog *Leiopelma*, lungfish *Neoceratodus*, monotreme and marsupial mammals, and ratite birds; and a variety of freshwater and terrestrial insects and other invertebrates (Crame 1992b). It is likely that many of these taxa had Triassic-Jurassic origins in tropical upland regions (Raven and Axelrod 1974). Their gradual dissemination southwards was undoubtedly aided by the presence of the Pangean supercontinent.

The second model (Fig. 5B) is almost exactly the opposite of the first in that it proposes that a number of major taxa arose in the polar regions; in time, some of these then moved out into the lower latitudes. At first sight such a proposition may seem much less likely, but it has to be borne in mind that the polar regions have not always been as they are today. Throughout much of the Mesozoic and Cenozoic eras climates were much milder and there was considerably more habitat space available for both terrestrial and marine taxa. This is particularly so in the Antarctic regions where, throughout the later Mesozoic era (150-65 Ma), remnants of the Gondwana supercontinent were still clustered (Fig. 4).

Strong support for the second model has come from the late Mesozoic-early Cenozoic fossil record of the Antarctic Peninsula region. Recent palaeobotanical discoveries, in particular, are now suggesting that a number of key elements in living southern hemisphere humid and perhumid forests may have originated there (Dettmann 1989). The micropalaeontological record of spores and pollen indicates that such elements include the fern

Lophosoria, the gymnosperms *Dacrydium* and *Lagarostrobus*, and angiosperms such as *Nothofagus*, together with representatives of the Myrtaceae, Proteaceae and Winteraceae (Dettmann 1989; Truswell 1989). These forms all appeared in the Campanian-Maastrichtian stages (83-65 Ma) and disseminated subsequently to lower-latitude localities. In the marine realm too, many of the beautifully preserved marine invertebrate taxa within the Eocene-(?)Oliogene La Meseta Formation of Seymour Island appear to be first occurrences. This essentially shallow-water, nearshore deposit has yielded examples of decapod crustaceans, bivalves, gastropods, echinoderms, brachiopods, barnacles and foraminifera that were only known previously from lower-latitude and deeper-water localities (Zinsmeister and Feldmann 1984). Further evidence that Antarctica was an important source for outer shelf and slope invertebrate taxa has been provided by biogeographical studies on certain living taxa. Iphimediid amphipods and serolid and arcturid isopods may have radiated from Antarctica through the later Cenozoic (Watling and Thurston 1989; Brandt 1991).

In essence, the first two models rely upon some form of dispersal mechanism from a putative centre of origin. As such, they are distrusted by a number of authorities who regard them as inherently untestable; each dispersal event involves a virtually unique set of organisms and circumstances. To such critics, a better way of approaching the problem of the origin of high-latitude biotas may be to assume that there were certain times in the past when the ranges of widespread taxa reached from one polar region to the other. These ranges were subsequently disrupted in the equatorial regions by the appearance of some form of barrier, be it topographic or climatic (or a combination of the two). The resultant disjunct distribution pattern is the classic bipolar one, and the differences between the first two models on the one hand (Fig. 5A,B), and the third on the other (Fig. 5C), reflects the current major dichotomy between the dispersal and vicariance schools of biogeography (Humphries and Parenti 1986; Crame 1992b).

Cladistic analyses of living terrestrial bipolar taxa, such as beech trees, chironomid midges and ratite birds, have been combined by Humphries and Parenti (1986, fig. 4.11a) into a consensus area cladogram. This shows that certain boreal taxa are more closely related to austral ones than those known from any other areas; postulated common ancestors were almost certainly forms that ranged from one hemisphere to the other across Pangea. Within the marine realm, striking bipolar patterns have now been established for the Jurassic (Pliensbachian and Tithonian stages) and Cretaceous (Aptian-Albian) periods, and it is likely that these, too, can be linked to the disintegration of the megacontinent (Crame 1993a). Such is the nature of these patterns as to suggest that climatic and sea level changes were also involved, at least to some extent.

Many living bipolar marine taxa have a fossil record stretching back to the late Oligocene-early Miocene, or even earlier. It is thought that the origin of these forms may be due not so much to tectonic as climatic vicariance. During a long period of latest Eocene to earliest Miocene (approx. 36-22 Ma) global cooling many ancestral taxa achieved widespread distributions, especially within the Pacific Ocean basin. Such distributions are then thought to have been disrupted in low-latitude regions by a late early to early middle Miocene (18-13 Ma) phase of global warming; at this time cool-temperate forms were effectively expelled from the tropics (Valentine 1984). Examples here would include representatives of the buccinid, naticid and littorinid gastropods, certain planktic foraminifera, fish, crustaceans and marine mammals (such as fur and elephant seals) (Lindberg 1991; Crame 1993a). A Pliocene-Pleistocene-Recent phase of bipolarity in both the marine and terrestrial realms may have been due to a combination of dispersal and vicariant events. Particularly important here would seem to have been the progressive closure of the isthmus of Panama (approx. 9.2-3.1 Ma), which not only blocked the flow of Atlantic and Caribbean waters into the Pacific, but also deflected the cool California Current southwards (Lindberg 1991; see also Crame 1993a,b for a fuller account of bipolar events through the Cenozoic).

3.2 Origin of Taxonomic Diversity Gradients

A concept closely related to the foregoing models is that of taxonomic diversity gradients from the equator to the polar regions. Clearly, in many groups of plants and animals, there are demonstrably more types in the low than the high latitudes (e.g. Arthur 1991). There is a general impression too that this may have been so for considerable periods of time, and latitudinal gradients can be demonstrated as far back as the Carboniferous and Permian periods (e.g. Stehli et al. 1969). To many observers there is a direct correlation between species richness and speciation rates; it has been logical to assume that there has been a greater evolutionary turnover in the tropics. The concept of major tropical centres of origin is still widely upheld (e.g. Briggs 1981).

Nevertheless, it is becoming apparent that both theoretical and empirical studies which purport to demonstrate faster evolutionary rates in the tropics are by no means unequivocal (Crame 1992b). It has also been shown that, as our knowledge of the high-latitude fossil record has increased, the temporal persistence of at least some meridional diversity gradients has been brought into question. For example, in the marine realm, living predatory prosobranch gastropods show a striking decrease in numbers of taxa from the low to high latitudes. However, in the mid-late Cretaceous, when the group diversified as part of the so-called Mesozoic marine revolution, a gradient in almost exactly the opposite direction existed (Taylor et al. 1980, text-fig. 11d)! Modern studies are showing too that some shallow, subtidal marine invertebrate assemblages from Antarctica can be just as diverse as those known from much lower latitudes. Taxa such as amphipods, polychaetes, sponges and bryozoans have proliferated in the highly seasonal (in terms of nutritional input) but otherwise extremely constant Antarctic marine environment (Clarke and Crame 1989).

Although there is obviously a need for much more data to substantiate fully these trends, a picture is beginning to emerge of an Earth upon which there may not have been preferred centres of origin. It may well be that, through time, there were equal opportunities for evolutionary innovation within any latitudinal belt. If this is indeed the case, how can we still account for tropical high diversity centres, probably one of the most important biogeographical enigmas on the face of the Earth today?

Of the many explanations of tropical high diversity foci (e.g. Arthur 1991), those of particular interest are the ones with a strong historical component (e.g. Vermeij 1978; Ricklefs 1987). Palaeontological studies are indicating increasingly that many taxa may have contracted into the tropics over considerable periods of time. In a way they may thus be regarded as refugia, although it is important to emphasize that this is in a very different sense from the concept of a "safe haven" for competitively and defensively inferior forms (Vermeij 1987). Repeated range retractions, perhaps in concert with climatic cycles, may also be at least a partial explanation for diversity foci in other latitudinal belts (Crame 1993b).

3.3 Mass Extinctions in the Polar Regions

The fossil record is beginning to provide evidence which suggests that there may also have been a latitudinal gradient in susceptibility to mass extinctions; tropical regions seem to have been much more vulnerable than polar ones (Crame 1992a, and references therein). This is particularly so for the Cretaceous-Tertiary mass extinction (65 Ma), with the James Ross Basin providing important evidence of phased transitions, rather than abrupt replacements, in the southern high latitudes (Zinsmeister et al. 1989). It may well be that higher-latitude communities were in some way buffered against abrupt environmental change; perhaps constant exposure to more severe, seasonal environments enabled them to cope better with sudden environmental stress. It is also likely that the Earth's higher-latitude regions played a fundamental role in supplying replacement taxa to fill ecological vacua left by mass extinctions in tropical communities (Crame 1992a).

4 Conclusions

1. One of the most significant developments in Antarctic Earth science over the last 30 years has been the improvement in our knowledge of the

continent's fossil record. It is now possible, for the first time, to distinguish the broad outlines of the history of life on our southernmost continent. Antarctica has become another valuable reference point for global palaeobiological syntheses.

2. Analysis of the prolific Antarctic fossil record over the last 600 million years shows that Early Palaeozoic (Cambrian) marine fossils have tropical/subtropical affinities, but almost all subsequent faunas and floras are of temperate aspect. Indeed it is becoming apparent that, ever since Antarctica achieved a mid- to high-latitude position in the southern hemisphere, it has been the site of differentiation of distinct austral biotas. These are not necessarily continuously present and it must be stressed that they do not necessarily reflect glacial climates, or anything approaching them. Nevertheless, it is striking how these austral biotas can be detected at regular intervals: the Early-Middle Devonian, Early Permian, Early-Middle Triassic, Jurassic (at least in part) and Cretaceous periods are all times when distinctive faunas and floras characterized Antarctica. The Late Cretaceous-Early Cenozoic Weddellian Province was, especially in the marine realm, a harbinger of today's distinctive living forms.
3. To account for the persistent development of distinct high-latitude biotas over considerable periods of geological time, three simple global models have been developed. In the first of these the polar regions are regarded as refugia; that is, they are sites of accumulation of taxa which have been displaced over time from lower latitudes. This is perhaps the common evolutionary perception of the poles. In the second model, the polar regions are regarded as major centres of origin. This will be an unfamiliar concept to many, but we must not let present day climatic conditions colour our perceptions of the past. In the Antarctic regions in particular, a combination of milder climates and greater habitat space within and around the component Gondwana continents may have acted as a powerful evolutionary stimulus. The third model emphasizes times when the ranges of formerly widespread ancestral taxa were split in the low-latitude regions to form bipolar patterns. A surprising variety of both living and fossil taxa is now known to be bipolar and there is an overriding impression that their formation has been a recurrent phenomenon through Earth history.
4. We are not yet in a position to rank the three evolutionary models in order of importance; there is strong anecdotal support for each of them from within the fossil record. What perhaps is of greater significance at the moment is to emphasize how each model incorporates the dynamic interplay of taxa over a considerable latitudinal range. No longer can we regard polar biotas as simply the product of progressive isolation at the top and bottom of the world. Study of the evolutionary history of the polar regions is shedding light on latitudinal range shifts which may also be of profound importance to our understanding of other global

ecological and biogeographical phenomena. These would include both the origin of latitudinal diversity gradients and regional responses to mass extinction events.

Acknowledgements. I am grateful to many colleagues at BAS for advice and information on a variety of topics included in the text.

References

Archangelsky S (1989) Plant distribution in Gondwana during the late Paleozoic. In: Taylor TN, Taylor EL (eds) Antarctic paleobiology. Springer, Berlin Heidelberg New York, pp 102–117

Arthur W (1991) The bulging biosphere. New Sci 1775:42–45

Bambach RK, Scotese CR, Ziegler AM (1980) Before Pangea: the geographies of the Paleozoic world. Am Sci 68:26–38

Benton MJ (1991) Polar dinosaurs and ancient climates. Trends Ecol Evol 6:28–30

Birkenmajer K, Zastawniak E (1989) Late Cretaceous-early Tertiary floras of King George Island, West Antarctica: their stratigraphic distribution and palaeoclimatic significance. In: Crame JA (ed) Origins and evolution of the Antarctic biota. Spec Publ Geol Soc Lond 47:227–240

Blodgett RB, Rohr DM, Boucot AJ (1990) Early and Middle Devonian gastropod biogeography. In: McKerrow WS, Scotese CR (eds) Palaeozoic palaeogeography and biogeography. Geol Soc Lond Mem 12:277–284

Bond GC, Nickeson PA, Kominz MA (1984) Breakup of a supercontinent between 625 Ma and 555 Ma: new evidence and implications for continental histories. Earth Planet Sci Lett 70:325–345

Bradshaw MA, McCartan L (1983) The depositional environment of the Lower Devonian Horlick Formation, Ohio Range. In: Oliver RL, James PR, Jago JB (eds) Antarctic earth science. Australian Academy of Science, Canberra and Cambridge University Press, Cambridge, pp 238–241

Brandt A (1991) Zur Besiedlungsgeschichte des antarktischen Schelfes am Beispiel der Isopoda (Crustacea, Malacostraca). Ber Polarforsch 98:1–240

Briggs JC (1974) Marine zoogeography. McGraw-Hill, New York

Briggs JC (1981) Do centers of origin have a center? Paleobiology 7:305–307

Buggisch W, Kleinschmidt G (1991) Stratigraphy and facies of the (meta-) sediments in the Shackleton Range, Antarctica. Abstr 6th Int Symp Ant Earth Sci. National Institute of Polar Research, Tokyo, p 91

Butterworth PJ, Macdonald DIM (1991) Basin shallowing from the Mesozoic Fossil Bluff Group of Alexander Island and its regional tectonic significance. In: Thomson MRA, Crame JA, Thomson JW (eds) Geological evolution of Antarctica. Cambridge University Press, Cambridge, pp 449–453

Calquist S (1987) Pliocene *Nothofagus* wood from the Transantarctic Mountains. Aliso 11:571–583

Case JA (1989) Antarctica: the effect of high latitude heterochroneity on the origin of the Australian marsupials. In: Crame JA (ed) Origins and evolution of the Antarctic biota. Spec Publ Geol Soc Lond 47:217–226

Chaloner WG, Lacey WS (1973) The distribution of Late Paleozoic floras. Spec Pap Palaeontol 12:271–289

Clarke A, Crame JA (1989) The origin of the Southern Ocean marine fauna. In: Crame JA (ed) Origins and evolution of the Antarctic biota. Spec Publ Geol Soc Lond 47:253–268

Clarkson PD, Hughes CP, Thomson MRA (1979) Geological significance of a Middle Cambrian fauna from Antarctica. Nature 279:791–792

Cocks LRM (1989) Antarctica's place within Cambrian to Devonian Gondwana. In: Crame JA (ed) Origins and evolution of the Antarctic biota. Spec Publ Geol Lond 47:9–14

Collinson JW (1991) The palaeo-Pacific margin as seen from East Antarctica. In: Thomson MRA, Crame JA, Thomson JW (eds) Geological evolution of Antarctica. Cambridge University Press, Cambridge, pp 199–204

Cox LR (1969) Family Buchiidae Cox, 1953. In: Moore RC (with Teichert C, McCormick L, Williams RB) (eds) Treatise on invertebrate paleontology, pt N, vol 1, Mollusca 6: Bivalvia. The Geological Society of America and University of Kansas, Lawrence, pp N374–N377

Crame JA (1986) Late Mesozoic bipolar bivalve faunas. Geol Mag 123:611–618

Crame JA (1992a) Late Cretaceous palaeoenvironments and biotas: an Antarctic perspective. Antarct Sci 4:371–382

Crame JA (1992b) Evolutionary history of the polar regions. Hist Biol 6:37–60

Crame JA (1993a) Bipolar molluscs and their evolutionary implications. J Biogeog 20: 145–161

Crame JA (1993b) Latitudinal range fluctuations in the marine realm through geological time. Trends Ecol Evol 8:162–166

Dalziel IWD (1991) Pacific margins of Laurentia and East Antarctica-Australia as a conjugate rift pair: Evidence and implications for an Eocambrian supercontinent. Geology 19:598–601

Dalziel IWD (1992) Antarctica; a tale of two supercontinents? Annu Rev Earth Planet Sci 20:501–526

Dalziel IWD, Elliot DH, Jones DL, Thomson JW, Thomson MRA, Wells NA, Zinsmeister WJ (1981) The geological significance of some Triassic microfossils from the South Orkney Islands, Scotia Ridge. Geol Mag 118:15–25

Damborenea SE, Mancenido MO (1992) A comparison of Jurassic benthonic faunas from South America and New Zealand. J R Soc NZ 22:131–152

Davidson G (1992) Piecing together the Pacific. New Sci 1804:25–29

Dettmann ME (1989) Antarctica: Cretaceous cradle of austral temperate rainforests. In: Crame JA (ed) Origins and evolution of the Antarctic biota. Spec Publ Geol Soc Lond 47:89–105

Doumani CA, Boardman RS, Rowell AJ, Boucot AJ, Johnson JG, McAlester AL, Saul J, Fisher DW, Miles RS (1965) Lower Devonian fauna of the Horlick Formation, Ohio Range, Antarctica. Antarct Res Ser 6:241–281

Edwards D (1989) Silurian-Devonian paleobotany: problems, progress and potential. In: Taylor TN, Taylor EL (eds) Antarctic paleobiology. Springer, Berlin Heidelberg New York, pp 89–101

Elliot DH (1991) Triassic-Early Cretaceous evolution of Antarctica. In: Thomson MRA, Crame JA, Thomson JW (eds) Geological evolution of Antarctica. Cambridge University Press, Cambridge, pp 541–548

Feldmann RM, Woodburne MO (eds) (1988) Geology and paleontology of Seymour Island, Antarctic Peninsula. Geol Soc Am Mem 169:1–566

Hammer WR (1989) Triassic terrestrial vertebrate faunas of Antarctica. In: Taylor TN, Taylor EL (eds) Antarctic paleobiology. Springer, Berlin Heidelberg New York, pp 42–50

Harland WB, Armstrong RL, Cox AV, Craig LE, Smith AG, Smith DG (1989) A geologic time scale 1989. Cambridge University Press, Cambridge

Hennig H (1911) Le conglomérat Pleistocène à Pecten de l'Ile Cockburn. Wiss Ergeb Schwed Südpol 3:1–76

Hill D (1965) Archaeocyatha from Antarctica and a review of the phylum. Sci Rep Transantarct Exp 10:1–151

Hoffman PF (1991) Did the breakout of Laurentia turn Gondwanaland inside-out? Science 252:1409–1412

Hooker JJ, Milner AC, Sequeira SEK (1991) An ornithopod dinosaur from the Late Cretaceous of West Antarctica. Antarct Sci 3:331–332.

Howarth MK (1981) Palaeogeography of the Mesozoic. In: Cocks LRM (ed) The evolving earth. British Museum (Natural History), London and Cambridge University Press, Cambridge, pp 197–220

Humphries CJ, Parenti LR (1986) Cladistic biogeography. Clarendon, Oxford

Hurley NF, Van der Voo R (1987) Paleomagnetism of Upper Devonian reefal limestones, Canning Basin, Western Australia. Geol Soc Am Bull 98:123–137

Jefferson TJ (1983) Palaeoclimatic significance of some Mesozoic Antarctic fossil forests. In: Oliver RL, James PR, Jago JB (eds) Antarctic earth science. Australian Academy of Science, Canberra and Cambridge University Press, Cambridge, pp 593–598

Kennett JP (1977) Cenozoic evolution of Antarctic glaciation, the Circum-Antarctic Ocean, and their impact on global paloceanography. J Geophys Res 82:3843–3860

Laird MG (1981) Lower Palaeozoic rocks of Antarctica. In: Holland CH (ed) Lower Palaeozoic rocks of the world, 3. John Wiley, Chichester, pp 257–314

Lindberg DR (1991) Marine biotic interchange between the northern and southern hemispheres. Paleobiology 17:308–324

Moores EM (1991) Southwest U.S. – East Antarctic (SWEAT) connection: a hypothesis. Geology 19:425–428

Pirrie D, Marshall JD (1990) High-paleolatitude Late Cretaceous paleotemperatures: new data from James Ross Island, Antarctica. Geology 18:31–34

Playford G (1989) Proterozoic and Paleozoic palynology of Antarctica: a review. In: Taylor TN, Taylor EL (eds) Antarctic paleobiology. Springer, Berlin Heidelberg New York, pp 51–70

Powell AWB (1951) Antarctic and subantarctic Mollusca: Pelecypoda and Gastropoda. Discovery Rep 26:47–196

Powell AWB (1965) Mollusca of the antarctic and subantarctic seas. In: van Mieghem J, van Oye P (eds) Biogeography and ecology in Antarctica. Dr W Junk, The Hague, pp 333–380

Powell C McA, Veevers JJ (1987) Namurian uplift in Australia and South America triggered the main Gondwanan glaciation. Nature 326:177–179

Quilty PG (1983) Bajocian bivalves from Ellsworth Land, Antarctica. NZ J Geol Geophys 26:395–418

Raven PH, Axelrod DI (1974) Angiosperm biogeography and past continental movements. Ann Mo Bot Gard 61:539–673

Raymond A, Kelley PH, Lutken CB (1989) Polar glaciers and life at the equator: The history of Dinantian and Namurian (Carboniferous) climate. Geology 17:408–411

Rees MN, Rowell AJ (1991) The pre-Devonian Palaeozoic clastics of the central Transantarctic Mountains: stratigraphy and depositional settings. In: Thomson MRA, Crame JA, Thomson JW (eds) Geological evolution of Antarctica. Cambridge University Press, Cambridge, pp 187–192

Rees MN, Pratt BR, Rowell AJ (1989) Early Cambrian reefs, reef complexes, and associated lithofacies of the Shackleton Limestone, Transantarctic Mountains. Sedimentology 36:341–361

Rich THV, Rich PV (1989) Polar dinosaurs and biotas of the early Cretaceous of southeastern Australia. Natl Geogr Res 5:15–53

Ricklefs RE (1987) Community diversity: Relative roles of local and regional processes. Science 235:167–171

Rowell AJ, Rees MN (1989) Early Palaeozoic history of the upper Beardmore Glacier area: implications for a major Antarctic structural boundary within the Transantarctic Mountains. Antarct Sci 1:249–260

Rowell AJ, Rees MN, Braddock P (1987) Silurian marine fauna not confirmed from Antarctica. Alcheringa 11:137

Runnegar B, Campbell KSW (1976) The Palaeozoic faunas of Australia. Earth Sci Rev 12:235–257

Schmidt DL, Rowley PD (1986) Continental rifting and transform faulting along the Jurassic Transantarctic Rift, Antarctica. Tectonics 5:279–291

Schopf JM, Askin RA (1980) Permian and Triassic floral biostratigraphic zones of southern land masses. In: Dilcher DL, Taylor TN (eds) Biostratigraphy of fossil plants. Dowden, Hutchinson & Ross, Stroudsberg, pp 119–151

Scotese CR, Barrett SF (1990) Gondwana's movement over the South Pole during the Palaeozoic: evidence from lithological indicators of climate. In: McKerrow WS, Scotese CR (eds) Palaeozoic palaeogeography and biogeography. Geol Soc Lond Mem 12:75–85

Scotese CR, McKerrow WS (1990) Revised world maps and introduction. In: McKerrow WS, Scotese CR (eds) Palaeozoic palaeogeography and biogeography. Geol Soc Lond Mem 12:1–21

Scotese CR, Bambach RK, Barton C, Van der Voo R, Ziegler AM (1979) Paleozoic base maps. J Geol 87:217–277

Shergold JH, Cooper RA, Mackinnon DI, Yochelson EL (1976) Late Cambrian Brachiopoda, Mollusca, and Trilobita from Northern Victoria Land, Antarctica. Palaeontology 19:247–291

Soloviev IA, Grikurov GE (1979) Newly obtained dates on the distribution of Cambrian trilobites in the Argentina and Shackleton Ranges. In: The Antarctic. The Committee Reports 18. Academy of Sciences USSR, Soviet Committee of Antarctic Research, Nauka, Moscow, pp 54–73 (in Russian)

Stehli FG, Douglas RG, Newell ND (1969) Generation and maintenance of gradients in taxonomic diversity. Science 164:947–949

Stevens GR (1989) The nature and timing of biotic links between New Zealand and Antarctica in Mesozoic and early Cenozoic times. In: Crame JA (ed) Origins and evolution of the Antarctic biota. Spec Publ Geol Soc Lond 47:141–166

Storey BC, Garrett SW (1985) Crustal growth of the Antarctic Peninsula by accretion, magmatism and extension. Geol Mag 122:5–14

Stump E (1991) Pre-Beacon tectonic development of the Transantarctic Mountains. Abstr 6th Int Symp Ant Earth Sci. National Institute of Polar Research, Tokyo, pp 551–556

Taylor JD, Morris NJ, Taylor CN (1980) Food specialization and the evolution of predatory gastropods. Palaeontology 23:375–409

Taylor TN, Taylor EL, Collinson JW (1986) Paleoenvironment of Lower Triassic plants from the Fremouw Formation. Antarct J US 21:26–27

Thomson MRA (1975) New palaeontological and lithological observations on the Legoupil Formation, north-west Antarctic Peninsula. Br Antarct Surv Bull 41 & 42:169–185

Thomson MRA (1983) Antarctica. In: Moullade M, Nairn AEM (eds) The Phanerozoic geology of the world II. The Mesozoic B. Elsevier, Amsterdam, pp 391–422

Thomson MRA, Tranter TH (1986) Early Jurassic fossils from central Alexander Island and their geological setting. Br Antarct Surv Bull 70:23–39

Trechmann CT (1917) Cretaceous Mollusca from New Zealand. Geol Mag 4:294–305

Truswell EM (1989) Cretaceous and Tertiary vegetation of Antarctica: a palynological perspective. In: Taylor TN, Taylor EL (eds) Antarctic paleobiology. Springer, Berlin Heidelberg New York, pp 71–88

Valentine JW (1984) Neogene marine climate trends: implications for biogeographic and evolution of shallow-sea biota. Geology 12:647–650

Veevers JJ (1988) Gondwana facies started when Gondwanaland merged in Pangea. Geology 16:732–734

Veevers JJ (1989) Middle/Late Triassic (230±5 Ma) singularity in the stratigraphic and magmatic history of the Pangean heat anomaly. Geology 17:784–787

Vermeij GJ (1978) Biogeography and adaptation. Harvard University Press, London
Vermeij GJ (1987) Evolution and escalation. Princeton University Press, Princeton, NJ
Watling L, Thurston MH (1989) Antarctica as an evolutionary incubator: evidence from the cladistic biogeography of the amphipod family Iphimediidae. In: Crame JA (ed) Origins and evolution of the Antarctic biota. Spec Publ Geol Soc Lond 47:297–313
Webb PN (1990) The Cenozoic history of Antarctica and its global impact. Antarct Sci 2:3–21
Webers GF (1982) Upper Cambrian molluscs from Northern Victoria Land and their stratigraphic implications. In: Craddock C (ed) Antarctic geoscience. University of Wisconsin Press, Madison, pp 635–638
Young GC (1989) The Aztec fish fauna (Devonian) of Southern Victoria Land: evolutionary and biogeographic significance. In: Crame JA (ed) Origins and evolution of the Antarctic biota. Spec Publ Geol Soc Lond 47:43–62
Young GC (1990) Devonian vertebrate distribution patterns and cladistic analysis of palaeogeographic hypotheses. In: McKerrow WS, Scotese CR (eds) Palaeozoic palaeogeography and biogeography. Geol Soc Lond Mem 12:243–255
Ziegler AM (1990) Phytogeographic patterns and continental configurations during the Permian period. In: McKerrow WS, Scotese CR (eds) Palaeozoic palaeogeography and biogeography. Geol Soc Lond Mem 12:363–379
Ziegler AM, Scotese CR, McKerrow WS, Johnson ME, Bamback RK (1979) Palaeozoic paleogeography. Annu Rev Earth Planet Sci 7:473–502
Zinsmeister WJ (1982) Late Cretaceous-Early Tertiary molluscan biogeography of the southern circum-Pacific. J Paleontol 56:84–102
Zinsmeister WJ, Feldmann RM (1984) Cenozoic high latitude heterochroneity of southern hemisphere marine faunas. Science 224:281–283
Zinsmeister WJ, Feldmann RM, Woodburne MO, Elliot D (1989) Latest Cretaceous/earliest Tertiary transition on Seymour Island, Antarctica. J Paleontol 63:731–738

Geological Contributions from Antarctica

Franz Tessensohn[1]

1 Introduction

This volume contains several other geoscientific papers. The main aspect of this contribution is a discussion of the role which geoscientific research in Antarctica plays in a global context:

- What *general* results does it yield – apart from the *local* inventory of geologic features?
- What does it contribute to the solution of global geoscientific problems, e.g. plate tectonics or climate research?
- What specific features of the Antarctic are relevant to the "outside"?
- What are the reasons that this continent is such an interesting and rewarding object for geoscientific studies?

Many geoscientific studies in Antarctica are based on some rather basic premises, e.g. that Antarctica

- forms
  - one of seven major lithospheric plates which at present cover the surface of the globe;
  - one of six continents of the Earth;
  - one of five major segments of the former supercontinent Gondwana;
- holds
  - localities of a rather unique geological significance, e.g.
  - the area of old crust in Enderby Land;
  - the fossil key locality of Seymour Island;
  - the glacial beds close to the South Pole which contain plant and wood relics or
  - the lake of lava in the active volcano of Mt. Erebus;
- provides
  - good opportunities to study plate tectonic features in certain key areas, e.g.

[1] Bundesanstalt für Geowissenschaften und Rohstoffe, Stilleweg 2, D-30655 Hannover, Germany

Hempel (Ed.) Antarctic Science

  - continental accretion in the cordilleran-type Andean chain or
  - continental split in the Ross Sea rift.
- contains
  - one of the best climatic archives of the past, preserved for different periods in different media with different time scales, e.g.
  - in inland ice and lake sediments for the last 200 000 years,
  - in marine sediments for the last several million years (comprising the whole "ice age") and
  - in the terrestrial Gondwana sediments for some hundreds of million years.

Opportunities to study these features within the ice-covered continent are found where rocks are accessible, e.g. along the rocky coasts or on inland mountains above the ice, sometimes in icefree oases surrounded by ice, and also in marine sediments offshore.

2 Antarctica in the Global Geo-Inventory

The Antarctic plate is one of seven major lithospheric plates (Fig. 1). It forms the end product in a long process of development which started when the present southern continents were united in a huge landmass, Gondwana. When this landmass disintegrated, the continental fragments moved away from each other. Simultaneously, new oceans formed between them. In fact, this process which eventually led to the formation of most of the present oceans is one of the fundamental steps in the evolution of the present world.

The Antarctic plate which comprises the Antarctic continent and the adjacent parts of the Southern Ocean (Fig. 2) is almost completely surrounded by mid-oceanic ridges and thus, together with Africa, forms a central stationary setting the study of which may give insights into the nature of the convection cells driving the plates.

Although at present largely covered by ice, Antarctica is a true continent as indicated by the presence of thick continental crust under the ice, or more obviously, by the presence of rocky coasts and interior mountain ranges. With a diameter of almost 4000 km Antarctica is larger in size than Australia. Knowledge of its geological structure and history is therefore an essential component for understanding the Earth's structure and history.

Antarctica was the central piece of the supercontinent Gondwana (Fig. 3) which contained all the present southern continents plus parts of India for at least 300–400 million years. Gondwana contains records of the early floral evolution and of the development of the first land animals. Antarctica is of prime importance for the spatial reconstruction of this huge landmass, for it formed the centre of Gondwana and was in contact with all other continental Gondwana fragments (Krynauw et al. 1991).

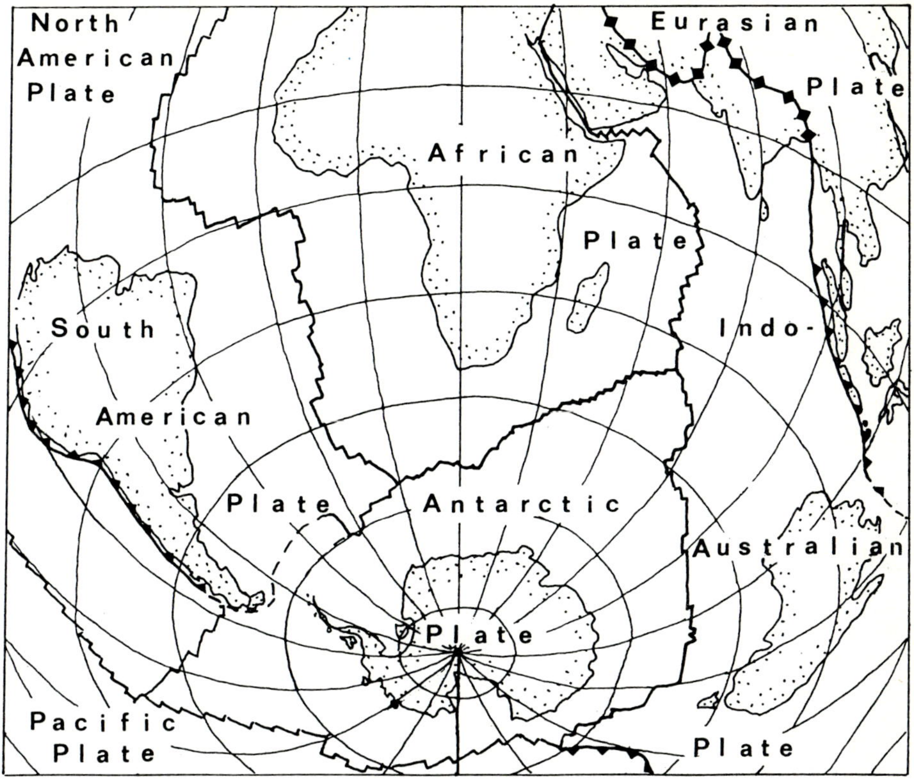

Fig. 1. Antarctic plate within the global framework

3 The Antarctic Climatic Record

But not only these features are of interest to the geosciences. Gondwana also contains some of the best climatic records of the past with evidence for an enormous glaciation within a major period of plant development and coal formation. Some *170 million years* are documented in the Gondwana sediments which are well preserved in Antarctica. This is an important data source for the study of long-term climatic developments associated with a major glaciation.

More recent climatic variations like the glaciations during the last *25 million years* (at least) and interglacial and postglacial changes are recorded in the youngest Antarctic sediments, onshore and offshore (Webb 1990, 1991; Barrett 1991). These sediments can best be studied in drill cores. The first cores from the Ross Sea (Hayes and Frakes 1975) have already provided interesting new information as there was ample evidence that the Antarctic

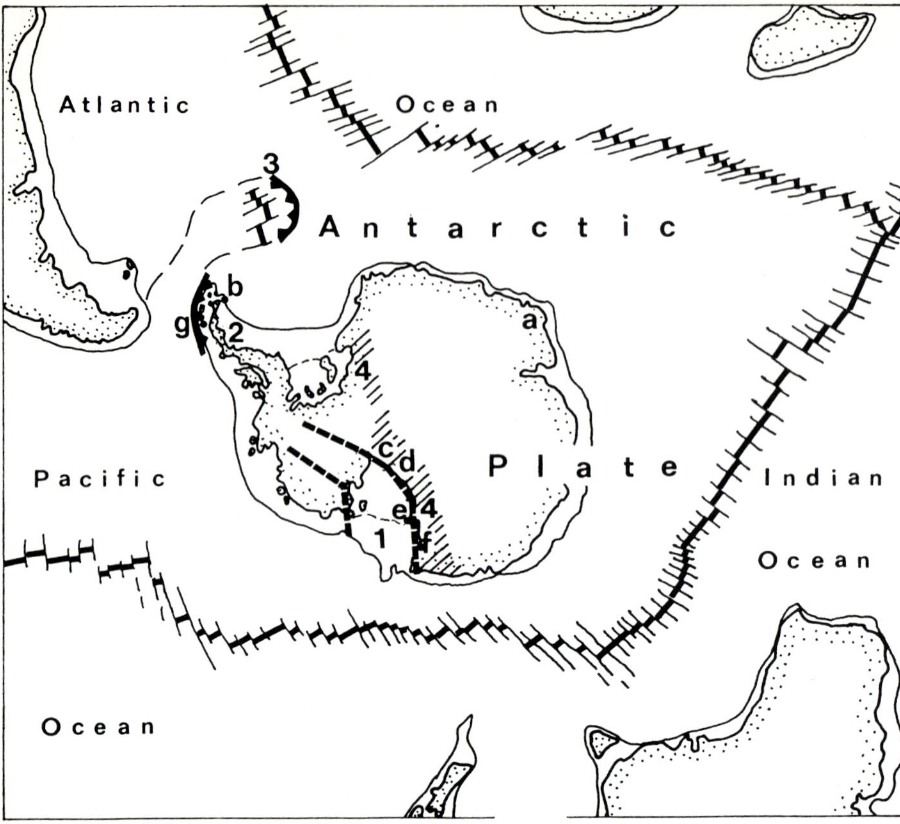

Fig. 2. Map of the Antarctic continent within the Antarctic plate, showing key localities and plate tectonic features discussed in the text; *1* Ross Sea Rift; *2* Antarctic Peninsula; *3* Scotia Arc; *4* Ross Foldbelt. *a* Enderby Land; *b* Seymor Island; *c* Central Transantarctic Mountains; *d* Beardmore Glacier; *e* Mt. Erebus; *f* Mt. Melbourne; *g* Deception Island

glaciation reached much further back in time than previously assumed, especially as compared to the northern hemisphere glaciations. Leaves of beech trees (*Nothofagus*) found in sediment cores drilled from the fast ice in the Ross Sea (Barrett 1991) provided new evidence, together with the sensational plant and wood discoveries of the Beardmore Glacier (Webb et al. 1984, 1987; Harwood 1985) for a fluctuating, discontinuous process of glaciation rather than a continuous one. In warmer intervals just before the last glaciation there were trees growing only 150 km north of the South Pole (see below).

On an even smaller scale, information on the last *several thousand years* is contained in Antarctic lake sediments (Lyons et al. 1985) and for a longer period in the ice of the interior of the continent (Oeschger, this Vol.). The study of the natural climatic processes documented here can provide a

Fig. 3. Gondwana, the vast continental landmass of the past, with Antarctica in the centre

background against which one may rate the changes induced by human impact.

4 Plate Tectonic Features of Antarctica

Plates are rather large and comparatively rigid segments forming the solid outer shell of the Earth. Plate tectonics explain the dynamic system of growth and destruction of these plates which is supposed to be driven by convection cells in the mantle. The present plates are explained in the first place but then also the geologically ancient ones. The main processes in this worldwide system take place at the plate margins as demonstrated, for example, by the concentration of earthquakes and volcanic eruptions along these boundaries.

Continents and oceans grow and become destroyed at these boundaries. All the major mountain chains are formed along plate boundaries. Although constructive and destructive processes take place contemporaneously in different regions, there is a clear cyclicity in the system. This is particularly manifest in birth, growth and destruction of the oceans. Such a cycle is conveniently divided into two phases, the first one being mainly extensional, the second one mainly compressional:

1. The split of a continent usually begins with a rift stage which is documented by the existence of a rather pronounced graben or graben system (example East African Rift) accompanied by thinning of the crust and a certain type of volcanism. This can be regarded as the embryonic stage in

the formation of a new ocean. In a second step, further extension leads to the complete split of continental crust and the upwelling of mantle material in the zone of the initial crack (example Red Sea), the juvenile stage of a new ocean.

Step three is usually a fairly long one and characterized by the continuous addition of new volcanic material along the central crack. Through this addition in the centre the ocean grows continuously on both sides of the mid-oceanic ridge (example Atlantic). This period is called the adult stage of an ocean.

Thus the youngest material of the new ocean floor is in or close to the centre, the oldest material close to its margins. These continental margins on either side were initially formed as the flanks of the rift graben. They persist more or less passively (passive margin) until the beginning of the second major phase.

2. At a certain stage of cooling the oldest formed oceanic crust begins to sink and becomes decoupled from the continental passive margin. It is moved under this margin and begins the process of subduction back into the mantle. This triggers a whole series of secondary geological events: The friction between the two plates causes numerous and deep-seated earthquakes. The cumulative thickness of the plates leads to melting of the crust, extensive magmatism and uplift in the form of high mountain ranges. The passive margin becomes active (example Pacific coast of the Americas).

In a typical section across an active margin the following features are encountered from ocean to continent: A deep sea trench where the oceanic plate bends down under the continent, a volcanic arc some distance inland above the zone of melting, a foreland basin between the two receiving a high amount of sediments from the rising mountain chain, and finally a back-arc basin behind this chain where there is local extension and related graben formation.

Eventually, all margins of the ocean become active (example Pacific). In some cases there is no adjacent continental plate. This may lead to subduction of one oceanic plate under another (example Philippines), also accompanied by strong volcanism and uplift in the form of island arcs.

Plate tectonics have become the leading theory for the geosciences, for it can explain most of the major dynamic and magmatic processes on Earth. Plate tectonics are also an important tool for the analysis of older cycles which cannot be directly observed any more. But one can derive the processes from the rock products which they have left behind during the different stages of formation. Plate tectonics finally help to understand the whole evolution of the planet (compare Dalziel, this Vol.).

The theory has been outlined here to some extent so that major research projects in Antarctica can be seen against this global background. Examples of ongoing plate tectonic processes as well as of older cycles can be studied in Antarctica.

4.1 Ross Sea Rift

The Ross Sea rift (Cooper and Davey 1987; Behrendt et al. 1991; Cooper et al. 1991; Tessensohn and Wörner 1991) forms a major tectonic structure (Fig. 2) which splits the stable Antarctic continent on a total length of 3000 km. The deep-reaching cracks in the crust allow magma to rise from the mantle as manifested by widespread and presently active volcanism (LeMasurier 1991), e.g. Mt. Erebus (Fig. 4).

The rift is comparable in size to similar structures on other continents, e.g. the East African rift with its active volcanoes in Kenya and Ethiopia or the Basin and Range province of the western United States. Compared to other rifts the Ross Sea rift is markedly asymmetric, both in the configuration of its up to 10-km deep sedimentary basins and in the shape of its flanks where a rift shoulder is developed on one side only. This rift shoulder, however, is a rather unique feature, for it forms the 3000-km-long and up to 4-km-high Transantarctic Mountains.

Most continental rifts are above sea level. The Ross Sea rift is drowned which in terms of research possibilities has the advantage that its internal structure can rather effectively be studied by marine seismic investigations. This has resulted in a truly international effort. Many countries have contributed to the very systematic investigations in a coordinated grid system of measurements.

Fig. 4. Mt. Erebus, an active volcano on Ross Island, is related to the Ross Sea rift, a major plate tectonic feature of Antarctica

The weak point for the interpretation of the evolution of this rift system during the last 100 to 150 million years is the lack of stratigraphic calibration for the sediments. There are efforts at present to overcome the situation by organizing an international scientific drill project in the Ross Sea.

4.2 Mid-Oceanic Ridges Around Antarctica

Apart from the small areas of the Antarctic Peninsula and the Scotia Sea the Antarctic plate is almost completely surrounded by mid-oceanic ridges (Fig. 2). One aspect of this situation is the fact that the plate cannot drift away from its present position. The only other plate in a similar stationary position is Africa. The stationary setting has kept Antarctica in a polar position for some time and thus may be one factor for the generation of the glaciation(s).

If one regards the development of the mid-oceanic ridge system around the Antarctic in time, it seems to reflect the eventual separation of the latter from the neighbouring continents in a clockwise spiral (Behrendt et al. 1991), beginning with the breakup of Gondwana in the Jurassic and ending with the rifting of the Ross Sea.

There are two triple points with the Indian and Atlantic ridges. The ridge between Australia and Antarctica forms a central symmetrical spreading axis between the mainland of the continents but changes with a pronounced shear component into an asymmetric position between Tasmania/New Zealand and Victoria Land/Marie Byrd Land.

4.3 Antartic Active Margins

As already pointed out, active margins are the sites of plate subduction and consequently of continental growth through subduction-related magmatism and mountain-building. In principle, the whole of West Antarctica has been formed by these processes during the past 500 million years. During several repetitions of the same basic process, the subduction of the heavier Pacific plate under the lighter Antarctic (and earlier Gondwana) plate, the Antarctic continent grew by the addition of several mountain belts. Similar processes occur all around the Pacific as part of the active circum-Pacific belt which is also called ring of fire because of the volcanism associated with it.

One of the best areas to study these processes is the *Antarctic Peninsula*. Some features like the well-developed deep sea trench on the Pacific side and the distribution of earthquake epicentres indicate that subduction may still be going on at present. At the same time, the products of processes which are only slightly older can be studied on the South Shetland Islands (Trouw and Gamboa 1991). The marine passages between the islands allow marine seismic investigations which portray the rock formations down to several kilometres below the sea floor.

Many typical subduction-related features occur in a relatively small region, e.g. a deep sea trench on the Pacific side of the South Shetland Islands, silicic sediments scratched off the seafloor (Elephant Island), blue schist indicating subduction-related high pressure (Elephant Island), the folded sedimentary wedge of the foot of the arc (Livingstone Island), the volcanic arc produced by melting processes in depth (King George Island), and a rifted back-arc basin (Bransfield Strait) in the brittle continental crust.

The volcanic island arc stage is also beautifully documented by the active volcanoes of the South Sandwich Islands. However, the setting is different here from the Peninsula, because two oceanic plates collide: The oceanic crust of the southern Atlantic is subducted under the small Scotia plate.

4.4 Fossil Pacific Margin

The products of former subduction processes are preserved in ancient mountain belts. Three successive episodes are documented in West Antarctica, the Andean, Gondwanide and Ross Foldbelts (Kleinschmidt and Bradshaw 1991). The *Ross Foldbelt* (Fig. 2) forms one of the longest preserved segments of the former active Gondwana margin and follows the Pacific margin of the East Antarctic continental core from the Ross Sea through to the Wedell Sea. Subduction related features preserved in the rock record include submarine volcanism of the volcanic island arc type, high pressure metamorphism, and a granitic magmatic arc (Kleinschmidt and Tessensohn 1987).

This belt was formed about 500 million years ago (Fig. 5). Although oceanic crust of such a high age does not exist in the present Pacific, the evidence derived from the fossil mountain belt allows the conclusion that in fact there must have been a predecessor, a proto-Pacific, at this time. The analysis of ancient or fossil plate tectonic products is thus an important prerequisite for the reconstruction of earlier stages in the evolution of the Earth.

5 Unique Geologic Localities in Antarctica

All continents contain one or the other locality of a rather unique geological significance, and so does Antarctica. The following examples are a rather subjective choice and the list is certainly not complete.

5.1 Old Crust

The early generation of continental crust can be studied in only a few areas of the world. One of these lies in *Enderby Land*, East Antarctica (Fig. 2), where very old crust is preserved (Sheraton et al. 1991). It has an age of

Fig. 5. The Ross Foldbelt of Antarctica. The dark vertical rocks in the lower part are the products of folding in an active margin setting at the ancient Gondwana margin. Light coloured horizontal cover rocks are plant-bearing terrestrial Gondwana sediments, overlain in turn by a dark volcanic sill. Locality: Unconformity Valley, Morozumi Range, North Victoria Land

about 3800 million years (Sobotovich 1976) which is rather high compared to an estimated total age of the Earth of about 4500 million years.

Studies of rather exotic minerals in these rocks indicate that the geochemical, atmospheric and marine environment at this juvenile phase of our planet may have been rather different from the one today. It is rather unique that, apart from this opportunity to study the early evolution of our planet, Antarctica also provides a means to study the early development of our solar system. The clue for these studies are the meteorites which are recovered from Antarctica in large quantities.

5.2 Fossils

The evolution of the Antarctic flora and fauna is discussed by Crame (this Vol.). Fossil evidence comes from strata all over Antarctica, but there are a few localities of quite exceptional importance which are worth mentioning here.

Seymour Island (Fig. 2) discovered by a Swedish expedition at the end of last century has proved to be one of the key fossil localities of Antarctica.

The Cretaceous and lower Tertiary formations exposed on the island are otherwise not well represented on the continent, although both eras may be documented in the offshore sedimentary record. The Seymour Island locality contains marine faunas like ammonites, fish, crabs, lobsters as well as the record of a landbridge between South America, Antarctica and, possibly, Australia as indicated particularly by the presence of dinosaurs and birds (Feldman and Woodburne 1988). The most spectacular discovery on the island, however, was the first mammal ever found in Antarctica, a small marsupial (Woodman and Zinsmeister 1984).

The Beacon sediments in the *Central Transantarctic Mountains* (Fig. 2) are an important source of information for terrestrial life and faunal and floral evolution on the Gondwana continent. These strata contain abundant plants of the Permian and Triassic periods (Schopf 1962; Taylor et al. 1986) and also rather large land animals, e.g. the first amphibians (Colbert 1982; Hammer et al. 1986).

5.3 Glaciations

Rather recently the upper *Beardmore Glacier* (Fig. 2) has become one of the key localities for the analysis of the development of the Antarctic ice cap. Here, about 2000 m above sea level and only 150 km from the South Pole, fossil wood and beech (*Nothofagus*) leaves were found in a glacial deposit, the Sirius Formation (Webb and Harwood 1987). The wood is not petrified, it is still burnable and floats on water. There is evidence that the beech vegetation grew here in situ close to some valley glaciers (Webb et al. 1987). This can only mean that there was not one long continuous glaciation but there were warmer intervals which allowed plants to grow so close to the South Pole.

Additional evidence was found in the fine matrix of the glacial material, for it contained marine microfossils which the glacial ice has transported from some unknown sea basins of interior Antarctica (Webb et al. 1984; Harwood 1985). The age of the fossils indicates that there were several periods of open marine conditions interrupted from other periods without fossil evidence which may mean periods of glacial cover.

Although the interpretation of these glacial deposits is by no means unanimous, the discovery of the site has led to a completely new approach to the analysis of the glacial history of the Southern Hemisphere. A continuous ice coverage of Antarctica from some starting point in the Tertiary to the present date can no longer be taken for granted.

5.4 Active Volcanism

Mt. Erebus (Figs. 2, 4), an active volcano on Ross Island in the Ross Sea, was originally discovered by James Clark Ross on one of the earliest expeditions to Antarctica. The mountain, which is almost entirely covered by ice, is permanently producing steam out of its crater. The crater is partly filled with a lake of molten lava, however no lava eruptions are recorded in historic times. The tectonic setting as well as petrography and geochemistry of the volcanic rocks are directly comparable to some of the famous volcanoes along the East African rift, e.g. Mt. Kilimandjaro or Mt. Kenya.

Another active volcano of the same province is *Mt. Melbourne*, some 400 km north of Erebus (Fig. 2). Mt. Melbourne is dormant at present, but there have been eruptions during the present period of glaciation, for there are layers of ash enclosed in the ice. In addition, there is hot steam coming out of several vents in the summit area. Around these vents there are areas of hot soil which are the habitat for a rather unique plant community, including mosses. This environment has therefore become a specially protected area within the Antarctic Treaty System.

Another, rather well-known, active volcano is *Deception Island* on the southern termination of the Bransfield Strait. Because there is a breach in the rim of the large crater the sea has entered the interior. Because it provides a good shelter and natural harbour, there were several research stations on the island until an eruption buried them under a thick layer of ash. Now the island is one of the main visiting points for tourist cruises.

Not a unique locality but a rather unique feature is the general absence of earthquake activities on the Antarctic continent. The reason for this phenomenon is not known at present.

6 Conclusions

There are three main reasons why Antarctica has to be considered in any global geo-inventory: (1) It is a true continent as indicated by a thick continental crust, a high average elevation, and rocky coasts and interior mountain ranges. (2) It has been the central piece of Gondwana, the supercontinent of the past which contained all present southern continents. (3) The Antarctic plate which comprises the Antarctic continent and the adjacent Southern Ocean is one of seven major plates which make up the solid shell of the Earth.

The sedimentary archives of the Antarctic contain three sets of important climatic informations with a different time-scale and resolution. A long-term climatic record (170 million years) is contained in the Gondwana sediments and includes information on a period of coal formation and on a major glaciation. The medium-term climatic record (25 million years) is preserved in the sediments of the marine basins and informs on the pre-glacial and early glacial climatic periods. Finally, the short-term record in the order of up to 200 000 years can be found in the inland ice and in lake sediments. It provides the history of the recent glacial period and the climatic background for possible human-induced changes.

Within the present setting of plates on the globe, the processes of plate construction and destruction can best be studied in certain key areas, mainly along the plate margins as outlined by earthquake and volcanic activities. Examples of these plate margins are present in Antarctica: continental split in the Ross Sea rift; the mid-oceanic ridges in the oceans surrounding Antarctica; an active island arc in the volcanic South Sandwich Islands; cordilleran-type subduction-related mountains in the Antarctic Peninsula (as part of the presently active circum-Pacific belt).

Apart from local results of scientific exploration, geoscientific studies in Antarctica thus contribute in many aspects to the solution of global geological problems. An important side effect is the fact that different geological "schools" meet, cooperate and compete in Antarctica. This fosters the scientific discussion and leads to the development of new ideas and hypotheses. Many important views in the past had their origin in the Antarctic or found a good reception there, facilitated by the opportunities provided by the Antarctic community.

References

Barrett PJ (1991) Antarctica and global climate change: a geological perspective. In: Harris CM, Stonehouse B (eds) Antarctica and global climate change, Belhaven, London, pp 35–49

Behrendt JC, Le Masurier WE, Cooper AK, Tessensohn F, Trehu A, Damaske D (1991) Geophysical studies of the West Antarctic Rift System. Tectonics 10, 6:1257–1273

Colbert EH (1982) Mesozoic vertebrates of Antarctica. In: Craddock C (ed) Antarctic geoscience. University of Wisconsin Press, Madison, pp 619–627

Cooper AK, Davey FJ (eds) (1987) The Antarctic continental margin: Geology and geophysics of the western Ross Sea. Circum-Pacific Council for Energy and Natural Resources, Earth Sci Ser V 5B Houston, 118 pp

Cooper AK, Behrendt J, Brancolini G, Davey F, Fitzgerald P, Hinz K, LeMasurier W, Ten Brink U, Tessensohn F, Wörner G (1991) Major rifting in the Ross Sea area. SCAR Conference on Antarctic science – Global Concerns, abstr invited poster A 6, abstr vol II, Bremen, p 62

Feldman RM, Woodburne MO (eds) (1988) Geology and paleontology of Seymor Island, Antarctica. Geol Soc Am Mem 169:566

Hammer WR, Defauw SL, Ryan WJ, Tamplin JW (1986) New vertebrates from the Fremouw Formation (Triassic), Beardmore Glacier region, Antarctica. Antarct J US 21-5:24–26

Harwood DM (1985) Late Neogene climatic fluctuations in the southern high latitudes: Implications of a warm Pliocene and deglaciated Antarctic continent. South Afr J Sci 81:239–241

Hayes DE, Frakes LA (eds) (1975) Leg 28. Init Rep Deep Sea Drilling Proj 28, 1017 pp

Kleinschmidt G, Bradshaw JD (1991) Active Pacific Gondwana margin. SCAR Conference on Antarctic Science – Global Concerns, abstr invited poster A 4, abstr vol II, Bremen, p 60

Kleinschmidt G, Tessensohn F (1987) Early Paleozoic westward directed subduction at the Pacific margin of Antarctica. In: McKenzie GD (ed) Gondwana six: structure, tectonics, and geophysics. Am Geophys Union Geophys Monogr 40:89–105

Krynauw JR, Bradshaw MA, Corner B, Pankhurst RJ, Storey BC, Thomson MRA (1991) Antarctica in Gondwana. SCAR Conference on Antarctic Science – Global Concerns, abstr invited poster A 3, abstr vol II, Bremen, p 59

LeMasurier W (1991) Antarctic volcanism in its global context. SCAR Conference on Antarctic Science – Global Concerns, abstr invited poster A 7, abstr vol II, Bremen, p 63

Lyons WB, Mayewski PA, Donahue P, Cassidy D (1985) A preliminary study of the sedimentary history of Lake Vanda, Antarctica: climatic implications. N Z Mar Freshwater Res 19:253–260

Schopf JM (1962) A preliminary report on plant remains and coal of the sedimentary section in the central range of the Horlick Mountains, Antarctica. Rep Inst Polar Stud, Ohio State University, Columbus, 2, 61 pp

Sheraton JW, Tingey RJ, Harley SL, Sandiford MA (1991) Old crust. SCAR Conference on Antarctic Science – Global Concerns, abstr invited poster A 8, abstr vol II, Bremen, p 64

Sobotovich EV (1976) The oldest rocks of Antarctica. Int Geol Rev 18-4:371–388

Taylor TN, Taylor EL, Collinson JW (1986) Paleoenvironment of Lower Triassic plants from the Fremouw Formation. Antarct J US 21-5:26–27

Tessensohn F, Wörner G (1991) The Ross Sea rift system (Antarctica) – structure, evolution and analogues. In: Thomson MRA, Crame CA, Thomson JW (eds) Geological evolution of Antarctica. Cambridge University Press, Cambridge, pp 273–277

Trouw RAJ, Gamboa LAP (1991) Plate interactions off the Antarctic Peninsula. SCAR Conference on Antarctic Science – Global Concerns, abstr invited poster A 5, abstr vol II, Bremen, p 61

Webb P-N (1990) The Cenozoic history of Antarctica and its global impact. Antarct Sci 2, 1:3–21

Webb P-N (1991) Global impacts of Antarctica's past climate and environment. SCAR Conference on Antarctic Science – Global Concerns, abstr invited poster A 2, abstr vol II, Bremen, p 58

Webb P-N, Harwood DM (1987) Terrestrial flora of the Sirius Formation: Its significance for Late Cenozoic glacial history. Antarct J US 22-4:7–9

Webb P-N, Harwood DM, McKelvey BC, Stott LD (1984) Cenozoic marine sedimentation and ice volume variation on the East Antarctic craton. Geology 12:287–291

Webb P-N, McKelvey BC, Harwood DM, Mabin MCG, Mercer JH (1987) Sirius Formation of the Beardmore Glacier region. Antarct J US 22-1/2:8–13

Woodburne MO, Zinsmeister WJ (1984) The first land mammal from Antarctica and its biogeographic implications. J Paleontol 58-4:913–948

Antarctica as a Space Laboratory

D. J. Lugg[1]

Antarctica is unique in that humans only began living there at the beginning of the twentieth century, and are still transient visitors with no "permanent civilization" present in the accepted sense. Those in the early Antarctic groups saw parallels between Antarctica and space. Such is reflected in an entry in Edward Wilson's diary for the 22 May, 1902 (Wilson 1966):

> "It was a wonderful night . . . Sounds carried an immense distance. The stillness was almost uncanny. One could imagine oneself in another dead planet. I could easily imagine we were standing not on the Earth but on the Moon's surface. Everything was so still and dead and cold and unearthly."

The hostile, dangerous and unfamiliar Antarctic environment with its total isolation, cold and photoperiodicity is arguably the most extreme and certainly the most isolated on Earth (Fig. 1). The majority of the small, confined, groups wintering for up to a year, have had to travel great distances to reach their destination, are totally self-sustaining and require complex operations for maintenance. They therefore provide an excellent analogue for groups in Space (Fig. 2), and have stimulated polar veterans such as Paul-Emile Victor to create his now famous drawing of the Polar Space connection (Fig. 3). Smith and Jones (1962) listed the similarities and differences between space and Antarctica and concluded

> "Nevertheless, the antarctic situation, taken as a whole, is about as similar to the astronaut's as we are likely to find on earth".

Those views of 30 years ago and reiterated by Stuhlinger (1969) are true today as, despite modern technology, the unforgiving natural polar environment has not changed and some groups are still totally isolated for nearly a year. The continuous presence changes annually or in some cases biannually. Without taking into account the fact that women are now in many Antarctic communities, the groups of today are quite different in composition from the early expeditions whose staff were from similar backgrounds and interests. More specialists now occupy the polar ranks. In the same 30-year period despite the total number of persons in space being less than an annual overwintering population for the whole Antarctic continent, space

[1] Australian Antarctic Division, Kingston, Tasmania 7050, Australia

Hempel (Ed.) Antarctic Science

Fig. 1. This small, isolated group in the Prince Charles Mountains is typical of those in Antarctica that are among the most isolated on Earth. (ANARE photo – D. Lugg)

personnel have also become more heterogeneous. Stereotyped pilot astronaut/cosmonauts have given way to mission specialists and scientists from a variety of disciplines. Construction workers and support crew are proposed in future space work. This further reinforces the Antarctic-space analogue.

Human studies in Antarctica have changed from the *heroic era* when they were few and limited. The development of these studies has been described by Wilson (1965). The International Geophysical Year (IGY) 1958/59 changed the nature of work in Antarctica from exploration to science. Although the scientific program of IGY did not officially include human biology or medicine, the increased scientific effort and additional populations stimulated research, much of which was still ad hoc. Some of the IGY research was reported in the comprehensive reviews of Edholm (1964), and Wilson (1965) who ended his with a plea for international co-ordination of human studies.

In 1972 a symposium was organized by the SCAR subcommittee on Human Biology and Medicine. In an introduction to the Proceedings, Edholm (Edholm and Gunderson 1973) commented on the

"well planned systematic professional studies in human biology" but felt that "there has not as yet been any general realisation of the significance or opportunities concerning human biology".

Fig. 2. Remote Antarctic populations provide an excellent analogue for groups in space. (ANARE photo – D. Lugg)

Fig. 3. Paul-Emile Victor's 1969 drawing of the Polar-Space connection

Fig. 4. Heat exchange with the environment was studied on the International Biomedical Expedition to the Antarctic (IBEA). (IBEA photo – D. Lugg)

He considered that there was a requirement for a "genuine international collaboration". In 1974 the SCAR subcommittee became a Working Group with the dual roles of research and medical practice. Gunderson's (1974) review shows the nexus between polar clinical medicine and research as well as the diversity and sophistication of the research.

Under the influence of the Working Group human studies have become well planned, systematic and multidisciplinary, and have increased in both volume and standard (Koerner 1982; Fifield 1987; Lugg 1987). One such successful international SCAR-sponsored programme was the International Biomedical Expedition to the Antarctic (IBEA) organized by the Working Group and carried out in austral summer 1980/81 (Fig. 4). This was the first Antarctic expedition organized solely for human biological studies (Rivolier et al. 1988).

What human studies have been performed in Antarctica? In order to answer this question all Antarctic Bibliography abstracts between 1951–1988 were reviewed (Lugg 1991a). Of the total, 1179 or 2.8% were considered to be medical science. When one considers that much of the field research has been carried out by physicians who are responsible for health care and carry out research as a spare-time endeavour, the number of papers produced compare more than favourably with other Antarctic disciplines. A survey of the fields of research of each paper together with the percentages of the total for each field, illustrates the multidisciplinary nature of the research.

Reviewing the topics with regard to time shows the changing nature of the research. Initially, there was basic research on the study of the effects of cold and isolation, with clinical and applied studies on psychology, behaviour, selection, nutrition, and dentistry. As technology and attitudes to research have changed, many more topics of research have been introduced with greatest emphasis now being placed on studies that facilitate living and working in Antarctica, especially as Antarctic environmental factors are not totally reproduceable in temperate laboratories. Recent work by Reed et al. (1990a,b) suggesting links between changing thyroid hormone economy and cold adaptation in personnel in Antarctica, the Polar T_3 Syndrome, may see a revival of cold adaptation studies in Antarctica.

A detailed appraisal of topics of research is beyond the scope of this paper. Reviews of long-term and current national programmes such as those of France (Rivolier et al. 1983), the United Kingdom (Norman 1989) and Australia (Lugg 1991b) give such details. Table 1 lists some of the research currently being performed. Much of this research is providing valuable data which have a relevance beyond Antarctica. A review of Antarctic bibliography (Lugg 1991a) also provides interesting data on countries carrying out medical science. Of the publications 86% has come from six countries and 96% from ten countries. The important factor is, however, that a further 11 countries, most of them just commencing Antarctic science, have produced papers. This augurs well for the future.

In 1987 The Division of Polar Programs of the National Science Foundation (NSF) and the National Aeronautics and Space Administration (NASA) cosponsored a conference entitled "The Human Experience in Antarctica: Applications to Life in Space". From the interest shown and the large attendance it would appear that the overall goal of the organizers, to revitalize and promote behavioural research in space flight, Antarctica,

Table 1. Current human studies in Antarctica

Thermal adaptation
Changes in immune response
Microbiology
Hormone adaptation to cold
Biorhythms
Fitness and health
Cardiovascular studies
Nutrition and energy balance
Photobiology
Epidemiology
Evaluation of stressors
Psycho-social and behavioural adaptation
Neuro and psychophysiological changes
Group dynamics
Sleep

and comparable settings, was achieved. The Proceedings of the Sunnyvale Conference (Harrison et al. 1991) is an important publication, as are the recommendations for agencies and researchers involved in isolated and confined environments. At the Conference the analogies between space and Antarctic research came to the fore and proposals were made for performing research in Antarctica which would be of value in the short term to Antarctic agencies, but in the longer term would be important for those in space. Studies such as those by Stuster (1986), Palinkas (1988), Pierce (1988), Rivolier and Bachelard (1988), and the Simulation Mission Study Group (SIMIS) for the European Space Agency (ESA 1991), confirm that there is an ongoing activity.

In order to facilitate all the disparate work being performed by both space and Antarctic agencies an ad hoc Group on Antarctic Space-related Human Factors Research was established at the Twentieth Scientific Committee on Antarctic Research (SCAR) Meeting in Hobart in September 1988 (SCAR 1989). The aims of this group were to identify and to study those mutual problems faced both in polar regions and space environments with the aim of enhancing the performance, health and safety of people in both settings. The SCAR ad hoc Group met on four occasions between 1989 and 1993. Following collaborative agreements between Antarctic and Space Agencies, Antarctic field research has commenced in austral winter 1993 with further research being planned for winter 1994 and beyond. Figure 5 illustrates the international network for Antarctic medicine and medical research.

By establishing active channels of communication including meetings, and continuing dialogue with NASA Life Sciences Division and ESA, it has been possible to identify the facilities and support and to determine methods and procedures for productive operational research. The projects may involve existing summer and winter station and traverse groups, and potential simulated space stations in Antarctica. Obviously, groups who are in complete isolation would be preferred.

A pilot microbiological study was performed in Antarctica (Fig. 6) by doctors from the United Kingdom, France and Australia. The project aimed to show whether the population structure of the human commensal, *Staphylococcus epidermidis*, is clonal and persistent or whether genetic drift occurs in isolation. Further proposals will be put to future SCAR meetings. This is in accord with NASA's space life sciences strategy for the twenty-first century (Nicogossian and Gaiser 1991) and the use of Antarctica to support the Space Exploration Initiative (SEI) (NASA-NSF 1990).

Having reviewed the substantial human research that has been done in Antarctica, it is relevant to look at the areas that could be studied using Antarctica as a space laboratory. Despite our inability to replicate the weightlessness in space, the approximately half-normal Earth gravity of Mars and other space-specific effects such as radiation, the prospects are excellent for performing human research with benefit to both Antarctic operators and space agencies. Additional benefits are the Mars-like charac-

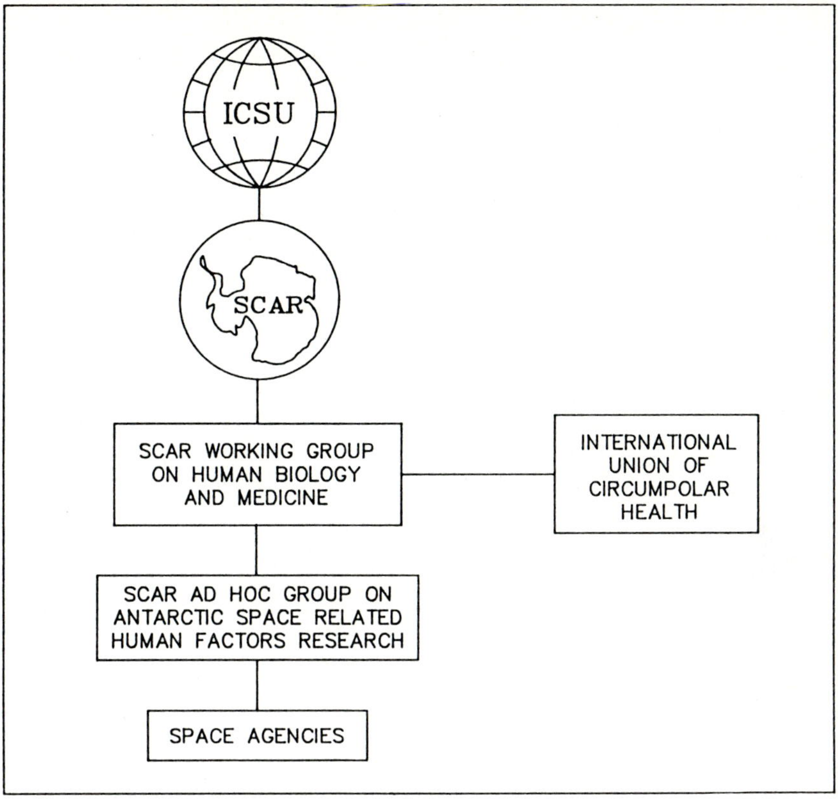

Fig. 5. International network for Antarctic medicine and medical research

teristics of certain Antarctic localities. These include the dry valleys and other areas which resemble the geology of the surface of the red planet as well as its meteorology. Exobiologists have already made significant studies of Antarctic lakes and rocky areas.

The most dramatic and certainly the most discussed areas of research in Antarctica that are seen as important for space are the behavioural and psycho-social adaptations to the isolated and confined living. The sameness of associates and tasks, the small group size, the enforced socialization, and the inability to escape from Antarctica result in the development of "cultures". These differ from group to group and year to year, and defy prediction. All Antarctic groups can describe unusual if not bizarre behaviour and practices (Fig. 7). Such behaviour has lessons for space planners. Will "space culture" mimic that of winterers in Antarctica? Will we see the development of folklore, customs and ceremonials?

The occurrence of psychological/psychiatric problems in Antarctica is relatively low. However, one major case can cause the group considerable difficulty. In space one case could be disastrous. Some Antarctic agencies have considerable experience in psychological screening. Tests for selection

Fig. 6. Microbiological studies in Antarctica. (IBEA photo – D. Lugg)

of space personnel could be used and validated on Antarctic wintering populations. If these were performed at a number of different national stations, there would be additional benefits of cross-cultural studies. This would be an additional benefit as it is likely that future long-term space probes will have participants from many nations. The experience of group dynamics in the Antarctic isolation and confinement, especially the change over the last 15 years from all-male to mixed gender groups, is most relevant. The "retribalization" of returning polar people is an excellent model for returning spacefarers.

The nature of the Antarctic groups make them ideally suited to a wide range of space-related human factors research. Work studies, use of leisure time, ergonomic and design studies, environmental monitoring of quarters, waste disposal, and preventive and occupational medicine could all be studied at existing Antarctic wintering stations. Research on thermal balance and nutrition and energy balance have been and are being studied in Antarctica. These are most pertinent for future groups in space. Critical space life-support systems, such as the growing of food and recycling of water, will need evaluation in isolation. A future prospect may include specific simulated space facilities to do this.

A serendipitous finding in a small immunology project of the International Biomedical Expedition to the Antarctic (IBEA) (Roberts-Thomson et al. 1985) led to follow up immunological studies on wintering personnel (Williams et al. 1986). Assessment of cell-mediated immunity indicates

Fig. 7. All Antarctic groups can describe unusual if not bizarre behaviour and practices. (ANARE photograph)

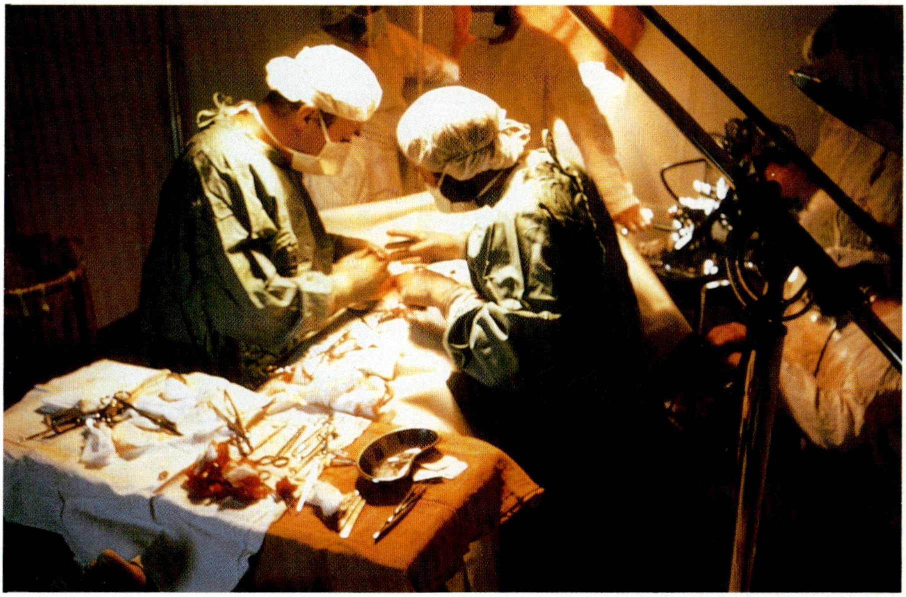

Fig. 8. An appendicectomy being performed at an Antarctic station. (ANARE photo – D. Parer)

decreased cutaneous responses with significant hypoergy and anergy. This may reflect altered immune activity induced by environmental and stress factors. Although no correlations have been found to date with endocrine or psychological measurements made on the same subjects, the interaction of brain, endocrinological functions, and immunological status in Antarctica remains of great importance. Further studies are progressing on the immune suppression, including lymphocyte function, virus reactivation and whether vitamin D levels in expeditioners have any correlation with depressed immune responses.

The vulnerability of the totally isolated polar groups, whose ability to resist infections may be influenced by anxiety, depression or other environmental stressors, is of great concern to those responsible for health care maintenance in Antarctica. This concern is equally true of those planning health care on long-term space flights. The ideal would be that all crew members could be guaranteed free of all medical problems for the duration of the flight. Valuable lessons can be learnt from Antarctica where epidemiological research, although revealing no Antarctic specific health disorder, has helped to identify possible medical processes and the possibility of the occurrence of conditions, such as appendicitis, that could prove troublesome in space. The research reinforces that no guarantees can be given for space with regard to health prediction. Additional to the Antarctic experience are the unknown medical effects of long-term weightlessness and the critical area for space travellers of radiation effects.

The presence of renal stones in Antarctic winterers previously free of such problems has been a great problem to many doctors and agencies alike. Research is needed on the incidence of and specific factors involved in this condition in Antarctica. This is of similar relevance to space as the crew of Skylab had a negative calcium balance resulting from alterations in bone mineral metabolism (Schneider et al. 1989). Such changes in space could lead to renal stone formation. Other pathology could include accelerated osteoporosis, impaired fracture healing and electrolyte changes.

Conditions such as renal stones highlight the decisions to be made on the level of services, equipment and specialist backup to be provided under the health maintenance facility on board the spacecraft. Much of this clinical, preventive and emergency medicine will be similar to that faced by the sole Antarctic medical practitioners and their advisers over the last 40 years in Antarctica. Doctors have diagnosed complex medical conditions, performed craniotomies, laparotomies (Fig. 8) and other surgical operations (Fig. 9) as well as carrying out dental, anaesthetic and laboratory procedures. One doctor (Rogozov 1964) removed his own appendix while he was wintering in Antarctica.

In space such procedures may be limited (Nicogossian and Pool 1989) and NASA Life Sciences Division have recognized that medical care systems on Moon and Mars would be similar to those in Antarctica (Nicogossian et al. 1990), but communications to Earth for diagnosis and discussion on manage-

Fig. 9. A patient with a fractured leg exercises at a nearby penguin rookery. (ANARE photo – D. Lugg)

ment should be superior to those of past years in Antarctica. "Radio blackouts" due to polar cap absorption (PCA) are well known to polar practitioners and fortunately now less of a problem where satellite communications give a more reliable service. It is the area of diagnosis and transmission of test results, despite long space lag time, that will be of paramount importance in space. Antarctic medical practice is seen as an excellent analogue for the development of the space health maintenance facility.

Throughout this conference we have seen Antarctica in the global scene. With changes in the polar ozone layer, monitoring of ultraviolet (UV) radiation in Antarctica and its effects on staff is being carried out by a number of nations. As Mars has no ozone layer UV wavelengths fall unimpeded on the surface. Antarctic monitoring therefore will be valuable as a predictor for humans on the surface of Mars as well as for current workers in Antarctica. Such specific research on the icy wastes of Antarctica may assist the success of space exploration of the Martian polar caps in the twenty-first century.

Today humans are still only transient visitors to Antarctica. In the short time that groups have lived in Antarctica polar technology has evolved along with social and cultural factors, and environmental attitudes; these now being debated worldwide by nations both involved and not involved in Antarctica. I hope that I have shown that human studies of excellence have been performed in Antarctica over a long period, and pointed out the role of SCAR in these, and above all the potential of Antarctica as a space laboratory. Gunderson (1991), the doyen of Antarctic behavioural scientists said in a foreword to the Conference on "The Human Experience in Antarctica: Applications to Life in Space":

> "The Antarctic experience has provided a useful model not only for difficult field behavioural research but also for international and multidisciplinary cooperation".

The statement accurately summarizes the contents of this chapter, as regards not only behavioural but all human research.

This SCAR conference has shown how Antarctic Science is important in relation to global problems. However, little has been said concerning the social impacts of global change. Antarctica as a space laboratory, with collaboration of international space and polar agencies, makes important contributions to social and other human research. Such research which is necessary for space will facilitate optimizing living and working conditions for scientists and others in Antarctica. But there are implications for populations at large, as research necessary for space may yield solutions to some of the health problems we face on Earth.

The use of Antarctica as an analogue is based on its uniquely isolated populations. If these populations become less isolated as was shown recently by the mid-winter flight of a US Hercules aircraft to McMurdo Station to repatriate a patient from the nearby Scott Base, then Antarctica as the best location to support the Space Exploration Initiative will be lost. It is therefore timely that SCAR-sponsored scientific studies on human biology and medicine be organized.

References

Edholm OG (1964) Man and the environment. In: Priestley R, Adie RJ, Robin GdeQ (eds) Antarctic research. Butterworth, London, pp 39–60

Edholm OG, Gunderson EKE (eds) (1973) Polar human biology. Proc SCAR IUPS IUBS Symp of Human biology and medicine in the Antarctic. Heinemann, London, pp 1–443

ESA (1991) Final report of Simulation Mission Study Group (SIMIS). ESA Rep LTPO-SR-91-01, Paris, pp 1–148

Fifield R (1987) International research in the Antarctic. Oxford University Press, Oxford, pp 120–126

Gunderson EKE (ed) (1974) Human adaptability to Antarctic conditions. Antarct Res Ser 22:1–131

Gunderson EKE (1991) Foreword. In: Harrison AA, Clearwater YA, McKay C (eds) From Antarctica to outer space – life in isolation and confinement. Springer, Berlin Heidelberg New York

Harrison AA, Clearwater YA, McKay C (eds) (1991) From Antarctica to outer space – life in isolation and confinement. Springer, Berlin Heidelberg New York, pp 1–410

Koerner FC (1982) Polar medicine – a literature review. Appendix to polar biomedical research – an assessment. National Academy Press, Washington, DC, pp 1–85

Lugg DJ (1987) International co-ordination of Antarctic medical research and medicine. Arct Med Res 46(1):29–34

Lugg DJ (1991a) Humans on Ice: a review of research of those living in Antarctica since IGY 1957–58. In: Postl BD, Gilbert P, Goodwill J et al. (eds) Circumpolar Health 90. University of Manitoba Press, Winnipeg, pp 499–501

Lugg DJ (1991b) Current international human factors research in Antarctica. In: Harrison AA, Clearwater YA, McKay C (eds) From Antarctica to outer space – life in isolation and confinement. Springer, Berlin Heidelberg New York, pp 31–42

NASA-NSF (1990) Use of Antarctic analogs to support the Space Exploration Initiative (SEI). Joint publication NASA-NSF, Washington, DC, pp 1–19

Nicogossian AE, Gaiser KK (1991) The space life sciences strategy for the 21st century. Presented at 9th IAA Man in Space Symposium, Cologne, Germany, June 1991

Nicogossian AE, Pool SL (1989) Medical care and health maintenance in flight. In: Nicogossian AE, Leach Huntoon C, Pool CL (eds) Space physiology and medicine, 2nd edn. Lea and Febiger, Philadelphia, pp 349–363

Nicogossian AE, Sulzman FM, Gaiser KK, Teeter RC (1990) Physiological adaptation to space flight. Presented to American Institute of Aeronautics and Astronautics (AIAA) Space Programs and Technologies Conference, Alabama, Sept 1990

Norman JN (1989) Medical care and human biological research in the British Antarctic Survey medical unit. Arct Med Res 48(3):103–116

Palinkas LA (1988) The human element in space: lessons from Antarctica. Engineering and construction, and operations in space. Aerospace Division ASCE, Albuquerque, pp 1044–1055

Pierce CM (1988) Mental health factors in spaceflight. Aviat Space Environ Med 2:99–101

Reed HL, Silverman ED, Shakir KMM, Dons R, Burman KD, O'Brian JT (1990a) Changes in serum triiodothyronine (T3) kinetics after prolonged Antarctic residence: the polar T3 syndrome. J Clin Endocrinol Metab 70:965–974

Reed HL, Brice D, Shakir KMM, Burman KD, D'Alesandro MM, O'Brian JT (1990b) Decreased free fraction of thyroid hormones after prolonged Antarctic residence. J Appl Physiol 69:1467–1472

Rivolier J, Bachelard C (1988) Study of analogies between living conditions at an Antarctic scientific base and on a space station. ESA Rep, Paris, pp 1–158

Rivolier J, Cazes G, Bachelard C (1983) Summary of the French research in medicine and psychology conducted with EPF and TAAF. TAAF, Paris, pp 1–58

Rivolier J, Goldsmith R, Lugg DJ, Taylor AJW (eds) (1988) Man in the Antarctic. Taylor and Francis, London, pp 1–223

Roberts-Thomson P, Lugg DJ, Vallverdu R, Bradley J (1985) Assessment of immunological responsiveness in members of the International Biomedical Expedition to the Antarctic 1980/81. J Clin Lab Immunol 17:115–118

Rogozov LI (1964) Self operation. Soviet Antarctic Expedition Information Bull 4:223 American Geophysical Union, Washington, DC (English translation)

SCAR (1989) Meeting of SCAR WGHB&M – summary report 20th Meet SCAR, Hobart, Australia, 12–16 Sept 1988. Polar Rec 245:86–87

Schneider VS, Leblanc A, Rambaut PC (1989) Bone and mineral metabolism. In: Nicogossian AE, Leach Huntoon C, Pool Cl (eds) Space physiology and medicine. 2nd edn. Lea and Febiger, Philadelphia, pp 214–221

Smith WM, Jones MB (1962) Astronauts, Antarctic scientists, and personal autonomy. Aerospace Med 33:162–166

Stuhlinger E (1969) Antarctic research, a prelude to space research. Antarct J US 4(1):1–7

Stuster JW (1986) Space station habitability recommendations based on a systematic comparative analysis of analogous conditions (NASA Contractor Report 3943). Anacapa Sciences, Santa Barbara, pp 1–209

Williams DL, Climie A, Muller HK, Lugg DJ (1986) Cell mediated immunity in healthy adults in Antarctica and the sub-Antarctic. J Clin Lab Immunol 20:43–49

Wilson EA (1966) Diary of the *Discovery* expedition to the Antarctic regions 1901–1904. Savours A (ed) Blandford, London, pp 143–144

Wilson O (1965) Human adaptation to life in Antarctica. In: van Mieghem J, van Oye P, Schell J (eds) Biogeography and ecology in Antarctica. Monographiae Biologicae, vol XV. W Junk, The Hague, pp 690–752

Antarctic Benthos: Present Position and Future Prospects*

W. E. Arntz[1] and V. A. Gallardo[2]

1 Introduction

Over a century ago, the first substantial benthic samples taken in Antarctic waters by the staff of the *Challenger* primarily served the purpose of completing the inventory of the world's ocean fauna. This first discovery phase, with a strong emphasis on taxonomy, continued during the first half of the twentieth century, before Antarctic stations were established ashore. The focus of research then shifted to studies of life history, behaviour, and physiology of the benthic fauna in shallow waters. Occasionally, as for example in McMurdo Sound, a further step was taken towards the study of ecological interactions in situ.

It is not the purpose of this chapter to summarize the history of benthic research in the Antarctic up to the present day. Excellent older reviews were provided by Hedgpeth (1969, 1971) and Dell (1972). We rather attempt to highlight recent progress against the background of research in the past, which has been considered in extenso in the majority of reviews to be cited in this paper. The important role played by the early explorers has recently been stressed by Davenport and Fogg (1989) and Dayton (1990).

It might seem doubtful whether another review of benthic research in Antarctica is needed. In fact, over 20 papers have reviewed the Antarctic benthos since Hedgpeth (1969, 1971) first called the public attention to the fact that the benthic fauna of the Southern Ocean is of special interest. At least 13 of these reviews appeared recently, i.e. in or after 1985. We will give a personal view of progress in benthic research from about that year although, of course, important achievements had been made before. In 1982, Lipps and Hickman had published their comparative report on the evolution of the Antarctic and deep sea faunas; in 1983, Clarke criticized the

* This paper is a short version of a more substantial review which is presently under revision. It summarizes the principal conclusions, however due to limited space in the ASC volume, the authors had to omit most details and discussions which will be included in the extended version.

[1] Sektion Biologie I, Alfred-Wegener-Institut für Polar- und Meeresforschung, Columbusstraße, D-27568 Bremerhaven, Germany

[2] Dpto. de Oceanografia, Universidad de Concepción, Apartado 2407, Concepción, Chile

Hempel (Ed.) Antarctic Science

anthropocentric view of the hardships of life in cold water, and in 1984/85 Picken, White and Arnaud published stimulating reviews on more recent results, mainly from shore-based Antarctic stations.

For the present review, we have considered a total of 280 papers related to benthic research, 145 of which have been published since 1984. We have excluded all purely taxonomical work (that is, without a connection to ecological questions). From the 145 recent papers, the following general conclusions can be drawn.

1. The bulk of the papers report work done in the Weddell Sea and in the Antarctic Peninsula/Scotia Arc region. This may reflect recent activities of RV *Polarstern* and shore-based work mainly on King George and Signy Islands, and the US stations of Palmer and McMurdo Sound. Shipboard sampling has, perhaps surprisingly, yielded more papers than work from the shore.
2. Relatively few papers have been published from areas north of the Antarctic Convergence, and nearly all work has been done in the high Antarctic.
3. Both deep- and shallow-water sampling have contributed a great deal to recent publications whereas scuba diving has not become as overwhelmingly important as was anticipated 10 years ago. Laboratory work in aquaria and cool containers is increasing in importance but still at an initial stage.
4. General descriptive biogeographical work and reports on community distribution have been the leading topics during the past 7 years followed by papers on species interactions (in a broad sense, including pelago-benthic coupling). Population dynamics and studies of reproduction and life histories have been catching up in importance but still contribute only a minority of the papers. Little work has been done on physiological and biochemical questions related to the Antarctic benthos.

In the following sections we will try and summarize the "state of the art" of different aspects of benthic research. Progress in the individual fields has been made at a different pace and in different ways. In some areas the data base has improved considerably, partly due to new facilities and techniques; in others old concepts have been examined critically and stimulating new ideas have been proposed. However, in many cases the available data are not yet sufficient to enable us to accept or reject certain hypotheses, or to produce definite statements as to the future development of certain areas of benthic research in Antarctica. The study of polar benthos, as any other field of polar research, is still developing.

2 Environmental Conditions in the Past

The questions of the duration and extent of continental glaciation, icebergs drifting and scouring in nearshore waters, whether the shelf was covered by or free of ice, whether the runoff from rivers was important, whether the environment was relatively constant or strongly fluctuating, and what the absolute seawater temperatures were, have been discussed for quite some time (see e.g. Lipps and Hickman 1982; Clarke and Crame 1989). Only in the past few years has there been a decisive improvement of our knowledge, based on more and much deeper drill holes in different parts of the continental shelf and its ice cover. Unfortunately, it is sometimes difficult to reconcile recent knowledge from these drill holes (Barrett 1989; Hambrey et al. 1989, 1991) with the record from marine sediments (Kennett 1985; Kennett and Barker 1990; Kennett and Stott 1990; Stott et al. 1990).

Since the Cretaceous, when Antarctica and Australia were still closely connected, water temperatures have decreased, however not linearly but with certain more rapid declines as in late Eocene (~38 Ma BP) and temporary warming at other times. Recent drilling data indicate that contrary to the record from marine sediments which provided evidence for full-scale glaciation only from mid-Miocene (15–20 Ma BP) there seems to have been a volume of ice on the Antarctic continent during the past 36 million years, i.e. at least from early Oligocene times, that was by no means less extensive than that of today. Furthermore, changes in the past, including variation in the extension of ice caps, advances and retreats of ice shelves, occurrence of icebergs and pack ice, and much warmer periods in between with vegetation growing in river valleys, occurred more commonly than was anticipated hitherto. Particularly during the Pliocene Antarctica was considerably warmer at times than at present (Kennett 1985; Haq et al. 1987; Quilty 1990).

This increased knowledge of the environmental history of the Southern Ocean is of utmost importance from an evolutionary point of view since both persistence of conditions (long-term "stability", see e.g. Sanders 1969) and frequency of disturbance or stress (see e.g. Dayton and Hessler 1972; Barrett and Rosenberg 1981; Bayne 1985) have been invoked when talking about faunal evolution in marine ecosystems (for a summary cf. Lipps and Hickman 1982).

3 Present Physical Factors Influencing the Benthos; Role of Physical Disturbance

Literature generally stresses the relative constancy of present physical conditions in the South Polar Sea.

Compared to other marine ecosystems, the relatively constant conditions are:

- *Low but stable temperatures*. The extreme continental case is McMurdo Sound where at 585 m depth an annual temperature range of $-1.8 \pm 0.2\,°C$ has been measured (Littlepage 1965). At some localities there are wider ranges due to the inflow of Warm Deep Water (Dunbar et al. 1985;

Bathmann et al. 1991; Arntz et al. 1992) or Cold Deep Water (Dayton 1990) moving up on the shelf. At Signy Island (S. Orkney I.) temperatures vary by 0.5 °C (Clarke et al. 1988), whereas at 17 m in Arthur Harbor they vary by 2.8 °C (Ayala and Valentine 1978).

- *Salinities that do not fluctuate much*. The normal range in the benthic realm is 34.6–34.9‰ (Lipps and Hickman 1982). Only locally – e.g. in the littoral of Subantarctic islands – may meltwater produce reduced values.
- *Less input from terrestrial sediments than in the Arctic*. Modern sediment input by meltwater from the Antarctic continent is considered to be minimal (Dunbar et al. 1985). However, dropstones from icebergs seem to make an exception (own obs.).
- *Isolation by deep sea, circum-Antarctic current systems*, and Antarctic Convergence. They all contribute to the constancy of conditions in the Southern Ocean (White 1984).

In contrast, certain other conditions fluctuate intensely:

- *The light regime* is highly seasonal. As a consequence, primary production is seasonal, too. Major fresh food input, and the vertical flux as a whole are restricted to a limited period of the year.
- *The sea-ice cover* varies in most areas (Clarke et al. 1988). Sea ice is important for life in the water column (Spindler 1990; Scharek 1991; Spindler and Dieckmann), whereas its importance is only hypothesized for the benthos underneath (Picken 1984; Arnaud 1985). Resuspended sediments may contain viable algal material (Berkman et al. 1986).
- *Anchor ice* is a major source of physical variation in shallow water communities (Dayton et al. 1974). It does not normally occur at depths >33 m (Dayton 1990). Anchor ice can encase plants and animals, tear them off the substrate, and is capable of lifting up to 25 kg (Picken 1984).
- *Fresh iceberg scours* have been found to be very common close to the ice edge (Lien et al. 1989; Galéron et al. 1992). Older scours have been found down to >400 m in the Weddell Sea but may mostly date from glacial periods when the water level was lower than today (Picken 1984).
- *Ice-shelves* seem to suppress the benthic fauna underneath and create unpredictable conditions due to temporary extension and the calving of icebergs. Most of the coastline and nearshore region of the Antarctic continent is covered by floating or grounded glacier ice, so beaches and true littoral areas are uncommon (Dunbar et al. 1985). A depauperate motile faunal element under the ice shelf was found as much as 430 km away from the ice edge (Bruchhausen et al. 1979; Dayton 1990; Lipps et al. 1979). Stations outside but close to the ice edge generally yield a reduced taxonomic richness in the south-eastern Weddell Sea which may be explained by the more frequent disturbance in this area from iceberg scouring and the shorter time of existence of the community (Galéron et al. 1992).

- *Variation of currents and circulation patterns* determines the grain size and composition of sediments (Dunbar et al. 1985). Coarser sediments are indicative of resuspending detritus, whereas on the soft bottoms of the trenches the slow currents do not resuspend particles, resulting in a meagre food supply for the epifauna (Voß 1988). Strong turbulent tidal currents, causing continuous lateral food advection, seem to favour a great number of species (Rauschert 1991). Circulation patterns are also responsible for dramatic differences in productivity in eastern McMurdo Sound as compared to the west (Barry 1988; Barry and Dayton 1988), accompanied by marked differences in density of the benthic faunal communities (Dayton and Oliver 1977; Dayton 1990).
- *Long-term/large scale modification of circulation patterns due to the 1982–83 ENSO event* may have caused heavy ice formation after a decade of low ice conditions (Dayton 1989).
- *Volcanic eruptions* may cause local benthic mortality (Gallardo et al. 1977; Gallardo 1987a). The infauna in these areas suffers recurrent and drastic alterations which result in an altered composition and scarcity of taxa (Gallardo et al. loc. cit.).

In summary, the Antarctic benthic environment is exposed to more physical variability and disturbance, both on a geological (see Sect. 2) and a recent time scale, than was thought in the past. It has never been a "stable", unchanging environment, as was already stated by Lipps and Hickman (1982; for a recent update see Clarke and Crame 1994), and it is not one today. Still and all, compared to other benthic marine ecosystems, it has some remarkably constant physical properties.

4 Evolution of the Benthic Fauna; Zoogeography

4.1 Retrospective: Which Factors Have Shaped Antarctic Benthos?

The Antarctic benthos in its present composition and diversity has evolved as a consequence of the long- and short-term abiotic environmental conditions in the past that have been referred to above, and of biotic interaction (cf. Sect. 9).

The factors that have been discussed in relation to the evolution of the benthic fauna include long-term stability, frequency of disturbance, low temperatures, temperature decrease, extreme seasonality in food supply, impact of various types of ice, and (lack of) terrestrial input. Any of them may have played an important part in the evolution of the Antarctic benthos. There have, however, been some recent arguments which may facilitate the discussion:

The Antarctic environment as a polar environment is older than anticipated and has not been as constant as we used to think (see Sect. 2). This means, a marked temperature gradient from the tropics to the pole developed earlier than anticipated, and the present polar fauna had more time to evolve under gradually changing conditions. Lipps and Hickman (1982) pointed out that long geologic periods are not required for speciation; however, is this true for the slow generation sequences in the Antarctic (see Sects. 8 and 11)?

Disturbance has been (and is) more frequent in the benthic environment than has been thought formerly (see Sects. 2 and 3). However, there has been a major discussion to what extent disturbance is rather favourable (Dayton and Hessler 1972) or unfavourable (Oliver and Slattery 1985) for the development of a diverse fauna; e.g. ice conditions and the advection of terrestrial material by ice and rivers have apparently been very variable in Antarctic history (see Sects. 2 and 3). Particularly, the advance and retreat of the ice cap on the continental shelf and above the upper slope (Grobe 1986) may have been detrimental for many shelf species, many others of which, however, survived (Brandt 1991). The resultant up-and-down movement of the fauna may explain the high degree of eurybathy in the Antarctic benthos (Klages 1991). On the other hand, obligate shallow benthic organisms such as macroalgae certainly were affected (Dayton 1990). Antarctic history may have led to periods in migration in and out of Antarctica, and may thus have contributed to speciation (Clarke and Crame 1994).

Low temperatures per se do not seem to present an insuperable problem for the evolution of a rich fauna (see Sect. 10), and speciation may proceed as effectively in cool as in warm waters (Clarke 1990a). (Again, the question is to what extent slow generation times may influence speciation!) Changes in temperature may have caused extinction on a geological time scale, but temperature decrease in Antarctica was slow and should have caused emigration rather than extinction (Clarke, loc. cit.).

The extremely seasonal input of food from the sea-ice and the water column is increasingly considered to be of great importance (Clarke 1988, 1990a; Clarke et al. 1988). Episodic availability of food requires particular adaptive responses which impose specific constraints on the types of organisms able to exploit such resources (Pearson and Rosenberg 1987). In the course of evolution this factor should have selected for organisms that sustain long starvation periods or live on food resources other than primary production and material sedimenting from the pelagic (see Sect. 9).

4.2 Antiquity and Origin of Present Fauna

Contrary to the view held some 30–40 years ago that the Antarctic fauna should not be very old, evidence has been accumulating in recent years that most of the Southern Ocean shallow water marine fauna evolved in situ since the Cretaceous or even earlier when the continents were still connected (Clarke and Crame 1989) and the fauna was still strikingly similar

(Menzies et al. 1973). More recent publications that support this view include the gastropods (Clarke 1990a; Clarke and Crame, loc. cit.) and the isopod families Serolidae and Arcturidae (Brandt 1991; Wägele and Brandt 1992). However, various other alternatives have also been discussed, e.g. immigration from the deep sea facilitated by similar conditions in that environment, or immigration from South America via the Scotia Arc (Watling and Thurston 1989). Particularly controversial has been the question whether (part of) the deep-sea fauna originated from the Antarctic or vice versa (Sieg 1988). On the other hand, the tanaidacean fauna of the Antarctic shelf is represented exclusively by "phylogenetically young" taxa (Sieg 1988, 1992) which, however, may also have an age of ≈30 million years. Almost the entire Antarctic tanaidacean fauna seems to have become extinct when temperatures dropped in Late Eocene, and cold-stenothermic eurybathic species then colonized the Antarctic shelf. Later this fauna was modified by Magellanian elements (Sieg 1988).

Decapoda Reptantia were rich in fossils in the Cretaceous (Pirrie 1989) and in the Eocene, with no break across the Cretaceous-Tertiary boundary (Feldmann and Tshudy 1989, and citations therein). Today there are no reptants at all in the high Antarctic although the few natant species that have made it are often very abundant (Arntz and Gorny 1991).

Other groups show a particular separation into closely related species which are almost certainly the product of radiations in situ (Clarke and Crame 1989 who give examples among pycnogonids, gastropods, echinoderms and ascidians). The Serolidae and Arcturidae both show patterns of radiation from stem taxa in the Antarctic (Brandt 1991; Wägele and Brandt 1992). However, most of the actual radiation is restricted to relatively few groups (White 1984; Dayton 1990). The ecological consequences of this fact have been discussed by the latter author.

Radiation is always connected with a high level of endemism which has been reported by several authors (Knox and Lowry 1977; Picken 1984; White 1984; Clarke and Crame 1989; Dayton 1990). Species endemism typically ranges between 57 and 95%, whereas at the genus level it is much lower (White 1984). Recent new data include isopods (87% endemic species of a total of 302, 20.7% endemic genera of a total of 121; Brandt 1991) and amphipods (about 95% endemic species of a total of >600; Jazdzewski et al. 1991). On the other hand, endemism of molluscs in the Weddell Sea is very low. Only two monoplacophorans, six gastropod and no bivalve species have been found exclusively in this area (Hain 1990).

4.3 Zoogeographic Affinities

The general value and the robustness of Hedgpeth's (1971) biogeographical conclusions as to the circum-Antarctic occurrence of many species plus smaller provinces or regions have been supported by Dell (1972), White

(1984) and Dayton (1990). A more recent division of areas based on amphipods and polychaetes by Sicinski (1986a,b) coincides in part with the regional division by Knox and Lowry (1977) but separates the Magellanic area and subdivides four Antarctic islands.

High affinities restricted to the high Antarctic (Davis Sea, Ross Sea, Adelie coast) have recently been shown for Weddell Sea gammaridean amphipods (Klages 1991) and gastropods (Hain 1990) whereas most Weddell Sea bivalves have a circum-Antarctic distribution which includes the South Shetlands, South Orkneys and Kerguelen (Hain, loc. cit.). For echinoderms similarities have been described between the Weddell Sea, the Scotia subregion and East Antarctica (Voß 1988; Gutt 1991a).

Links with other continents are strongest with South America, e.g. for sponges (with the Magellanic region; Sarà 1992), isopods (Brandt 1991; Wägele and Brandt 1992), bivalves and gastropods (Clarke and Crame 1989), but in most cases the palaeontological record is not good enough to decide whether immigration occurred in the one or other direction (Clarke and Crame, loc. cit.). Particularly low affinities have been encountered between sub-Antarctic amphipods and those of the Weddell Sea (only 18 of 101 spp. in common; Klages 1991).

In some cases a deep-sea connection has been established (tanaidaceans: Sieg 1992; molluscs: Arnaud and Bandel 1976 fide Clarke and Crame 1989; Dell 1972; Hain 1990 with earlier citations; a few isopod species: Hessler and Thistle 1975; amphipods: Watling and Thurston 1989; both latter citations fide Clarke and Crame 1989).

5 Distributional and Zonational Patterns (Relevance of Depth, Substrate, Distance from Ice Edge)

Distributional work on Antarctic benthos started with the investigation of Bullivant and Dearborn (1967). Since then types and zonation patterns of most Antarctic epifaunal assemblages have been described as essentially circumpolar (Hedgpeth 1971; Knox and Lowry 1977; Richardson and Hedgpeth 1977; Gallardo 1987a; Voß 1988; Dayton 1990).

Where there are ice shelves, the macrobenthos underneath is poor. There is no local primary production from algae, but bacterial densities and organic carbon are equivalent to deep-sea values. No benthic infauna has been collected, whereas motile epifauna (crustaceans and fish) was found several hundred kilometres from the ice edge (Bruchhausen et al. 1979; Lipps et al. 1979). Life under the ice shelf in the vicinity of the edge was rich and varied in the Ross Sea (Oliver et al. 1976; Dayton et al. 1984; Oliver and Slattery 1985) but sessile elements were scarce in the Weddell Sea in an area formerly covered by the ice shelf (Gerdes and Klages, pers. comm.).

In areas where there are no ice shelves, shallow-water benthic life is less rich due to the impact of anchor ice and iceberg scour. Faunal richness and biomass increase with depth (Zamorano 1983; Rauschert 1991; Arnaud 1992). Where anchor ice is absent, substratum type is the dominant factor in determining distribution and abundance of macrobenthic species (Kirkwood and Burton 1988). Zonal patterns have been reported for epibenthos on hard and cobble substrates (Dayton et al. 1974; Picken 1985; Smith and Simpson 1985; Tucker 1988; Dayton 1990; Rauschert 1991), however they are less distinct or non-existent for soft-bottom benthos (Gallardo 1987a; Mühlenhardt-Siegel 1988; Dayton 1990). Lower sublittoral and bathyal assemblages consist predominantly of suspension feeders which occur over a wide depth range (Picken 1984). Eurybathy has been reported for many groups including molluscs (Powell 1973 fide Picken 1980a; Davenport 1988; Hain 1990), gammaridean amphipods (Klages 1991), arcturid isopods (Wägele 1987b), caridean decapods (Arntz and Gorny 1991), holothurians (Gutt 1991c), echinoids (Brey and Gutt 1991), and sponges (Barthel and Gutt 1992).

Recent work on benthic communities in the Weddell Sea (Voß 1988; Galéron et al. 1992) separates a rich Eastern Shelf Community dominated by suspension feeders and a poorer Southern Shelf Community with a higher trophic diversity. Stations close to the shelf ice had "replicates" in similar depths at some distance from the ice-shelf (Galéron et al. loc. cit.). No relation between benthic biomass distribution and depth has been detected on the Weddell Sea shelf, but values were drastically lower on the slope and towards the deep sea (Gerdes et al. 1992).

Knowledge of Antarctic meiobenthos is very limited at any depth (Arnaud 1992). During the EPOS expedition, a near shelf ice, a slope and a deep-sea meiobenthic community were discriminated in the Weddell Sea (Herman and Dahms 1992).

6 Densities, Biomasses, Productivity

A traditional view is that densities and biomasses of macrobenthos are high in the Antarctic both on hard and soft bottoms (White 1984; Clarke and Crame 1989; Clarke 1990a), especially in comparison with the Arctic (Dayton 1990). Only in one case (McMurdo East Sound) have exceptionally high densities between 118 712 and 155 572 individuals m^{-2} been reported (Dayton and Oliver 1977); most other abundances were 1–2 orders of magnitude lower (Gallardo and Castillo 1969; Lowry 1975; Mühlenhardt-Siegel 1988, 1989; Gerdes et al. 1992). Biomass (wet wt.) values ranged between 0.12 g and 1644.20 g m^{-2} in the Weddell Sea (Gerdes et al. 1992) and between 9.06 g and 57.13 g m^{-2} in the Antarctic Peninsula/Scotia Arc area (Mühlenhardt-Siegel 1988). Antarctic biomass values are higher than those of Arctic macrobenthos at similar depth (Brey and Clarke 1994). Numbers and biomasses of individual taxa or species differ greatly. Many species occur in particularly low numbers. Relatively high densities and biomasses have been reported for amphipods (Jazdzewski et al. 1991), limpets (Picken 1980a) and scallops (Berkman 1990). Low mean values, but locally dense patches have been observed with amphipods (Klages 1991), caridean shrimps (Gutt et al. 1991), holothurians (Gutt and Piepenburg 1991) and sponges (Barthel and Gutt 1992). Low values have been found for epi- and endofauna in Admiralty Bay (Wägele and Brito 1990), shelled molluscs in the Weddell Sea (Hain 1990), echinoderm predators in McMurdo Sound (Dayton et al. 1974) and the sea urchin *Sterechinus* spp. (Brey and Gutt 1991).

Production: mean biomass ratios on a yearly basis (P/B year^{-1}) of Antarctic benthic invertebrates are considerably lower than those of their temperate counterparts, a fact which must result in a comparatively low overall production of macrobenthic communities. The lowest P/B found hitherto is that of *Sterechinus antarcticus* (0.12; Brey 1991). Other values have been published on molluscs (Picken 1979; Brey et al. 1993; Brey and Hain 1992), isopods (Luxmoore 1985; Wägele 1990) and mysids (Siegel and Mühlenhardt-Siegel 1988). For a recent compilation and further calculations see Brey and Clarke (1994).

Benthic meiofauna has been studied very little in terms of densities and biomasses, and no productivity estimates are available for this group from the Antarctic. Densities encountered at Kerguelen Islands (Soyer and de Bovée 1977) and in the south-east Weddell Sea (Herman and Dahms 1992) were low compared to data from temperate latitudes.

7 Species Richness, Diversity, Equitability

Species numbers, diversity and evenness of the Antarctic benthos are difficult to compare with other marine areas because the basic requirement for comparison (use of the same samplers, screen sizes and area extensions, and mathematics) is seldom met. In addition, breakdown to species level continues to be a major problem for many areas and taxa. Where species numbers are cited, they often refer to the whole Southern Ocean due to the circum-Antarctic occurrence of many species.

Traditionally, Antarctic benthos has been considered rich in species and diverse, particularly with regard to the epifauna (Dearborn 1968; Dell 1972). Diversity was found to be mostly high also for the infauna, with some areas such as Arthur Harbor approaching maximum diversity values (Lowry 1975), but others such as Chile Bay revealing comparatively low diversity (Gallardo and Castillo 1969). The older literature has been summarized in Richardson and Hedgpeth (1977) who also reported seasonal differences in diversity due to the reproduction of infaunal species. More recently, some reviewers have cautioned against overgeneralization (White 1984; Clarke and Crame 1989; Clarke 1990a; Gutt 1991c). Some groups, among them sessile suspension feeders (sponges, bryozoans), motile epibenthos (amphipods) and taxa which cover a wide range in terms of motility or trophic function (polychaetes), are rich in species; others seem to occupy an intermediate level (bivalves, gastropods, isopods); and some are missing altogether (stomatopods) or restricted to a few representatives (cirripedes and decapods where reptants are totally absent from the high Antarctic). In the latter case species richness was much higher in Cretaceous and Early Tertiary times (Clarke 1990a). Figure 1 presents some examples of present-day soft bottom communities in the Weddell Sea.

Within-group diversity is particularly high in amphipods (Klages 1991) and in isopods where three families account for at least one-half of the Antarctic species (White 1984) which now total 346 (Brandt 1990). All eight species of caridean shrimps in the Weddell Sea belong to different genera (Arntz and Gorny 1991). Holothurians (Gutt 1991c) and shelled gastropods and bivalves in this area (Hain 1990) also reveal a high taxonomic diversity, but in the latter group species richness in Antarctica is much less than in the tropics (Clarke 1994).

Little has been published recently on the evenness of Antarctic benthos. An extraordinary evenness was found in the dense macrofaunal assemblage in McMurdo Sound (Dayton and Oliver 1977) whereas Richardson and Hedgpeth (1977) recorded low equitability values at Arthur Harbor (Antarctic Peninsula). The dominance of a few nematod species prevented high values in the meiobenthic community studied in the Weddell Sea (Herman and Dahms 1992).

8 Dynamics of Antarctic Benthic Communities (Including Recolonization/Succession)

The slow rates of population growth in Antarctic benthic communities necessitate long intervals between periods of observation to determine the persistence of benthic communities and, particularly, the resilience of the fauna after catastrophes induced e.g. by anchor ice or iceberg scouring.

Long-term studies on Antarctic benthos are virtually non-existent although the US studies in McMurdo Sound provide some insight into the long-term development of different Antarctic shallow-water communities. The principal result seems to be that in most cases changes do not occur rapidly, however, there are exceptions (Dayton 1989). Very few colonization experiments have been carried out in the Antarctic both on hard and soft bottoms, and the results are contradictory. Buoys, sediment traps and other equipment deployed in the high Antarctic are normally not colonized by the fauna as is the case in other oceans. Plates exposed in McMurdo Sound in 1974

Fig. 1. Photographs of Antarctic soft bottoms revealing distinct differences in colonization. *Upper left* (still camera, ca. $0.5\,m^2$, 572 m depth, 78°06.3′ S/36°27.7′ W): Impoverished fauna in an area formerly covered by the ice shelf. Hexactinellid sponge: *Rossella racovitzae*, fish: *Trematomus loennbergii*. *Upper right* (still camera, ca. $0.3\,m^2$, 295 m, 71°38.0′ S/12°15.2′ W): Bryozoan debris, crinoid: *Promachocrinus kerguelensis*; *left* prostomium of echiuroid. *Lower left* (ROV, 196 m, 71°14.8′ S/12°59.3′ W): Rich three-dimensional association with several *R. racovitzae*, *Isodyctia* sp. (hexactinellids), another

hexactinellid with a *P. kerguelensis*, various bryozoan species, and the white ophiuroid *Astrotoma agassizii*. *Lower right* (ROV, same depth and position as preceding photo): Hexactinellids (*left Cinachyra barbata*; *centre* and *right R. racovitzae*) with epizoans such as ophiuroids, the crinoid *P. kerguelensis*, the holothuroid *Staurocucumis liouvillei* and the gastropod Margarella sp. (*white dots*). (All photos reproduced with kind permission by Julian Gutt, AWI Bremerhaven)

collected only two serpulid polychaetes in the first 3 years and were reported to be still bare after 5 years. However, in 1984 all the settling surfaces were heavily covered with benthic organisms (Dayton 1989). A settling plate array deployed at 670 m depth in the Weddell Sea for 1 year was not colonized at all (Gerdes, pers. comm.). In shallow water of Maxwell Bay, King George Island, asbestos cement plates taken up after 3 years revealed a rich sessile fauna including solitary ascidians up to 30 cm in size (Rauschert 1991).

In the dense infaunal shallow-water community in McMurdo Sound (cf. Sect. 6) early succession of benthos started in a similar way as in temperate communities whereas the subsequent successional stages lasted about three times longer (Oliver et al. 1976). No further information seems to be available on soft-bottom trays in the Antarctic.

9 Biotic Interactions/Trophic Dynamics

9.1 Pelagobenthic Coupling: the Input from Above

Older studies (e.g. El-Sayed 1971) have sometimes overestimated the amount of pelagic production in the Southern Ocean. More recent investigations have stressed the extremely seasonal nature and short duration of high productivity conditions in the water column (Barry 1988; Clarke 1988; El-Sayed 1988; Howard-Williams et al. 1990; Bathmann et al. 1991). Some recent information has become available on the amount and composition of particulate organic matter that contributes to sedimentation during this period and the time scales of sedimentation events (Dayton 1990; Bathmann et al. 1991).

The role of sea ice for the benthos may have been underestimated in the past. The number of organisms living in the pack ice may exceed that in the water column by several orders of magnitude (Spindler 1990), and detrital fallout, particularly during melting processes at the receding ice edge, may contribute significantly to the food input at the seafloor (Dayton 1990). Ice-associated productivity extends into the winter (Dieckmann 1987), a substantial proportion of diatoms found in sediment traps may be representative of ice algal species (Leventer and Dunbar 1987), and ice algae may contribute as much as 30% to primary production in the South Polar Sea (Spindler and Dieckmann 1991).

Much of the material "from above" seems to be utilized immediately by the large stocks of epibenthic suspension feeders; despite high inputs at times and supposed slow bacterial activity, carbon is not accumulated in the sediments (Dayton 1990). Living "on the second floor", e.g. on large sponges (Bullivant 1967; Voß 1988), may have developed as a response to food competition at the water–sediment interface. Epizoic life has also been found on motile fauna, e.g. foraminiferans living on scallops (Mullineaux and de Laca 1984) or gooseneck barnacles (*Scalpellum*) living on the stone crab *Paralomis* (own obs.). Lateral advection and resuspension may contribute substantially to improve food conditions during times of low primary productivity (Berkman et al. 1986; Brey et al. 1993).

9.2 Feeding Habits of Benthic Fauna

The study of food and feeding habits of Antarctic benthos has received considerable attention since it may explain part of the apparent discrepancy between seasonally limited food resources and the existence of a rich benthic life. Aquarium observations, especially the use of cool containers, have greatly contributed to the understanding of trophic interactions. The role of phytoplankton and the even smaller fractions of plankton for the epibenthic suspension feeders has received little attention. Among many groups of the macrobenthos, a predilection for necrophagy has been observed (Arnaud 1970, 1977, 1992; Presler 1986). Amphipods are the best recently investigated group, including herbivores where macroalgae exist (Knox and Lowry 1977), detritivores such as *Ampelisca* and a large number of scavengers, predators and omnivores which can utilize food resources year-round (Klages 1991). Some of the predators are highly specialized on certain taxa and have undergone adaptations of their mouthparts (Slattery and Oliver 1986; Coleman 1989a,b,c, 1990a,b, 1991; Klages and Gutt 1990a,b; Klages 1991; de Broyer and Klages 1991). The more developed isopod family Arcturidae filter phytoplankton and detritus (Wägele 1987a). Most isopods, however, belong to higher trophic levels and seem to be similarly independent of fluctuating primary production as are most amphipods (Luxmoore 1985; Juilfs and Wägele 1987; Wägele and Brandt 1992; Klages 1994).

Other recent benthic food studies include sponges (McClintock 1987), caridean shrimps (Perissonotto and McQuaid 1990), stone crabs (Arnaud and Miquel 1985), molluscs (Castilla and Rozbaczylo 1985; H. Wägele 1988; Hain 1990), holothurians (Gutt 1991a; Gutt et al. 1992), and ophiuroids (Kellogg et al. 1982; Fratt and Dearborn 1984; Dearborn et al. 1986). Already the classic study by Dearborn (1977) had revealed that asteroids have a wide range of feeding habits.

It can be concluded that a large proportion of Antarctic benthic organisms has developed a high degree of independence from fluctuating food conditions.

9.3 Benthos as Food for Other Organisms

Benthic fauna has been recorded from the stomachs of warm-blooded animals and fish in the Antarctic. Shallow-water limpets are preyed upon by gulls (Castilla and Rozbaczylo 1985). Different seal species and penguins take caridean shrimps, mysids and octopodid cephalopods (Dearborn 1965; Clarke and Macleod 1982a,b fide Kühl 1988; Green and Williams 1986; Green and Burton 1987; Perissonotto and McQuaid 1990 and references therein). In deeper water off the continental ice edge in the Weddell Sea the only benthic food of Weddell seals consisted of octopodid cephalopods (Plötz 1986; Plötz et al. 1991), and emperor penguins did not feed on benthos at all (Klages 1989).

Fish mostly prefer epibenthos, especially the more motile crustacean groups, to the – mostly sparse – infauna (Moreno 1971; Duarte and Moreno 1981; Wyanski and Targett 1981; Daniels 1982; Schwarzbach 1988; Kock 1992). The latter author found resource partitioning to be the rule and stated that (epi)benthos feeding fish are more numerous in the high Antarctic as compared to the Scotia Arc area.

9.4 Other Interactions and Interaction Experiments

Although there may be some degree of biotic disturbance locally (Zamorano et al. 1986; Gallardo 1987a), the Antartic benthos lacks "the persistent overwhelming (biotic) disturbances characterizing the Arctic" (Dayton 1990).

The only substantial in situ interaction studies on Antarctic benthos have been done in McMurdo Sound (Dayton et al. 1974; Dayton 1979, 1989; Oliver and Slattery 1985). The principal goal of these studies was to learn about interactions in the shallow-water sponge community and in a dense soft-bottom assemblage.

Predation was found to be the cardinal factor in preventing space monopolization by a single species in the sponge community and in maintaining the dense soft-bottom assemblage in McMurdo Sound, whereas competition for space seemed to be of minor importance (Dayton et al. 1974; Dayton 1990; Oliver and Slattery 1985). Some species have developed avoidance and escape responses (McClintock 1985). Under special circumstances, however, competition for space may also be important. Arborescent growth may represent a morphological adaptation which allows certain sponges to grow on relatively narrow bases, thus ameliorating competition with the more prostrate forms (Dayton et al. 1974). The upper spines of sea urchins (*Notocidaris*) are preferred by the small bivalve *Lissarca* and other species for settlement, and there is evidence of intra- and interspecific competition for space among these bivalves (Brey et al. 1993). The same may be true for the "second floor" localities on sponges.

10 Physiology/Autecology

Polar organisms pay for their ability of living at very low temperatures with a limited thermal tolerance (Clarke 1990a). This stenothermy seems to be more restrictive for Antarctic invertebrates than for their Arctic counterparts (George 1977; Arnaud 1985; Peck 1989).

While short-term and seasonal changes of seawater temperature are small compared with other marine ecosystems, the temperature changes from the Early Tertiary until today were equivalent to only 0.003 °C in a thousand years, and should not have posed a major challenge to the marine fauna on an evolutionary time-scale (Clarke 1990a). Moreover, most of the benthic fauna in the world ocean lives at relatively low temperatures. Cold water is not an unusual environment for marine organisms (Clarke 1988). However, polar organisms must be able to grow, reproduce, feed and evade predators at temperatures close to or even below 0 °C. From the start of physiological studies in the Antarctic attempts have been made to explain how this may have been achieved.

The concept of "cold adaptation" as a metabolic adaptation of cold-water poikilotherms (for references see Clarke 1983) was increasingly doubted from the mid-1970s when the first well-controlled experiments with polar benthic invertebrates were published (White 1975; Everson 1977; Maxwell 1977; Ralph and Maxwell 1977b,c; Ivleva 1980; Houlihan and Allan 1982; Maxwell and Ralph 1985; Davenport 1988; Peck 1989). All these experiments indicated that former findings in favour of metabolic "cold adaptation" were apparently based on experimental artifacts.

Low metabolic activity has usually been measured as oxygen consumption rates, and data referring to benthic species have been published for isopods (Luxmoore 1984), amphipods (Klekowski et al. 1973; Aarset and Torres 1989; Klages 1991), caridean shrimps (Maxwell and Ralph 1985), prosobranch gastropods (Houlihan and Allan 1982; Clarke, 1990b) and bivalves (Davenport 1988). They all coincide in that no elevation of oxygen consumption attributable to evolutionary history or zoogeographical position of polar animals is detectable. Recently, Clarke (1990a) has argued, furthermore, that respiration is a particularly misleading indicator of temperature compensation and should therefore be abandoned.

From an energy budget point of view, there is no reason to compensate for low temperatures except for calcification, a metabolic cost that indeed increases at lower temperatures (Nicol 1967; Clarke 1990a). Dwarfing in Antarctic (and deep-sea) organisms often occurs in species that secrete calcium carbonate (Lipps and Hickman 1982). As to other metabolic costs, there is even a distinct energetic advantage of living in cold water (Clarke 1983, 1987). Why, then, do polar invertebrates (and fish) not grow more rapidly, reproduce earlier and more frequently, move faster and more often than their temperate or tropical counterparts? There seem to be two possible answers: incomplete adaptation at the molecular level and seasonal food limitation.

The first alternative is difficult to confirm or reject because of the almost complete lack of biochemical studies on Antarctic benthic invertebrates (Clarke 1991). Recently, Dittrich (1990, 1992a,b) has shown that caridean shrimps and isopods from the Weddell Sea require a significantly lower activation energy of proteolytic enzymes at polar temperatures than related species from lower latitudes. Where the growth of an Antarctic organism is much faster than that of related species (see Dayton et al. 1974; Rauschert 1991), we can infer that growth has evolved compensation for temperature, but in the normal case where growth rates in polar organisms are slow we cannot simply conclude that this is due necessarily to direct limitation by temperature (Clarke 1991). Sluggishness of most Antarctic invertebrates kept in aquaria is obvious (Arnaud 1977), however, even extremely inactive species such as isopod parasites can be quite active when searching for their hosts (Wägele 1988, 1990). Amphipod scavengers, especially Lysianassidae, often rest motionless for long periods but become hyperactive (and quite fast!) when there is a food stimulus (Klages 1991). Temperature compensation on a molecular level for the Antarctic fauna during its evolution may have been only partial for many animals (Clarke 1987).

The second alternative indicates that in many cases temperature – at least where its seasonal oscillations are small – is not the primary factor at all. Instead it is the seasonally restricted availability of food resources resulting from short summer periods of phytoplankton production (Picken 1984; Clarke 1988; Bathmann et al. 1991; Arntz et al. 1992) that seems to keep Antarctic organisms from growing faster, reproducing earlier and more often and exhibiting a greater locomotor activity. However, most Antarctic animals may be able to survive seasonally fluctuating but predictable levels of food precisely because the water is cold (Clarke 1980), i.e. as a secondary factor temperature plays quite an important part.

The importance of food availability can best be seen in species where growth is strongly seasonal. On an annual basis growth is very slow in Antarctic benthos even in species which have been described as "relatively fast growing"; however, when growth is actually in progress, it may proceed relatively fast (Clarke 1988). This means that the (biochemical) capacity for faster growth exists, at least in these cases.

Reproduction of Antarctic animals, too, is closely tied to the seasonal cycle of food availability in some cases, whereas it is totally uncoupled in others. The latter is especially the case in animals which live at higher levels of the food web (White 1977; Clarke 1988). Benthic organisms do not appear to need seasonal lipid stores, possibly because their metabolic rates are so low that they do not require such reserves (Clarke 1977, 1982, 1988).

11 Life History Patterns and Strategies

Traditionally, a number of features have been assigned to Antarctic poikilotherms many of which were either deduced from animals which had been studied in the Arctic, or were hypothesized against the background of the specific environmental conditions in polar seas (White 1984). These features comprise a number of properties such as low fecundity, non-pelagic development, brooding, slow embryonic development etc. part of which have been quoted as "Thorson's Rule" (cf. Thorson 1950; Pearse et al. 1991). Numerous examples have been given in the literature (see, e.g. the reviews of White 1984, Picken 1984; and papers by Arnaud 1978; Picken 1979, 1980a,b).

More recent literature, especially from the high Antarctic, confirms that the above patterns, which are largely consistent with K strategies, are characteristic of many benthic species in Antarctica while revealing, at the same time, that there are important differences between taxa.

11.1 Duration of Embryonic Development; Egg Sizes

There are large variations in the degree of retardation of embryonic development and maturation from species to species (Wägele 1990). Very slow rates of embryonic development have been found in Antarctic isopods (Wägele 1987b, 1988, 1990) and gammaridean amphipods where they increase exponentially with decreasing habitat temperature (Klages 1994; de Broyer and Klages 1991; Arntz et al. 1992). Benthic caridean shrimps from the Weddell Sea become sexually mature only after 4–6 years (Gorny 1989; Arntz et al. 1992), whereas females of the shrimp *Chorismus antarcticus* at South Georgia mature in their third post-metamorphic summer (Clarke 1985). Embryonic development in these cases lasts about 18 and 14 months, respectively.

Very long developmental times have also been reported for Antarctic prosobranch gastropods (about 2 years till hatching; Hain 1992) and nudibranchs (H. Wägele 1988 and references therein). Solenogastres and polyplacophorans have a three to seven times longer development than their temperate counterparts (Hain, loc. cit.; Arntz et al. 1992). Long-lasting metamorphosis in echinoderms has been reported by Bosch et al. (1987), by Gutt (1991b), and Gutt et al. (1992) for holothurians.

In most cases long developmental times have been found to be connected with large egg sizes and, consequently, low fecundity. Examples include mysids (Siegel and Mühlenhardt-Siegel 1988); amphipods (Klages 1991); octopodid cephalopods (Kühl 1988) and other molluscs, with extremes in solenogastres and polyplacophorans (Hain and Arnaud 1992) and nudibranchs (H. Wägele 1988). Large eggs have also been reported for holothurians (Gutt et al. 1992), and small brood sizes have been found for brittle stars (Dayton and Oliver 1978).

While latitudinal clines within higher taxonomic categories have been well known for some time (for amphipods, see e.g. Bone 1972; Bregazzi 1972), recently also differences in the same crustacean species occurring in the low and high Antarctic have been detected. Caridean shrimps and the isopod *Ceratoserolis* bear fewer and larger eggs in the SE Weddell Sea compared with lower Antarctic regions (Wägele 1987b; Clarke and Gore 1992; Gorny et al. 1992).

11.2 Brooding, (Lack of) Pelagic Larvae

The predominance of brooding species in the Antarctic benthos and the "relative absence" of pelagic larvae have been discussed extensively in the literature (e.g. White 1984; Picken 1979, 1980b, 1984). Already Thorson (1950) hypothesized – mainly from evidence collected in the Arctic – that there is a strong general trend towards non-pelagic development and brood protection in polar waters due to the unsecurity of prolonged larval life in these waters where primary production is restricted to a short period each year.

The general validity of "Thorson's rule" has been challenged recently by Pearse et al. (1986, 1991) and Berkman et al. (1991) because of the increasing number of benthic species found recently with pelagic larvae. Some of these even have planktotrophic larvae.

However, as Pearse et al. (1991) state themselves, even excluding those groups such as the peracarids which are brooders in other oceans as well, and which are particularly dominant in the Antarctic, some taxa do indeed display unusually high incidences of brooding. This is particularly true with echinoderms. Nearly all (43) species of echinoids are brooders. One species even has special brood pouches (David and Mooi 1990). *Sterechinus neumayeri* and three asteroids with their planktotrophic larvae seem to be exceptions. A few asteroids have pelagic lecithotrophic larvae (Pearse et al., loc. cit.).

Brooding is assumed to be a common feature for Antarctic bivalves. Interestingly, it is just the common three larger bivalve species in McMurdo Sound (*Adamussium colbecki*, *Laternula elliptica*, *Limatula hodgsoni*) that have pelagic developmental stages. The other, >60, species of Antarctic bivalves are assumed to have non-planktonic development (Pearse et al. 1986). Twelve SE Weddell Sea bivalves have been shown to be brooders so far (Hain and Arnaud 1992). Even large neogastropods and opisthobranchs have a non-pelagic development (H. Wägele 1988; Pearse et al. 1991), and intra-capsular metamorphosis is the rule with Antarctic gastropods (Hain and Arnaud 1992). Pelagic lecithotrophic development as in *Nacella con-*

cinna (Picken 1980a) and the – facultatively planktotrophic – larvae of the Capulidae and Lamellariidae (Bandel et al. 1994) are exceptions. "Thorson's rule" seems to hold at least for most gastropods (Pearse et al. 1991).

Chorismus antarcticus and *Notocrangon antarcticus* have pelagic larval stages (Boysen-Ennen 1987; Piatkowski 1987), whereas such stages have never been found of the third common benthic shrimp species in the Weddell Sea, *Nematocarcinus lanceopes* (Arntz and Gorny 1991). A spectacular recent detection was balanomorph barnacle larvae in the plankton at McMurdo Sound (Foster 1989).

From data of Pearse et al. (1991) it looks as though it is mostly large species living in shallow water that account for the most notable exceptions to "Thorson's rule". Although non-pelagic development may not be as overwhelmingly prevalent among Antarctic invertebrates as believed until a short time ago, the number of pelagic, and particularly of planktotrophic, larvae that have been found is still very small compared with the immense total number of species in the Antarctic benthos. New sampling designs may improve the record to some extent, but the total number of species with pelagic larvae is likely to remain low. In fact, the selective conditions in the Antarctic pelagic rather encourage the evolution of benthic or short-term, non-feeding drifting stages (Pearse et al. 1991).

11.3 Growth, Final Size, Mortality, Longevity

Low annual growth, (often) large final size, low mortality and prolonged longevity have often been mentioned as another facet of living under polar conditions. There is an empirical inverse relationship between low annual growth rates and maximal size. Growth studies on Antarctic benthos in aquaria have so far been rather a failure for molluscs and amphipods (Klages 1991), but they have provided some useful data for isopods (Wägele 1990) and caridean shrimps (Maxwell and Ralph 1985; Bruns 1992). Field data on Antarctic bivalves (Ralph and Maxwell 1977a), nudibranchs (H. Wägele 1988), shrimps (Clarke and Lakhani 1979), mysids (Siegel and Mühlenhardt-Siegel 1988), echinoids (Brey 1991) and most sponges (Dayton et al. 1974) support the idea that annual growth rates are normally very low in the South Polar Sea. However, growth is often limited to the summer months and may be fairly rapid if averaged over these months only (Richardson 1979; Sagar 1980; Clarke 1991). Some new evidence for extended longevity and large final size has been presented for the amphipod *Eusirus perdentatus* (Klages 1994), and for caridean shrimps, which also reveal a latitudinal cline in maximum lengths (Arntz and Gorny 1991; Gorny et al. 1992).

Assuming that growth rings on Aristotle's lantern are annual, Brey (1991) concluded that *Sterechinus antarcticus* needs 50 years to grow to a diameter of 40 mm and may reach a maximum age of 75 years. Total annual mortality of this species was found to be extremely low. However, certain sponges (Dayton et al. 1974; Dayton 1989) and ascidians (Rauschert 1991) revealed remarkably fast growth in comparison with other species in the same environment.

11.4 Seasonality vs Non-Seasonality of Reproduction and Growth

Many different strategies have been realized in the Antarctic, from strict coupling to total decoupling of reproduction and growth to the short pulses of food input in summer (Thurston 1970; White 1970; Bregazzi 1972; Rakusa-Suszczewski 1972; Hardy 1977; Clarke and Lakhani 1979; Picken 1979, 1980a,b, 1984; White 1984; Clarke 1990b). The differing early development modes and release periods of Antarctic bivalves with pelagic larvae are influenced by the dependence or non-dependence of the larvae on seasonally produced food sources (Berkman et al. 1991). A further step, which has been developed by the majority of bivalves, is total avoidance of the dangerous pelagic zone by brood protection.

Recruitment of juvenile *Sterechinus neumayeri* most likely occurs in synchrony with the subsequent period of high level of benthic chlorophyll *a* concentrations (Bosch et al. 1987). In shallow water off Davis Station, small infaunal amphipods and tanaidaceans protected their brood throughout the winter and released their juveniles at times that coincided with the period of high primary productivity (Tucker 1988).

Conditions for the south-eastern Weddell Sea, where the whole range of strategies exists, have been summarized in Arntz et al. (1992). The shrimps *C. antarcticus* and *N. antarcticus* and the holothurian *E. steineni* make use of the improved food conditions for their offspring in Antarctic summer; this is also true for the deep-water shrimp *Nematocarcinus lanceopes*. Disconnection from the seasonal production cycle has been achieved by the holothurian *Psolus dubiosus* and a number of bivalves which produce fewer, larger eggs and protect their brood. An almost complete disconnection from the high Antarctic seasonal cycle is the case for most amphipods as they are scavengers, predators, or detritus feeders which brood their young and release their juveniles as fully developed organisms throughout the year. Decoupling sems to occur more often in the high Antarctic, in deeper water and with trophic generalists and scavengers which appear to be more common here than in shallow areas at lower latitudes, where primary food limitation is presumably less severe.

12 Conservation Aspects

Early protection measures in the Antarctic such as the Convention for the Regulation of Whaling were species-directed rather than environment-directed (Bonner 1989). A few benthic Crustacean species are used commercially in the sub-Antarctic, but no Antarctic benthic invertebrate species has been exploited despite the fact that high Antarctic scallops (*Adamussium colbecki*) and shrimps may be both abundant and large in size in some areas. However, their stocks would presumably not stand any serious exploitation except at extremely low levels of fishing mortality (Berkman 1990; Arntz and Gorny 1991). Furthermore, harvesting the scallops with conventional dredges would disrupt the apparently stable population and perturb the oligotrophic habitat (Berkman, loc. cit.).

The latter is particularly important because it stresses that species should not only be protected for their own sake but also as members of an ecosystem. The Convention (later Commission) on Conservation of Antarctic Marine Living Resources (CCAMLR), signed in 1980, was an improvement compared with the Whaling Convention in that it underlines the responsibility to maintain the balance of ecological relationships between harvested, dependent and related populations.

This allows the Commission to take a wider view than conventional fisheries protection measures (Bonner 1989).

Some benthic species are significant as food for demersal fish or other organisms (Sect. 9), but a much greater importance must be assigned to the fragile three-dimensional structure of epibenthic communities providing protection and many niches for fish and motile benthos. The structure of these communities has become known better in recent years, partly due to the increased use of underwater video and camera transects. Habitat destruction and other adverse impacts of bottom trawling have been reviewed by Kock (1990) and Constable (1991). The latter author also stressed the importance of protected areas as recovery sites for affected species in trawled areas.

Pollution of most kinds is still much lower in the Southern Ocean than in other seas (Cripps and Priddle 1991) although certain substances, e.g. organochlorines, may occur in similar concentrations. Since Antarctic invertebrates live much longer than temperate species and are an important food for fish, an accumulation of these substances in higher trophic levels of the Southern Ocean has to be assumed (Ernst and Klages 1991). Furthermore, the growing number of scientific stations, as well as increasing ship traffic connected with these stations and also with tourism, present a menace to some areas. Sewage from various Antarctic stations has led to eutrophication and greater turbidity of sublittoral waters of Maxwell Bay (King George Island), with negative impacts on the flora and fauna in this area (Rauschert 1991). As detrimental effects last longer than at lower latitudes due to the low resilience and slow recovery of benthic assemblages in the Antarctic (see Sect. 8), disturbance of any kind is more problematic.

From all these arguments it is obvious that there exists a strong requirement for benthic conservation sites in the Antarctic (Chittleborough 1987; Gallardo 1987b, 1991). A breakthrough for the protection of Antarctica was reached by the Antarctic Treaty Parties during the 1991 Madrid XI Special Consultative Meeting in the shape of a "Protocol on Environmental Protection within the Antarctic Treaty" (PEPAT) and Annexes I–IV (Antarctic Treaty 1991). In essence PEPAT designates Antarctica as a natural reserve, devoted to peace and science, prohibiting any activity relating to mineral resources other than scientific research, and, in general, committing all parties to a comprehensive protection of the Antarctic environment.

Up to now several (Marine) Sites of Special Scientific Interest [(M)SSSI] have been created in the Antarctic Peninsula/Scotia Arc area. In the future protection may be provided to any previously designated conservation area which includes intertidal and sublittoral zones within its boundaries, and also to marine belts surrounding protected islands (for examples, see Gallardo 1991).

13 Methodological Aspects

Benthic marine biologists are comparatively conservative in their methods. With relatively little modification, Petersen's grab (Petersen and Boysen-Jensen 1911) or Reineck's box corer (Reineck 1963) have been used for decades. Even nowadays, a large amount of samples is taken by this type of equipment or with trawls and dredges of different kinds. Especially the latter cannot be used for quantitative assessments of the fauna; they are, however, still quite useful for a general large-scale overview and for collecting large amounts of material.

In the high Antarctic sampling, which has to be done at considerable water depth, is often hampered by dropstones from melting icebergs. This leads to a high percentage of failures in grab and corer casts which may results quite expensive in terms of ship time. However, in recent years, several multiple corers have been developed which take a greater number of samples simultaneously. These corers include the multiple corer for meiofauna sampling (Barnett et al. 1984) and the "multibox corer" for sampling macrobenthos (Gerdes 1990), which is supplied with a video camera. The multibox corer, which is capable of indicating differences in small-scale distribution of the benthic fauna in the square-metre range, has performed well even in very coarse sediment of the eastern Weddell Sea shelf and slope (Galéron et al. 1992; Gerdes et al. 1992).

Underwater still (including stereographic) cameras and remotely operated vehicles (ROVs) with video cameras have gained much in importance. They are particularly useful in the Antarctic where the greater part of the benthic fauna lives either at or above the seafloor. UW cameras have been used for studies of the density and distribution of larger organisms (Hamada et al. 1986), and have been shown to be of great use for comparison with trawl catch data (Barthel et al. 1991; Brey and Gutt 1991; Gutt et al. 1991). The use of video cameras and ROVs for quantitative assessments still requires considerable improvement. Larger underwater vehicles of the "Alvin" or "Nautile" type have, to our knowledge, never been used in the Antarctic.

The icegoing research vessel *Polarstern* has enabled European Antarctic research to explore and investigate the permanent pack ice zone of the Weddell Sea. It represents a new research ship type which combines icebreaking abilities with facilities to use all kinds of gear to be deployed in the water column and at the seafloor, ample laboratory space and cool containers where live organisms can be kept at ambient temperature. This combined approach of studying preserved material from plankton nets, trawls, dredges and cores, and live organisms in the cool containers and lateron in the cool lab at the institute has provided many good scientific results in recent years.

14 Future Benthic Research in Antarctica

Both ship and shore-based work can contribute to the future development of Antarctic benthic research. Vessels of the *Polarstern* type and perhaps also manned or automatic underwater vehicles may explore the last unknown areas on the Antarctic map such as the permanent and seasonal pack ice zones around the continent, or certain areas beneath the ice shelves. These shipboard studies should focus on certain areas rather than continue the large-scale approach that had to be taken by many interdisciplinary cruises due to the restricted time available for individual fields of research. The study of processes such as pelago-benthic and bentho-pelagic coupling, including the role of the sea ice, will increase in importance and may open new interesting areas of research in the context of global warming. Increasing ozone depletion and resulting higher UV-B levels may have minor effects on the deeper zoobenthos, which mostly lacks pelagic larvae, but are likely to be important for shallow-water benthos (Karentz 1991). Major changes also have to be expected at the shelf-ice edges. Small-scale resolution of bottom topography, which will be helpful to explain colonization and distribution patterns of benthic fauna, can be obtained using hydrosweep/parasound techniques and may be particularly rewarding in areas of heavy iceberg scour. Hopefully, additional equipment will be developed to sample short-lived larval stages or drift stages close to the seafloor. It may be desirable to define sublittoral research areas close to the Antarctic continent which are monitored over several decades. For this purpose recently acquired knowledge from scientifically oriented trawling and dredging, bottom photography and video should be used.

At the same time there may be a revival and increased use of shore-based stations, principally because they can serve for certain types of studies which are difficult or even impossible to carry out from research vessels. These approaches include continuous sampling year-round; direct observation by divers; interaction experiments (predator ex- and inclusion, colonization and succession) in situ; and aquarium studies with a natural food supply from running sea water. Enclosed shallow-water areas of the "Benthosgarten" type (Arntz and Rumohr 1982; Gallardo and Retamal 1986), similar to the studies at McMurdo, may provide further insight into the population dynamics and production of individual species, their interactions and community dynamics. Experimental work on these items requires some patience in the Antarctic but may turn out to be particularly rewarding in the long term. As in deep water, the monitoring of specific areas may be useful. Some crowded areas such as the South Shetland Islands or the Palmer region where human interference by pollution from the shore, oil spills, etc. has become severe, should also serve for studies of the human impact on the benthic ecosystem.

Both in deep and shallow water, certain neglected fields of research should be intensified. Physiological and biochemical studies on polar poikilotherms almost exclusively refer to fish, whereas very little has been published on invertebrates despite several decades of discussion on "cold adaptation". The population dynamics and production biology of all but a handful of benthic species are virtually unknown, and interaction research has almost exclusively been done at McMurdo Sound. Both meio- and microfauna have received little attention at any depth, and macrofaunal studies in the deep sea have been extremely scarce. Finally, benthic taxonomy must not be further reduced since "all evolutionary, biogeographical, and ecological research absolutely depends on competent systematic research" (Dayton 1990).

Acknowledgements. We would like to thank Prof. Dr. A. Clarke (BAS, Cambridge) and Drs. T. Brey, D. Gerdes, J. Gutt, S. Hain and M. Klages for their very useful comments on former versions of this review. C. Klages, M. Romboy and A. Klepatz patiently retyped the various stages of the manuscript.

References

Note: To reduce the volume of the list of references, the authors have deleted all references before 1980 from this list although they appear in the text. The complete list of references is available from the authors on request.

Aarset AV, Torres JJ (1989) Cold resistance and metabolic responses to salinity variations in the amphipod *Eusirus antarcticus* and the krill *Euphausia superba*. Polar Biol 9:491–497

Antarctic Treaty (1991) Protocol on environmental protection to the Antarctic Treaty and Annexes I–IV. XIth Spec Consult Meet, Madrid, 3–4 Oct 1991

Arnaud PM (1985) Essai de synthèse des particularités éco-bio-logiques (adaptations) des invertébrés benthiques antarctiques. Oceanis 11(2):117–124

Arnaud PM (1987) Les écosystèmes benthiques des plateaux péri-insulaires des Iles françaises du sud de l'Océan Indien. Actes Colloq Rech Franç Terres Australes, Strasbourg, pp 129–138

Arnaud PM (1992) The state of the art in Antarctic benthic research. In: Gallardo VA, Ferretti O, Moyano HI (eds) Actas del Semin Int Oceanografia in Antartide. Centro EULA, Universidad de Concepción/Chile, pp 341–346

Arnaud PM, Hain S (1992) Quantitative distribution of the shelf and slope molluscan fauna (Gastropoda, Bivalvia) of the eastern Weddell Sea (Antarctica). In: Hempel G (ed) Weddell Sea ecology. Results of EPOS European *Polarstern* study. Springer, Berlin 1993. Polar Biol 12:103–109

Arnaud PM, Miquel JC (1985) The trophic role of the stone crab, *Lithodes murrayi*, in the benthic ecosystem of the Crozet Islands. In: Siegfried WR, Condy PR, Laws RM (eds) Antarctic nutrient cycles and food webs. Springer, Berlin Heidelberg New York, pp 381–388

Arntz WE, Gorny M (1991) Shrimp (Decapoda, Natantia) occurrence and distribution in the eastern Weddell Sea, Antarctica. Polar Biol 11:169–177

Arntz WE, Rumohr H (1982) An experimental study of macrobenthic colonization and succession, and the importance of seasonal variation in temperate latitudes. J Exp Mar Biol Ecol 64:17–45

Arntz WE, Brey T, Gerdes D, Gorny M, Gutt J, Hain S, Klages M (1992) Patterns of life history and population dynamics of benthic invertebrates under the high Antarctic conditions of the Weddell Sea. In: Colombo G, Ferrari I, Ceccherelli VU, Rossi R (eds) Marine eutrophication and population dynamics. Proc 25th EMBS, Ferrara/Italy. Olsen & Olsen, Fredensborg, pp 221–230

Bandel K, Hain S, Riedel F, Thiemann H (1994) *Limacosphaera*, an unusual mesogastropod larva of the Weddell Sea. Nautilus (in press)

Barnett PRO, Watson J, Connelly D (1984) A multiple corer for taking virtually undisturbed samples from shelf, bathyal and abyssal sediments. Oceanol Acta 7:399–408

Barrett G, Rosenberg R (eds) (1981) Stress effects on natural ecosystems. John Wiley, New York, 305 pp

Barrett PJ (ed) (1989) Antarctic Cenozoic history from the CIROS-1 drillhole, McMurdo Sound. DSIR, Wellington, 251 pp

Barry JP (1988) Hydrographic patterns in McMurdo Sound, Antarctica, and their relationship to local benthic communities. Polar Biol 8:377–391

Barry JP, Dayton PK (1988) Current patterns in McMurdo Sound, Antarctica and their relationship to local biotic communities. Polar Biol 8:367–376

Barthel D, Gutt J (1992) Sponge associations in the eastern Weddell Sea. Antarct Sci 4:137–150

Barthel D, Gutt J, Tendal OS (1991) New information on the biology of Antarctic deepwater sponges derived from underwater photography. Mar Ecol Prog Ser 69:303–307

Bathmann U, Fischer G, Müller PJ, Gerdes D (1991) Short-term variations in particulate matter sedimentation off Kapp Norvegia, Weddell Sea, Antarctica: relation to water mass advection, ice cover, plankton biomass and feeding activity. Polar Biol 11:185–195

Bayne BL (1985) Responses to environmental stress: tolerance, resistance and adaptation. In: Gray JS, Christiansen ME (eds) Marine biology of polar regions and effects of stress on marine organisms. John Wiley, New York, pp 331–349

Berkman PA (1990) The population biology of the Antarctic scallop, *Adamussium colbecki* (Smith 1902) at New Harbor, Ross Sea. In: Kerry KR, Hempel G (eds) Antarctic ecosystems. Ecological change and conservation. Springer, Berlin Heidelberg New York, pp 282–288

Berkman PA, Marks DS, Shreve GP (1986) Winter sediment resuspension in McMurdo Sound, Antarctica, and its ecological implications. Polar Biol 6:1–3

Berkman PA, Waller TR, Alexander SP (1991) Unprotected larval development in the Antarctic scallop *Adamussium colbecki* (Mollusca: Bivalvia: Pectinidae). Antarct Sci 3(2):151–157

Bone DG (1972) Aspects of the biology of the Antarctic amphipod *Bovallia gigantea* Pfeffer at Signy Island, South Orkney Islands. Br Antarct Surv Bull 27:105–122

Bonner WN (1989) Environmental assessment in the Antarctic. Ambio 18(1):83–89

Bosch I, Beauchamp KA, Steele ME, Pearse JS (1987) Development, metamorphosis, and seasonal abundance of embryos and larvae of the Antarctic sea urchin *Sterechinus neumayeri*. Biol Bull 173:126–135

Boysen-Ennen E (1987) Zur Verbreitung des Meso- und Makrozooplanktons im Oberflächenwasser der Weddellsee (Antarktis). Ber Polarforsch 35:126

Brandt A (1990) The deep sea isopod genus *Echinozone* Sars, 1897 and its occurrence on the continental shelf of Antarctica. Antarct Sci 2(3):215–219

Brandt A (1991) Zur Besiedlungsgeschichte des antarktischen Schelfes am Beispiel der Isopoda (Crustacea, Malacostraca). Ber Polarforsch 98:240

Brey T (1991) Population dynamcis of *Sterechinus antarcticus* (Echinodermata: Echinoidea) on the Weddell Sea shelf and slope, Antarctica. Antarct Sci 3(3):251–256

Brey T, Clarke A (1994) Population dynamics of marine benthic invertebrates in Antarctic and sub-Antarctic environments: Are there unique adaptions? Antarct Sci (in press)

Brey T, Gutt J (1991) The genus *Sterechinus* (Echinodermata: Echinoidea) on the Weddell Sea shelf and slope (Antarctica): distribution, abundance and biomass. Polar Biol 11:227–232

Brey T, Hain S (1992) Growth, reproduction and production of *Lissarca notorcadensis* (Bivalvia: Philobryidae) in the Weddell Sea, Antarctica. Mar Ecol Prog Ser 82:219–226

Brey T, Starmans A, Magiera U, Hain S (1993) *Lissarca notorcadensis* (Bivalvia: Philobryidae) living on *Notocidaris* sp. (Echinoida: Cidaroidea): population dynamics in limited space. Polar Biol 13:89–95

Bruns T (1992) Experimentelle Untersuchungen zur Larvalentwicklung antarktischer Garnelen (Decapoda, Natantia). Dipl Thesis, Universität Osnabrück, 129 pp

Castilla JC, Rozbaczylo N (1985) Rocky intertidal assemblages and predation on the gastropod *Nacella* (*Patinigera*) *concinna* at Robert Island, South Shetlands, Antarctica (1). Ser Cient INACH 32:65–73

Chittleborough RG (1987) A rationale for conservation areas within Antarctic waters. Select Sci Pap SC-CAMLR-SSP/4:513–535

Clarke A (1980) A reappraisal of the concept of metabolic cold adaptation in polar marine invertebrates. In: Bonner WN, Berry RJ (eds) Ecology in the Antarctic. Academic Press, London, pp 77–92

Clarke A (1982) Lipid synthesis and reproduction in the polar shrimp *Chorismus antarcticus*. Mar Ecol Prog Ser 9:81–90

Clarke A (1983) Life in cold water: the physiological ecology of polar marine ectotherms. Oceanogr Mar Biol Annu Rev 21:341–453

Clarke A (1985) The reproductive biology of the polar hippolytid shrimp *Chorismus antarcticus* as South Georgia. In: Gray JS, Christiansen ME (eds) Marine biology of polar regions and effects of stress on marine organisms. John Wiley, New York, pp 237–246

Clarke A (1987) Temperature, latitude and reproductive effort. Mar Ecol Prog Ser 38:89–99

Clarke A (1988) Seasonality in the Antarctic marine environment. Comp Biochem Physiol 90B(3):461–473

Clarke A (1990a) Temperature and evolution. Southern Ocean cooling and the Antarctic marine fauna. In: Kerry KR, Hempel G (eds) Antarctic ecosystems. Ecological change and conservation. Springer, Berlin Heidelberg New York, pp 9–22

Clarke A (1990b) Faecal egestion and ammonia excretion in the Antarctic limpet *Nacella concinna* (Strebel 1908). J Exp Mar Biol Ecol 128:227–246

Clarke A (1991) What is cold adaptation and how should we measure it? Am Zool 31:81–92

Clarke A (1994) Is there a latitudinal diversity cline in the sea? Trends Ecol Evol (in press)

Clarke A, Crame JA (1989) The origin of the Southern Ocean marine fauna. In: Crame JA (ed) Origins and evolution of the Antarctic biota. Geol Soc Spec Publ 47:253–268

Clarke A, Crame JA (1994) The Southern Ocean benthic fauna and climatic change: a historical perspective. Philos Trans R Soc B (in press)

Clarke A, Gore DJ (1992) Egg size and composition in *Ceratoserolis* (Crustacea: Isopoda) from the Weddell Sea. Polar Biol 12:129–134

Clarke MR, Macleod N (1982a) Cephalopods in the diet of elephant seals at Signy Island, South Orkney Islands. Br Antarct Surv Bull 57:27–31

Clarke MR, Macleod N (1982b) Cephalopod remains in the stomachs of eight Weddell seals. Br Antarct Surv Bull 57:33–40

Clarke A, Holmes LJ, White MG (1988) The annual cycle of temperature, chlorophyll and major nutrients at Signy Island, South Orkney Islands 1969–82. Br Antarct Surv Bull 80:65–86

Coleman CO (1989a) Burrowing, grooming, and feeding behaviour of *Paraceradocus*, an Antarctic amphipod genus (Crustacea). Polar Biol 10:43–48

Coleman CO (1989b) *Gnathiphimedia mandibularis* K.H. Barnard 1930, an Antarctic amphipod (Acanthonotozomatidae, Crustacea) feeding on Bryozoa. Antarct Sci 1(4): 343–344

Coleman CO (1989c) On the nutrition of two Antarctic Acanthonotozomatidae (Crustacea, Amphipoda). Gut contents and functional morphology of mouthparts. Polar Biol 9:287–294

Coleman CO (1990a) *Bathypanoploea schellenbergi* Holamn & Watling, 1983, an Antarctic amphipod (Crustacea) feeding on Holothuroidea. Ophelia 31(3):197–205

Coleman CO (1990b) Anatomy of the alimentary canal of *Parandania boecki* (Stebbing 1888) (Crustacea, Amphipoda, Stegocephalidae) from the Antarctic Ocean. J Nat Hist 24:1573–1585

Coleman CO (1991) Comparative fore-gut morphology of Antarctic Amphipoda (Crustacea) adapted to different food sources. Hydrobiologia 223:1–9

Constable AJ (1991) Potential impacts of bottom trawling on benthic communities in Prydz Bay, Antarctica. SC-CAMLR-X/BG/19:5

Cripps GC, Priddle J (1991) Review: Hydrocarbons in the Antarctic marine environment. Antarct Sci 3:233–250

Daniels RA (1982) Feeding ecology of some fishes of the Antarctic Peninsula. Fish Bull 80:575–588

Davenport J (1988) Oxygen consumption and ventilation rate at low temperatures in the Antarctic protobranch bivalve mollusc *Yoldia* (= *Aequiyoldia*) *eightsi* (Courthony). Comp Biochem Physiol 90A(3):511–513

Davenport J, Fogg GE (1989) The invertebrate collections of the *Erebus* and *Terror* Antarctic expedition; a missed opportunity. Polar Rec 25(155):323–327

David B, Mooi R (1990) An echinoid that "gives birth": morphology and systematics of a new Antarctic species, *Urechinus mortenseni* (Echinodermata, Holasteroida). Zoomorphology 110:75–89

Dayton PK (1989) Interdecadal variation in an Antarctic sponge and its predators from oceanographic climate shifts. Science 245:1484–1486

Dayton PK (1990) Polar benthos. In: Smith WO (ed) Polar oceanography, Part B: Chemistry, biology, and geology. Academic Press, London, pp 631–685

Dayton PK, Kooyman GL, Barry JP (1984) Benthic life under thick ice. Antarct J US 19:128

Dearborn JH, Ferrari FD, Edwards KC (1986) Can pelagic aggregations cause benthic satiation? Feeding biology of the Antarctic brittle star *Astrotoma agassizii* (Echinodermata: Ophiuroidea). Biology of the Antarctic Seas XVII. Antarct Res Ser 44:1–28

De Broyer C, Klages M (1991) A new *Epimeria* (Crustacea, Amphipoda, Paramphithoidae) from the Weddell Sea. Antarct Sci 3(2):159–166

Dieckmann G (1987) High phytoplankton biomass at the advancing ice edge in the northern Weddell Sea during winter (Abstr). EOS Trans AGU 68:1765

Dittrich B (1990) Temperature dependence of the activities of trypsin-like proteases in decapod crustaceans from different habitats. Naturwissenschaften 7:491f

Dittrich B (1992a) Thermal acclimation and kinetics of a trypsin-like protease in eucarid crustaceans. J Comp Physiol B 162:38–46

Dittrich B (1992b) Life under extreme conditions: aspects of evolutionary adaptation to temperature in Crustacean proteases. In: Hempel G (ed) Weddell Sea ecology. Results of EPOS European *Polarstern* study. Springer, Berlin 1993. Polar Biol 12:269–274

Duarte WE, Moreno CA (1981) The specialized diet of *Harpagifer bispinis*: its effects on the diversity of Antarctic intertidal amphipods. Hydrobiologia 80:241–250

Dunbar RB, Anderson JB, Domack EW (1985) Oceanographic influences on sedimentation along the Antarctic continental shelf. Antarct Res Ser 43:291–311

El-Sayed SZ (1988) Seasonal and interannual variabilities in Antarctic phytoplankton with reference to krill distribution. In: Sahrhage D (ed) Antarctic ocean and resources variability. Springer, Berlin Heidelberg New York, pp 101–119

Ernst W, Klages M (1991) Bioconcentration and biotransformation of ^{14}C-γ-hexachlorocyclohexane and ^{14}C-γ-hexachlorobenzene in the Antarctic amphipod *Orchomene plebs* (Hurley, 1965). Polar Biol 11:249–252

Feldmann RM, Tshudy DM (1989) Evolutionary patterns in macrurous decapod crustaceans from Cretaceous to early Cenozoic rocks of the James Ross Island region, Antarctica. In: Crame JA (ed) Origins and evolution of the Antarctic biota. Geol Soc Spec Publ 47:183–195

Foster BA (1989) Balanomorph barnacle larvae in the plankton at McMurdo Sound, Antarctica. Polar Biol 10:175–177

Fratt DB, Dearborn JH (1984) Feeding biology of the Antarctic brittle star *Ophionotus victoriae* (Echinodermata: Ophiuroidea). Polar Biol 3:127–139

Galéron J, Herman RL, Arnaud PM, Arntz WE, Hain S, Klages M (1992) Macrofaunal communities on the continental shelf and slope of the southeastern Weddell Sea, Antarctica. In: Hempel G (ed) Weddell Sea ecology. Results of EPOS European *Polarstern* study. Springer, Berlin 1993. Polar Biol 12:283–290

Gallardo VA (1987a) The sublittoral macrofaunal benthos of the Antarctic shelf. Environ Int 13:71–81

Gallardo VA (1987b) Analysis of Antarctic conservation areas with emphasis on marine areas. Ser Cient INACH 36:177–184

Gallardo VA (1991) The benthic realm and Antarctic conservation. Int Rep Univ Concepción, mimeo, 4 pp

Gallardo VA, Retamal MA (1986) Establecimiento de un "Bentosgarten" en la Antártica. Bol Antarct Chil 6(1):40–42

Gerdes D (1990) Antarctic trials with the multi-box corer, a new device for benthos sampling. Polar Rec 26(156):35–38

Gerdes D, Klages M, Arntz WE, Herman RL, J, Hain S (1992) Quantitative investigations on macrobenthos communities of the southeastern Weddell Sea shelf based on multibox corer samples. In: Hempel G (ed) Weddell Sea ecology. Results of EPOS European *Polarstern* study. Springer, Berlin 1993. Polar Biol 12:291–301

Gorny M (1989) Entwicklung und Wachstum der Garnelen (Decapoda, Natantia) *Nematocarcinus longirostris*, Bate (Nematocarcinidae) und *Chorismus antarcticus*, Pfeffer (Hippolytidae) im Weddellmeer (Hochantarktis). Dipl Thesis, Universität Bremen, 104 pp

Gorny M, Arntz WE, Clarke A, Gore DJ (1992) Reproductive biology of caridean decapods from the Weddell Sea. In: Hempel G. (ed) Weddell Sea ecology. Results of EPOS European *Polarstern* study. Springer, Berlin 1993. Polar Biol 12:111–120

Green K, Burton HR (1987) Seasonal and geographical variation in the food of Weddell seals, *Leptonychotes weddellii*, in Antarctica. Aust Wildl Res 14:475–489

Green K, Williams R (1986) Observations on food remains in faeces of Elephant, Leopard and Crabeater Seals. Polar Biol 6:43–45

Grobe H (1986) Spätpleistozäne Sedimentationsprozesse am antarktischen Kontinentalhang vor Kapp Norvegia, östliche Weddellsee. Ber Polarforsch 27:121

Gutt J (1991a) On the distribution and ecology of holothurians in the Weddell Sea (Antarctica). Polar Biol 11:145–155

Gutt J (1991b) Investigations on brood protection in *Psolus dubiosus* (Echinodermata: Holothuroidea) from Antarctica in spring and autumn. Mar Biol 111:281–286

Gutt J (1991c) Are Weddell Sea holothurians typical representatives of the Antarctic benthos? Meeresforschung 33:312–329

Gutt J, Piepenburg D (1991) Dense aggregations of three deep-sea holothurians in the southern Weddell Sea, Antarctica. Mar Ecol Prog Ser 68:277–285

Gutt J, Gorny M, Arntz WE (1991) Spatial distribution of Antarctic shrimps (Crustacea: Decapoda) by underwater photography. Antarct Sci 3(4):363–369

Gutt J, Gerdes D, Klages M (1992) Seasonality and spatial variability in the reproduction of two Antarctic holothurians (Echinodermata). Polar Biol 11:533–544

Hain S (1990) Die beschalten benthischen Mollusken (Gastropoda und Bivalvia) des Weddellmeeres, Antarktis. Ber Polarforsch 70:181

Hain S (1992) Maintenance and culture of living benthic molluscs from high-Antarctic shelf areas. Aquacult Fish Managem 23:1–11

Hain S, Arnaud P (1992) Notes on the reproduction of high Antarctic molluscs from the Weddell Sea. In: Hempel G (ed) Weddell Sea ecology. Results of EPOS European *Polarstern* study. Springer, Berlin 1993. Polar Biol 12:303–312

Hamada E, Numanami H, Naito Y, Taniguchi A (1986) Observation of the marine benthic organisms at Syowa Station in Antarctica using a remotely operated vehicle. Mem Natl Inst Polar Res Spec Issue 40:289–298

Hambrey MJ, Larsen B, Ehrmann WU, ODP Leg 119 Shipboard Scientific Party (1989) Forty million years of Antarctic glacial history yielded by Leg 119 of the Ocean Drilling Program. Polar Rec 25(153):99–106

Hambrey MJ, Ehrmann WU, Larsen B (1991) Cenozoic glacial record of the Prydz Bay continental shelf, East Antarctica. In: Barron J, Larsen B et al. (eds) Proc ODP Sci Results, College Station, TX, 119:72–132

Haq BU, Hardenbol J, Vail PR (1987) Chronology of fluctuating sea levels since the Triassic. Science 235:1156–1167

Herman RL, Dahms HU (1992) Meiofauna communities along a depth transect off Halley Bay (Weddell Sea – Antarctica). In: Hempel G (ed) Weddell Sea ecology. Results of EPOS European *Polarstern* study. Springer, Berlin 1993. Polar Biol 12:313–320

Houlihan DF, Allan D (1982) Oxygen consumption of some Antarctic and British gastropods: an evaluation of cold adaptation. Comp Biochem Physiol 73A(3):383–387

Howard-Williams C, Pridmore RD, Broady PA, Vincent WF (1990) Environmental and biological variability in the McMurdo ice shelf ecosystem. In: Kerry KR, Hempel G (eds) Antarctic ecosystems. Ecological change and conservation. Springer, Berlin Heidelberg New York, pp 23–31

Ivleva IV (1980) The dependence of crustacean respiration rate on body mass and habitat temperature. Int Rev Gesamten Hydrobiol 65(1):19–47

Jazdzewski K, Teodorczyk W, Sicinski J, Kontek B (1991) Amphipod crustaceans as an important component of zoobenthos of the shallow Antarctic sublittoral. Hydrobiologia 223:105–117

Juilfs HB, Wägele JW (1987) Symbiontic bacteria in the gut of the blood-sucking Antarctic fish parasite *Gnathia calva* (Crustacea: Isopoda). Mar Biol 95:493–499

Karentz D (1991) Ecological considerations of Antarctic ozone depletion. Antarct Sci 3:3–11

Kellogg DE, Kellog TB, Dearborn JH, Edwards KC, Fratt DB (1982) Diatoms from brittle star stomach contents: implications for sediment reworking. Antarct J US 17:167–169

Kennett JP (1985) Neogene paleoceanography and plankton evolution. Suid-Afr Tydskr Wetenskap 81:251–253

Kennett JP, Barker BF (1990) Latest Cretaceous to Cenozoic climate and oceanographic developments in the Weddell Sea, Antarctica: an ocean-drilling perspective. In: Kennett JP, Barker P et al. (eds) Proc Ocean Drilling Project Sci Results 113:937–960

Kennett JP, Stott LD (1990) 49. Proteus and proto-oceanus: ancestral Paleogen oceans as revealed from Antarctic stable isotopic results; ODP leg 113. In: Kennett JP, Barker P et al. (eds) Proc Ocean Drilling Project Sci Results 113:865–880

Kirkwood JM, Burton HR (1988) Macrobenthic species assemblages in Ellis Fjord, Vestfold Hills, Antarctica. Mar Biol 97:445–457

Klages M (1991) Biologische und populationsdynamische Untersuchungen an ausgewählten Gammariden (Crustacea: Amphipoda) des südöstlichen Weddellmeeres, Antarktis. Diss Universität Bremen, 240 pp

Klages M (1994) Distribution, reproduction and population dynamics of the Antarctic gammaridean amphipod *Eusirus perdentatus* Chevreux, 1912 (Crustacea). Antarct Sci (in press)

Klages M, Gutt J (1990a) Observations on the feeding behaviour of the Antarctic gammarid *Eusirus perdentatus* Chevreux, 1912 (Crustacea: Amphipoda) in aquaria. Polar Biol 10:359–364

Klages M, Gutt J (1990b) Comparative studies on the feeding behaviour of high Antarctic amphipods (Crustacea) in the laboratory. Polar Biol 11:79–83

Klages N (1989) Food and feeding of emperor penguins in the eastern Weddell Sea. Polar Biol 9:385–390

Kock KH (1990) The effect of bottom trawling on benthic assemblages. SC-CAMLR-IX/BG/15:6

Kock KH (1992) Antarctic fish and fisheries. Studies in polar research. Cambridge University Press, Cambridge, 359 pp

Kühl S (1988) A contribution to the reproductive biology and geographical distribution of Antarctic Octopodidae (Cephalopoda). Malacologia 29(1):89–100

Leventer A, Dunbar RB (1987) Diatom flux in McMurdo Sound, Antarctica. Mar Micropalaeontol 12:49–64

Lien R, Solheim A, Elverhøi A, Rokoengen K (1989) Iceberg scouring and sea bed morphology on the eastern Weddell Sea shelf, Antarctica. Polar Res 7:43–57

Lipps JH, Hickman CS (1982) Origin, age, and evolution of Antarctic and deep-sea faunas. In: Ernst WG, Morin JG (eds) The environment of the deep sea. Prentice-Hall, Englewood Cliffs, pp 324–356

Luxmoore RA (1984) A comparison of the respiration rate of some Antarctic isopods with species from lower latitudes. Br Antarct Surv Bull 62:53–65

Luxmoore RA (1985) The energy budget of a population of the Antarctic isopod *Serolis polita*. In: Siegfried WR, Condy PR, Laws RM (eds) Antarctic nutrient cycles and food webs. Springer, Berlin Heidelberg New York, pp 389–396

Maxwell JGH, Ralph R (1985) Non-cold-adapted metabolism in the decapod *Chorismus antarcticus* and other sub-Antarctic marine crustaceans. In: Siegfried WR, Condy PR, Laws RM (eds) Antarctic nutrient cycles and food webs. Springer, Berlin Heidelberg New York, pp 397–406

McClintock JB (1985) Avoidance and escape responses of the Subantarctic limpet *Nacella edgari* (Powell) (Mollusca: Gastropoda) to the sea star *Anasterias perrieri* (Smith) (Echinodermata: Asteroidea). Polar Biol 4:95–98

McClintock JB (1987) Investigation of the relationship between invertebrate predation and biochemical composition, energy content, spicule armament and toxicity of benthic sponges at McMurdo Sound, Antarctica. Mar Biol 94:479–487

Mühlenhardt-Siegel U (1988) Some results on quantitative investigations of macro-zoobenthos in the Scotia Arc (Antarctica). Polar'Biol 8:241–248

Mühlenhardt-Siegel U (1989) Quantitative investigations of Antarctic zoobenthos communities in winter (May/June) 1986 with special reference to the sediment structure. Arch Fischereiwiss 39(1):123–141

Mullineaux LS, de Laca TE (1984) Distribution of Antarctic benthic foraminifers settling on the pecten *Adamussium colbecki*. Polar Biol 3:185–189

Oliver JS, Slattery PN (1985) Effects of crustacean predators on species composition and population structure of soft-bodied infauna from McMurdo Sound, Antarctica. Ophelia 24(3):155–175

Pearse JS, Bosch I, McClintock JB, Marinovic B, Britton R (1986) Contrasting tempos of reproduction by shallow-water animals in McMurdo Sound, Antarctica. Antarct J US 31:182–184

Pearse JS, McClintock JB, Bosch I (1991) Reproduction of Antarctic benthic marine invertebrates: tempos, modes, and timing. Am Zool 31:65–80

Pearson TH, Rosenberg R (1987) Feast and famine: structuring factors in marine benthic communities. In: Gee JHR, Giller PS (eds) Organization of communities past and present. Blackwell, Oxford, pp 373–395

Peck LS (1989) Temperature and basal metabolism in two Antarctic marine herbivores. J Exp Mar Biol Ecol 127:1–12

Perissinotto R, McQuaid C (1990) Role of the sub-Antarctic shrimp *Nauticaris marionis* in coupling benthic and pelagic food-webs. Mar Ecol Prog Ser 64:81–87

Piatkowski U (1987) Zoogeographische Untersuchungen und Gemeinschaftsanalysen am antarktischen Makroplankton. Ber Polarforsch 34:150

Picken GB (1980a) The distribution, growth, and reproduction of the Antarctic limpet *Nacella* (*Patinigera*) *concinna* (Strebel, 1908). J Exp Mar Biol Ecol 42:71–85

Picken GB (1980b) Reproductive adaptations of Antarctic benthic invertebrates. In: Bonner WN, Berry RJ (eds) Ecology in the Antarctic. Academic Press, London, pp 67–75

Picken GB (1984) Marine habitats-benthos. In: Bonner WN, Walton DWH (eds) Key environments: Antarctica. Pergamon, Oxford, pp 154–172

Picken GB (1985) Benthic research in Antarctica: past, present and future. In: Gray JS, Christiansen ME (eds) Marine biology of polar regions and effects of stress on marine organisms. John Wiley, Chichester, pp 167–183

Pirrie D (1989) Shallow marine sedimentation within an active margin basin, James Ross Island, Antarctica. Sediment Geol 63:61–82

Plötz J (1986) Summer diet of Weddell seals (*Leptonychotes weddellii*) in the eastern and southern Weddell Sea, Antarctica. Polar Biol 6:97–102

Plötz J, Ekau W, Reijnders PJH (1991) Diet of Weddell seals *Leptonychotes weddellii* at Vestkapp, eastern Weddell Sea (Antarctica), in relation to local food supply. Mar Mammal Sci 7:136–144

Presler P (1986) Necrophagous invertebrates of the Admiralty Bay of King George Island (South Shetland Islands, Antarctica). Pol Polar Res 7(1-2):25–61

Quilty PG (1990) Significance of evidence for changes in the Antarctic marine environment over the last 5 million years. In: Kerry KR, Hempel G (eds) Antarctic ecosystems. Ecological change and conservation. Springer, Berlin Heidelberg New York, pp 4–8

Rauschert M (1991) Ergebnisse der faunistischen Arbeiten im Benthal von King George Island (Südshetlandinseln, Antarktis). Ber Polarforsch 76:75

Sagar PM (1980) Life cycle and growth of the Antarctic gammarid amphipod *Paramoera walkeri* Stebbing, 1906. J R Soc NZ 10:259–270

Sarà M (1992) I poriferi nell'ecosistema antartico: la provincia magellanica. In: Gallardo VA, Ferretti O, Moyano HI (eds) Actas del Semin Int Oceanografia in Antartide. Centro EULA, Universidad de Concepción/Chile, 7–9 March 1991, pp 517–522

Scharek R (1991) Die Entwicklung des Phytoplanktons im östlichen Weddellmeer (Antarktis) beim Übergang vom Spätwinter zum Frühjahr. Ber Polarforsch 94:195

Schwarzbach W (1988) Die Fischfauna des östlichen und südlichen Weddellmeeres: geographische Verbreitung, Nahrung und trophische Stellung der Fischarten. Ber Polarforsch 54:94

Sicinski J (1986a) Benthic assemblages of Polychaeta in chosen regions of the Admiralty Bay (King George Island, South Shetland Islands). Pol Polar Res 7(1-2):63–78

Sicinski J (1986b) Application of the dendrite analysis in the discussion on the biogeography of the Antarctic. Pol Polar Res 7(3):305–317

Sieg J (1988) Das phylogenetische System der Tanaidacea und die Frage nach Alter und Herkunft der Crustaceenfauna des antarktischen Festlandsockels. Z Zool Syst Evolutionsforsch 26:363–379

Sieg J (1992) On the origin and age of the Antarctic Tanaidacean fauna. In: Gallardo VA, Ferretti O, Moyano HI (eds) Actas del Semin Int Oceanografia in Antartide. Centro EULA, Universidad de Concepción/Chile, 7–9 March 1991, pp 421–430

Siegel V, Mühlenhardt-Siegel U (1988) On the occurrence and biology of some Antarctic Mysidacea (Crustacea). Polar Biol 8:181–190

Slattery PN, Oliver JS (1986) Scavenging and other feeding habits of Lysianassid amphipods (*Orchomene* spp.) from McMurdo Sound, Antarctica. Polar Biol 6:171–177

Smith JMB, Simpson RD (1985) Biotic zonation on rocky shores of Heard Island. Polar Biol 4:89–94

Spindler M (1990) A comparison of Arctic and Antarctic sea ice and the effects of different properties on sea ice biota. In: Bleil U, Thiede J (eds) Geological history of the Polar Oceans: Arctic versus Antarctic. Kluwer, Dordrecht, pp 173–186

Spindler M, Dieckmann GS (1991) Das Meereis als Lebensraum. Spektrum Wiss Febr 1991:48–57

Stott LD, Kennett JP, Shackleton NJ, Corfield RM (1990) 48. The evolution of Antarctic surface waters during the Paleogene: inferences from the stable isotopic composition of planktonic foraminifers, ODP leg 113. In: Kennett JP, Barker P et al. (eds) Proc Ocean Drilling Project Sci Results 113:849–863

Tucker MJ (1988) Temporal distribution and brooding behaviour of selected benthic species from the shallow marine waters off the Vestfold Hills, Antarctica. In: Ferris JM, Burton HR, Johnstone GW, Bayly IAE (eds) Biology of the Vestfold Hills, Antarctica. Hydrobiologia 165:151–159

Voß J (1988) Zoogeographie und Gemeinschaftsanalyse des Makrozoobenthos des Weddellmeeres (Antarktis). Ber Polarforsch 45:145

Wägele H (1988) Riesenwuchs contra spektakuläre Farbenpracht–Nudibranchia der Antarktis. Nat Mus 118(2):46–53

Wägele JW (1987a) The feeding mechanism of *Antarcturus* and a redescription of *Spinacoronatus* Schultz, 1978 (Crustacea: Isopoda: Valvifera). Philos Trans R Soc Lond B 316:429–458

Wägele JW (1987b) On the reproductive biology of *Ceratoserolis trilobitoides* (Crustacea: Isopoda): latitudinal variation of fecundity and embryonic development. Polar Biol 7:11–24

Wägele JW (1988) Aspects of the life-cycle of the Antarctic fish parasite *Gnathia calva* Vanhöffen (Crustacea: Isopoda). Polar Biol 8:287–291

Wägele JW (1990) Growth in captivity and aspects of reproductive biology of the Antarctic fish parasite *Aega antarctica* (Crustacea, Isopoda). Polar Biol 10:521–527

Wägele JW, Brandt A (1992) Aspects of the biogeography and biology of Antarctic isopods (Crustacea). In: Gallardo VA, Ferretti O, Moyano HI (eds) Actas del Semin

Int Oceanografia in Antartide. Centro EULA, Universidad de Concepción/Chile, 7–9 March 1991, pp 417–420

Wägele JW, Brito TAS (1990) Die sublitorale Fauna der maritimen Antarktis. Erste Unterwasserbeobachtungen in der Admiralitätsbucht. Nat Mus 120:269–282

Watling L, Thurston MH (1989) Antarctica as an evolutionary incubator: evidence from the cladistic biogeography of the amphipod family Iphimediidae. In: Crame JA (ed) Origins and evolution of the Antarctic biota. Geol Soc Spec Publ 47:297–313

White MG (1984) Marine benthos. In: Laws RM (ed) Antarctic ecology, vol 2. Academic Press, London, pp 421–461

Wyanski DM, Targett TE (1981) Feeding biology of fishes in the endemic Antarctic Harpagiferidae. Copeia 3:686–696

Zamorano JH (1983) Zonación y biomasa de la macrofauna bentónica en Bahía South, Archipiélago de Palmer, Antártica. Ser Cient INACH 30:27–38

Zamorano JH, Duarte WE, Moreno CA (1986) Predation upon *Laternula elliptica* (Bivalvia, Anatinidae): a field manipulation in South Bay, Antarctica. Polar Biol 6:139–143

Future Ocean-Atmosphere Research in the Antarctic Region

Ernst Augstein[1]

1 General Remarks

In the framework of this consideration we will primarily address research topics of the Antarctic region which are more or less interrelated with global climate. Earlier as well as recent field and model investigations have provided convincing evidence that certain physical processes in the polar atmosphere, ocean and cryosphere may support, enhance or even initiate global climate changes. Since the main components of the Earth's climate are strongly coupled through non-linear interaction mechanisms, any perturbations excited somewhere in the system may suffer from significant changes in amplitude and frequency during their migration through the interrelated physical domain. Therefore, investigations of Antarctic geophysical processes should ideally cover a wide range of time and space scales and should include all important interacting climate elements. But in reality individual field observations and model studies have to be limited with respect to space and time due to technical, logistic and financial restrictions as well as to the lack of experienced manpower. Consequently, international programmes are required to create joint research projects in such a way that various single contributions may provide a satisfactory basis for reasonable syntheses.

The geography of the south polar region is predominantly characterized by the ice-covered Antarctic continent which is more or less centred around the South Pole. The land mass is surrounded by a water belt which contains the only closed oceanic circulation around the Earth's axis. This Antarctic Circumpolar Current (ACC) is primarily driven by a likewise closed belt of rather steady westerly winds. South of these circumpolar circulations both the atmospheric and the oceanic flows are governed by three cyclonic gyres which occupy the Weddell Sea, the Ross Sea and the less pronounced Prydz Bay. Large ice shelves cover the southerly parts of the first two ocean basins. They form major outflows of the continental ice sheet. The seasonal variations of the solar radiation and the given atmospheric and oceanic flow

[1] Alfred-Wegener-Institut für Polar-und Meeresforschung, Postfach 120161, D-27515 Bremerhaven, Germany

Hempel (Ed.) Antarctic Science

patterns adjacent to the continent favour a strong annual cycle of the extent and the volume of sea ice around Antarctica.

These various observational facts provide already distinct hints that the Antarctic climate system must be significantly controlled by interacting oceanic, atmospheric and cryospheric processes. A closer look into the different climate components would reveal that biological and chemical mechanisms are as well participating in the determination of the actual state and in the generation of changes of the environmental conditions. Therefore, both field studies and modelling efforts have to take into account a rather complex interacting natural system in order to diagnose and to predict the influence of the Antarctic region on global climate. Consequently, all relevant processes of and coupling mechanisms between the above mentioned subsystems have to be investigated in great detail.

2 The Atmosphere

The lower boundary of the atmospheric circulation in the southern latitudes is primarily controlled by the sloping ice sheet surface over the Antarctic continent and by a strongly variable sea ice cover of the surrounding ocean. A recent numerical experiment of Egger (1991) suggests that even the dynamics of the large-scale stratospheric flow seem to be remarkably affected by the low-level cold air drainage from the centre of the continent to the coast. Studies of Simmonds and Budd (1990) with an atmospheric global circulation model reveal that minor changes of sea ice concentration lead to substantial variations of the local air – sea heat exchanges and to significant modifications of the atmospheric temperature and wind velocity distributions in the southern hemisphere. Although these model results are quite convincing, they have still to be proven by observations. In addition, more refined numerical experiments are required to study all important impacts of the Antarctic region on global climate.

The influence of clouds through diabatic heating of the air column and surface precipitation on the one hand, and through their radiative properties on the other hand, is qualitatively more or less understood but considerable uncertainties still exist with regard to the quantitative cloud effects. Therefore, intensive research is still required to develop satisfactory concepts for the model treatment of clouds and its consequences on the atmospheric heat budget, the surface radiation balance and the snow deposition. Another topic which needs intensive investigations is the interaction between the troposhere and the stratosphere in high southern latitudes. It is rather likely that besides the lateral stratospheric exchanges between high and mid latitudes the vertical fluxes across the tropopause have also a considerable impact on the air temperature and the concentration of chemical constituents in the stratosphere. Thus, both the horizontal and the vertical transports may affect the recently discovered stratospheric ozone depletions during spring time each year.

3 The Ocean

The large scale circulation of the Southern Ocean is qualitatively well known from observations and can more or less be described by oceanic general circulation models (Wolff et al. 1991). But neither of the present models nor the existing data are sufficient to properly analyze and to quantitatively reproduce the actual thermohaline state and the current field of the Southern Ocean in desirable detail. The present uncertainties with regard to the oceanic conditions south of about 50° S are primarily due to the sparsity and inhomogeneity of observations particularly in the sea ice belt area, the coarse resolution of numerical models, the poor treatment or even neglect of sea ice and ice shelves and the application of unproven algorithms for vertical small-scale mixing and deep convection in model computations.

Some of these deficiencies can nowadays be overcome with the aid of improved observational techniques. Ice-breaking research vessels and new highly capable satellites with various types of sensors form a profound basis for extended data acquisitions. And the rapid development of computational power enables us to carry out numerical studies of the relevant processes and of the large-scale circulation of the Southern Ocean with a hierarchy of models.

4 Sea Ice and Ice Shelves

The sea ice cover of the Antarctic area is predominantly characterized by a strong annual cycle. Satellite observations show that it is subject to remarkable interannual variations and to significant short-term alterations in specific areas, as e.g. the coastal zone and the ice margin (Zwally et al. 1983). Modelling of sea ice has been carried out with various numerical concepts which differ mainly with respect to their treatment of the relevant physical processes (see e.g. Hibler and Ackley 1983). According to our present knowledge both factors, the thermodynamics and the dynamics of sea ice – including internal forces, need to be considered to reproduce the important features of the observed ice cover (Hibler and Bryan 1987). Furthermore, sufficient simulation of the natural conditions requires the application of models in which at least the sea ice and the upper ocean are coupled (Lemke et al. 1990). The final goal, however, must be to include also the atmosphere into the models as demonstrated e.g. by Koch (1988). The improvement of models depends to a large extent also on field programmes which provide information on melting and freezing, the extent, concentration, thickness, motion and deformation of sea-ice, as well as on the atmospheric and oceanic thermal and dynamical forcings at the sea surface. Additionally, the vertical and horizontal structures of the oceanic mixed layer and of the atmospheric boundary layer have to be monitored for a successful investigation of coupled ocean – sea ice – atmosphere system. On the modelling side particular efforts must be made to achieve a better treatment of the interaction between the ice-covered ocean and the atmosphere.

Recent model studies of Hellmer and Olbers (1991), and earlier observations of Foldvik et al. (1985) indicate that thermohaline processes at the lower boundary of the large ice shelves seem to be of primary importance for the formation of Antarctic Bottom Water. Simultaneously, the bottom melting and freezing at the ice shelves contribute to the mass budget of e.g. the Filchner-Ronne Ice Shelf (Oerter et al. 1991). Therefore, the ocean – ice shelf interaction processes have to be investigated observationally to provide a realistic basis for their treatment in ice shelf, ocean and climate models.

Besides the thermal, radiative and dynamical influences, the sea ice cover also acts as a barrier for the air – sea gas exchange. Notably the CO_2 and oxygen fluxes between the atmosphere and the ocean practically cease in the presence of sea ice. Since respiration processes in the water column continue during winter, the CO_2 partial pressure should increase and the oxygen concentration should decrease in the upper water column of the Antarctic sea ice belt from fall to spring. In addition, CO_2 diffuses from the Warm Deep Water into the mixed layer so that there is a net transport of CO_2 from the ocean to the atmosphere in the sea ice regime during the melting season (Weiss 1987). In the northerly ice-free areas, however, the ocean is primarily a CO_2 sink. Such particularities of the air – sea gas exchanges need still more systematic research and require an appropriate mathematical description in climate models.

5 Research Activities

Most of the aforementioned research topics will be addressed during the last decade of this century by ongoing and anticipated national or international projects. The latter are mainly coordinated in the framework of the World Climate Research Programme and the International Geosphere-Biosphere Programme. Most of the work will be concerned with process studies which will lead to important model improvements. But it is also highly necessary to considerably extend the long-term observational network in the Antarctic region. Valuable data are presently gathered by in situ measurements at the Antarctic wintering stations and through remote sensing techniques by visible, infrared and microwave satellite imagery. Both of these observational methods need not only to be maintained but they must be supplemented by e.g. automatic surface stations on the continent and by drifting buoys on the ocean which transfer their data via satellite to landstations in Europe or on other continents. Furthermore, ocean moorings, equipped with current meters, thermistors, conductivity cells, sediment traps and upward looking sonars must be deployed and maintained for at least one sunspot cycle at specified locations of the Southern Ocean through multinational support. The full set of measurements together with the already existing data would help to detect climate changes and it would also provide satisfactory observational time series for improving ocean – atmosphere circulation models and for testing of climate simulations. As a final result, one would then obtain more reliable scenario calculations and perhaps even climate predictions.

References

Egger J (1991) On the mean atmospheric circulation over Antarctica. Geophys Astrophys Fluid Dynam 58:75–90

Foldvik A, Gammelsroed T, Slotsvik N, Toerrensen T (1985) Oceanographic conditions on the Weddell Sea shelf during the German Antarctic expedition 1979/80. Polar Res 3:209–226

Hellmer HH, Olbers D (1991) On the thermohaline circulation under the Filchner-Ronne ice shelves. Antarct Sci 3:433–442

Hibler WD III, Ackley SF (1983) Numerical simulations of the Weddell Sea pack-ice. J Geophys Res 88:2873–2887

Hibler WD III, Bryan K (1987) A diagnostic ice-ocean model. J Phys Oceanogr 17: 987–1015

Koch C (1988) A coupled sea-ice-atmospheric boundary layer model. Part I: Description of the model and 1979 standard run. Beitr Phys Atmos 61:344–354

Lemke P, Owens WB, Hibler WD III (1990) A coupled sea ice-mixed layer pycnocline model for the Weddell Sea. J Geophys Res 95:9513–9525

Oerter H, Eicken H, Miller H (1991) The physical properties of the marine ice under the Filchner-Ronne shelf ice, Filch.-Ronne Ice Shelf Progr Rep 5. Alfred-Wegener-Inst, Bremerhaven, pp 13–18

Simmonds I, Budd WF (1990) A simple parameterization of ice leads in a general circulation model, and the sensitivity of climate to change in Antarctic ice concentration. Ann Glaciol 14:266–269

Weiss RF (1987) Winter Weddell Sea Project 1986: Trace gas studies duing legs ANT V/2 and V/3 of *Polarstern*. Antarct J US 22:99–100

Wolff J-O, Maier-Raimer E, Olbers D (1991) Wind-driven flow over topography in a zonal β-plane channel: a quasigeostrophic model of the Antarctic Circumpolar Current. J Phys Oceanogr 21:236–264

Zwally HJ, Comiso JC, Parkinson C, Campbell WJ, Carsey FD, Gloersen P (1983) Antarctic sea ice, 1973–1976: Satellite passive-microwave observations. NASA, Washington, DC, 206 pp

Future of Antarctic Science – Biosphere

Gerd Hubold[1]

1 Introduction

Due to its climatic conditions, ice cover, rocky soils and bare mountains, continental Antarctica is a hostile place for life. Only specialized species of bacteria, fungi and lichens are adapted to live in niches on and in stones, ice and soil. The few species of insects and higher plants are restricted to the circum-Antarctic islands.

Warm-blooded top predators use the coasts of Antarctica as breeding and resting sites. These species, however, are not connected to the terrestrial ecosystems, but are parts of essentially marine food chains. In contrast to the sparse terrestrial life, the sea surrounding Antarctica hosts a wealth of species and large stocks of phytoplankton, zooplankton, krill, squids, fish, mammals and birds. When harvested, the stocks of fur seals, elephant seals, whales and (most recently) fish rapidly declined.

The surprising contrast between the vast former seal and whale concentrations on one hand, and their sensitivity to exploitation on the other, has prompted rather antagonistic approaches to the future handling of the Antarctic environment. These approaches range from the still persisting model of a highly productive Antarctic ocean, which contains large protein resources for future generations, to the conservationist approach, which claims the necessity to protect Antarctica from virtually any human activity, possibly even scientific investigations.

It is not within the scope of this short presentation to evaluate these different approaches and to cover both terrestrial and marine Antarctic systems. My focus is on some biological features which are of basic importance for the understanding of Antarctic marine biology and which may contribute to a rational development of future Antarctic biological research as an input to the developing global environmental research under the International Geosphere Biosphere Program (IGBP).

[1] Institut für Seefischerei, Palmaille 9, D-22767 Hamburg, Germany

Hempel (Ed.) Antarctic Science

2 The Antarctic Marine Ecosystem

As in other seas, life in the Southern Ocean is based on sunlight and nutrients. Nutrient concentrations are generally high, so light and, to a lesser degree, temperature, are the limiting factors for primary production. In sea ice areas, the water column is shaded by the ice cover. In ice-free areas, wind-induced turbulence causes down-mixing of the plant cells leading to light limitation of their photosynthesis.

High phytoplankton productivity is regularly found in areas where both shading by ice cover and turbulence are at their minimum, i.e. near ice edges and in meltwater areas. In and under the ice itself, dark-adapted plants can grow and specialized food chains are based on this production.

As a result of the light and turbulence conditions, overall average productivity of Antarctic waters is rather low. Productivity is widely determined by the extent and dynamics of sea ice and stabilized meltwater areas. Compared with other seas, we cannot expect therefore exceptional high levels of animal production based on this overall low primary production.

These average calculations are of interest to global budget calculations. In biological terms, however, they are not meaningful, because biological systems do not operate on global scales and on averages. Even the smallest functional unit of life, the cell, creates concentration gradients against the environment. On different scales, life is organized in clones, clusters, patches, schools, accumulations, colonies, communities, etc. Emperor penguins would not survive the harsh Antarctic winter if they were evenly distributed over the sea ice. Thus, biological systems often operate with high concentrations on rather small spatial scales and make a global average approach both difficult and meaningless.

To create and maintain biological concentrations in the sea, reliable physical or biological mechanisms must exist. Only large animals are capable of determining their aggregations by active movement. Smaller organisms, or those with small larvae as e.g. fishes, are temporally planktonic, i.e. they are subject to a passive drift with currents in the ocean. Their life cycles and populations are strongly coupled to hydrography in so-called retention systems formed by gyres, fronts and currents (Sinclair 1988).

The sea ice area is probably one of the most prominent retention systems in the Antarctic, which works on a seasonal cycle of sea-ice formation and melting, providing the habitat for a very well-developed surface ice-and-meltwater community. Other communities exist in the Antarctic, e.g. in the circum-Antarctic West Wind Drift, in the large gyre systems as e.g. Ross Sea and Weddell Sea, and in the stable coastal currents of the East Wind Drift.

The productive zones of the sea-ice community and the connected pelagic food chains move across the Southern Ocean with the melting zones and with the surface currents such as the coastal current. By this movement, pelagic biomass originating from larger areas is regularly transported along the coasts. At key sites, sedentary local predator colonies can harvest the production which is brought to them by the hydrographic conveyor belts. The Weddell-Scotia Confluence zone, and the islands herein, are such hot spots in the Atlantic Sector, where hydrographic conveyor belts pass by loaded with pelagic biomass from elsewhere.

In these areas the large penguin and seal colonies of many thousands of individuals rely on a regular advection of their prey at the right time and the right place. In a crowded penguin rookery the chicks cannot be brought up on local krill production of about $5\,\mathrm{g\,m^{-2}\,year^{-1}}$ around the colony site. Recent detail analyses of commercial krill catches show that 90% of the fleet's catches were reported from within 100 km around the predator colonies in the Atlantic sector (SC-CAMLR 1991). Evidently, the fishery exploits the same concentration mechanisms that enable the survival of the predator colonies.

Thus, many Antarctic species (and a potential fishery) depend on regular oceanographic current patterns and advection of organic matter and plankton to the key sites. This makes the Antarctic food webs (and a potential fishery) highly sensitive to possible changes to those environmental factors which govern the advection and retention systems. This sensitivity may be more accentuated in the Southern Ocean than in most other marine systems.

3 Future Research

Modellers tell us that global warming would almost certainly produce most marked effects in polar latitudes. This would almost certainly result in drastic changes in the distribution, character and thickness of sea ice and in terrestrial ice cover. Crucial alterations in the determining physical factors on land and in the Southern Ocean dynamics, i.e. the currents and the sea ice, can be expected from this. What will then be the effect on the biota, and what might be mediating influences of the biota themselves on such a change in an ecosystem which is highly sensitive to environmental factors?

Future biological science will have to address these questions on various levels and by multiple approaches:

- Terrestrial microbial systems will be studied under the international BIOTAS programme coordinated by SCAR. BIOTAS promotes standardized observations on certain terrestrial habitats to understand key factors determining the colonization processes of plants and microbes in Antarctic land biotopes. It will also observe possible effects of environmental changes on these processes.
- Ongoing marine programmes, such as the CCAMLR ecosystem monitoring of plankton, krill and predators in open water areas shall provide information about possible changes in the key components of marine

ecosystems. These investigations, however, do not tell much about the key processes involved, especially those related to sea ice and its variability.

- With improving satellite technique better synoptic pictures of Antarctic surface patterns will become available. The expectation, however, that looking at Antarctica from space can provide qualitative progress in our understanding of biological processes seems unrealistic. Although detailed analyses of satellite images of surface chlorophyll distribution can provide a synoptic picture of near-surface accumulations of plant pigments at the sea-ice margins, the underlying processes are not understood. Detailed ground truth data from ships and shore based stations are essential to interprete the colourful pictures from space.
- The investigation of global carbon fluxes is an approach which is taken within the Southern Ocean JGOFS programme. This approach at present lacks profound knowledge on biological processes determining these fluxes on scales of patches and retention systems. Extrapolations of local sedimentation to the global scale have to consider the patchiness problem, and also the plasticity of organisms to react individually and in a non-linear fashion to their environment.

One example of future biological research in the Antarctic is the forthcoming EASIZ programme. Considering the principal ongoing and planned Antarctic marine research approaches, the SCAR/SCOR Group of Specialists on Southern Ocean Ecology (GOS-SOE) has developed a catalogue of biological research objectives and strategies named EASIZ (Ecology of the Antarctic Sea Ice Zone) as a coordinated ecological research initiative focussed on the Antarctic Sea Ice Zone and its relevance to global change (Anonymous 1992a,b).

The overall objective of the EASIZ programme is "to determine the role of the Antarctic sea-ice zone on Antarctic marine systems and in the control of global biogeochemical and energy exchanges".

Priority attention will be given to:

- the factors controlling the life cycles and survival of the biota (organism-related biological approach);
- the impact of sea ice and biota on ocean-atmosphere exchanges (system-oriented approach);
- the nature of biogeochemical cycles in water column and benthos (organism-, carbon-, and system-oriented approach).

This proposed research requires a major long-term programme of inter- and multidisciplinary research (10 years minimum) at a variety of temporal and spatial scales on an annual basis.

In order to achieve this, two broad areas of research activities are identified:

- experimentally based studies of key processes and geographical areas (considering the concept of patchiness and hot spots in the ecosystem);

– acquisition of data through remote sensing (to achieve synoptic information on the scales of key processes and sites).

Such activities are interdependent through their common need for information gathered from ship- and shore-based research.

The implementation of such a research programme requires international cooperation, coordination and direction. Part of these scientific questions are covered by the existing or developing Southern Ocean JGOFS and Southern Ocean GLOBEC programmes. The scientific framework for EASIZ will provide a biological input into the physical-meteorological programmes under IGBP, and it will also focus on biological processes necessary to develop an ecological understanding of the functioning of Antarctic sea ice zone biocenoses. The scope of this programme includes processes on and in the sea ice itself, as well as in the water column and on the seafloor beneath, especially in Antarctic shelf and near-shore areas (Anonymous 1993).

References

Anonymous (1992a) SCAR/SCOR group of specialists on Southern Ocean ecology. Rep Meet held in Trondheim, Norway, 21–23 May 1990. SCAR Rep 8:1–22

Anonymous (1992b) SCAR/SCOR Group of specialists on Southern Ocean Ecology. Rep Meet held in Bremerhaven, Germany, 17–18 Sept 1991. SCAR Rep 8:23–40

Anonymous (1993) SCAR/SCOR Group of specialists on Southern Ocean Ecology. Rep Meet held in Cambridge, 7–9 Sept 1992. SCAR Rep (in press)

SC-CAMLR (1991) Report of the Tenth Meeting of the Scientific Committee (SC-CAMLR-X), CCAMLR, Hobart, Australia, 427 pp

Sinclair M (1988) Marine Populations. An Essay on population regulation and speciation. Books in recruitment fishery oceanography. Wash Sea Grant Prog, University of Washington Press, Seattle, pp 1–252

Printing: Saladruck, Berlin
Binding: Buchbinderei Lüderitz & Bauer, Berlin